Hypergroup Theory and Algebrization of Incidence Structures

Hypergroup Theory and Algebrization of Incidence Structures

Editors

Dario Fasino
Domenico Freni

Basel • Beijing • Wuhan • Barcelona • Belgrade • Novi Sad • Cluj • Manchester

Editors
Dario Fasino
University of Udine
Udine, Italy

Domenico Freni
Università di Udine
Udine, Italy

Editorial Office
MDPI
St. Alban-Anlage 66
4052 Basel, Switzerland

This is a reprint of articles from the Special Issue published online in the open access journal *Mathematics* (ISSN 2227-7390) (available at: https://www.mdpi.com/si/mathematics/Hypergroup_Theory).

For citation purposes, cite each article independently as indicated on the article page online and as indicated below:

Lastname, A.A.; Lastname, B.B. Article Title. *Journal Name* **Year**, *Volume Number*, Page Range.

ISBN 978-3-0365-8568-0 (Hbk)
ISBN 978-3-0365-8569-7 (PDF)
doi.org/10.3390/books978-3-0365-8569-7

Cover image courtesy of Dario Fasino and Domenico Freni

© 2023 by the authors. Articles in this book are Open Access and distributed under the Creative Commons Attribution (CC BY) license. The book as a whole is distributed by MDPI under the terms and conditions of the Creative Commons Attribution-NonCommercial-NoDerivs (CC BY-NC-ND) license.

Contents

About the Editors vii

Dario Fasino and Domenico Freni
Preface to the Special Issue on "Hypergroup Theory and Algebrization of Incidence Structures"
Reprinted from: *Mathematics* **2023**, *11*, 3424, doi:10.3390/math11153424 1

Christos G. Massouros and Gerasimos G. Massouros
On the Borderline of Fields and Hyperfields
Reprinted from: *Mathematics* **2023**, *11*, 1289, doi:10.3390/math11061289 5

Osman Kazancı, Sarka Hoskova-Mayerova and Bijan Davvaz
Algebraic Hyperstructure of Multi-Fuzzy Soft Sets Related to Polygroups
Reprinted from: *Mathematics* **2022**, *10*, 2178, doi:10.3390/math10132178 41

Stefano Innamorati and Fulvio Zuanni
A Combinatorial Characterization of $H(4, q^2)$
Reprinted from: *Mathematics* **2022**, *10*, 1707, doi:10.3390/math10101707 57

Jan Chvalina, Bedřich Smetana, Jana Vyroubalová
Construction of an Infinite Cyclic Group Formed by Artificial Differential Neurons
Reprinted from: *Mathematics* **2022**, *10*, 1571, doi:10.3390/math10091571 67

Mario De Salvo, Dario Fasino, Domenico Freni and Giovanni Lo Faro
Commutativity and Completeness Degreesof Weakly Complete Hypergroups
Reprinted from: *Mathematics* **2022**, *10*, 981, doi:10.3390/math10060981 81

Giovanni Lo Faro, Salvatore Milici and Antoinette Tripodi
Uniform (C_k, P_{k+1})-Factorizations of $K_n - I$ When k Is Even
Reprinted from: *Mathematics* **2022**, *10*, 936, doi:10.3390/math10060936 99

Mario De Salvo, Dario Fasino, Domenico Freni and Giovanni Lo Faro
G-Hypergroups: Hypergroups with a Group-Isomorphic Heart
Reprinted from: *Mathematics* **2022**, *10*, 240, doi:10.3390/math10020240 107

Anak Nongmanee and Sorasak Leeratanavalee
v-Regular Ternary Menger Algebras and Left Translations of Ternary Menger Algebras
Reprinted from: *Mathematics* **2021**, *9*, 2691, doi:10.3390/math9212691 125

Metod Saniga, Henri de Boutray, Frédéric Holweck and Alain Giorgetti
Taxonomy of Polar Subspaces of Multi-Qubit Symplectic Polar Spaces of Small Rank
Reprinted from: *Mathematics* **2021**, *9*, 2272, doi:10.3390/math9182272 137

Irina Cristea and Milica Kankaraš
The Reducibility Concept in General Hyperrings
Reprinted from: *Mathematics* **2021**, *9*, 2037, doi:10.3390/math9172037 155

About the Editors

Dario Fasino

Dario Fasino is an Associate Professor in Numerical Analysis in the Department of Mathematics, Computer Science and Physics of the University of Udine, Italy. The majority of his work concerns the spectral properties and computational properties of structured matrices arising in a variety of subjects. His most current research interests lie at the intersection of network science and numerical linear algebra. Around 2000, he started a fruitful collaboration with D. Freni in semihypergroup and hypergroup theory.

Domenico Freni

Domenico Freni is an Associate Professor at the University of Udine, Italy, and works in the Department of Mathematics, Computer Science and Physics. He received a PhD degree from the Blaise Pascal University of Clermont-Ferrand II and worked with Y. Sureau and M. Gutan. He is the author of more than 50 papers on hypercompositional algebra and group theory. He has served as a reviewer in several WOS-indexed journals and has participated in research projects with national and departmental funding. Since the end of the 1990s, he has set up a research group that, together with M. De Salvo, D. Fasino and G. Lo Faro, has made a conspicuous contribution to the field of semihypergroup and hypergroup theory.

Editorial

Preface to the Special Issue on "Hypergroup Theory and Algebrization of Incidence Structures"

Dario Fasino * and Domenico Freni *

Dipartimento di Scienze Matematiche, Informatiche e Fisiche, Università di Udine, 33100 Udine, Italy
* Correspondence: dario.fasino@uniud.it (D.F.); domenico.freni@uniud.it (D.F.)

This work contains the accepted papers of a Special Issue of the MDPI journal *Mathematics* entitled "Hypergroup Theory and Algebrization of Incidence Structure". As Guest Editors of this Special Issue, we have invited significant and original contributions dealing with algebraic hyperstructures in a broad sense, of which hypergroups and adjacency structures are two prominent representatives.

Algebraic hyperstructures are natural generalizations of ordinary algebraic structures when the composition operator is multivalued. For this reason, the study of algebraic hyperstructures is also called hypercompositional algebra. The first results of this theory appeared in the 1940s. Since then, it has undergone a lively development, beginning in the 1970s with the work of various research groups in France, Greece, and Italy on the theory of hypergroups, hyperrings, and hyperfields. At present, hypercompositional algebra is being studied by many researchers on almost all continents, and is characterized by a great diversity of style and subject matter, as evidenced by the contributions collected in this work.

Since its origin, hypercompositional algebra has had a close relationship with classical algebra, borrowing concepts and methods from it. For example, an important class of hypergroups is constructed from the quotient structure of a classical group with respect to a non-normal subgroup. As studies progressed, the theory found new and profound relationships with Galois theory, geometry, and, in particular, incidence structures. Today, hypercompositional algebra has a fruitful variety of connections with other areas of mathematics, such as fuzzy set theory, combinatorics, and probability, with applications in various other sciences such as computer science, artificial intelligence and physics, as well as other natural and social sciences.

The Special Issue received 24 distinct submissions of which 10 (42%) were published after peer review; 13 contributions (54%) were rejected by the MDPI Editorial Board and only one (4%) after peer review. The published papers were written by 24 different authors from 10 different countries, with an average number of 2.9 authors per paper, and address theoretical aspects, applications, and related computational issues of algebraic hyperstructures. The names of the authors in alphabetical order are: Jan Chvalina, Irina Cristea, Bijan Davvaz, Henri De Boutray, Mario De Salvo, Dario Fasino, Domenico Freni, Alain Giorgetti, Frédéric Holweck, Šárka Hošková-Mayerová, Stefano Innamorati, Milica Kankaraš, Osman Kazancı, Sorasak Leeratanavalee, Giovanni Lo Faro, Christos Massouros, Gerasimos Massouros, Salvatore Milici, Anak Nongmanee, Metod Saniga, Bedrich Smetana, Antoinette Tripodi, Jana Vyroubalová, and Fulvio Zuanni. We provide hereafter a brief overview of their contributions. The reader will find here a plethora of different problems, focuses and methods that provide insight into the variegated features of hypercompositional algebra.

In [1], the authors define new classes of hyperfields and hyperrings. They classify finite hyperfields as quotient hyperfields or non-quotient hyperfields, and analyze structures resulting from the subtraction of a multiplicative subgroup from a field. This paper includes an extensive bibliography on the subject, which not only provides the interested reader with a detailed roadmap of hyperfield theory but also opens up further investigation into the boundary between classical and hypercompositional algebra.

The authors of [2] analyze the relations between multi-fuzzy soft sets and polygroups. Moreover, they extend some algebraic properties of fuzzy soft polygroups and soft polygroups to multi-fuzzy soft polygroups. Finally, they define new operations on a multi-fuzzy soft set and present some of their algebraic properties.

The main result in [3] is the establishment of a novel combinatorial characterization of $H(4, q^2)$, a Hermitian variety embedded in the projective space $PG(4, q^2)$. This characterization makes it possible to remove an unnecessary hypothesis that is present in a previously known analogous characterization, except for a few small cases.

The authors of [4] consider a class of linear differential operators with time-dependent coefficients inspired by artificial neurons, called differential neurons. With these objects, they first define an infinite cyclic group isomorphic to $(\mathbb{Z}, +)$, extending the monoid of differential neurons with their negative powers. This construction is then extended by successive steps until a non-commutative join space is defined.

The paper [5] introduces a family of hypergroups, here called weakly complete, that generalizes the construction of complete hypergroups. Furthermore, the authors define the degree of completeness of finite hypergroups which, in some sense, quantifies the extent to which the hypergroup is close to being complete. For weakly complete hypergroups, this degree can be computed by explicit formulas.

In the context of graph factorizations, the paper [6] provides a complete solution to an unsolved existence problem for uniform factorizations of complete simple graphs in terms of cycles and paths with certain specified sizes. The solution crucially relies on two constructions that allow to derive factorizations of larger graphs from the knowledge of simpler cases.

The objective of [7] is to take the first step in the classification of G-hypergroups, that is, hypergroups whose heart is a non-trivial group. This research has an emphasis on G-hypergroups whose heart is a torsion group. In particular, the authors characterize G-hypergroups which are of type U on the right or right cogroups. The paper also includes the hyperproduct tables of all G-hypergroups with sizes up to 5, up to isomorphisms.

In [8], the authors introduce the notion of v-regular ternary Menger algebras, which generalizes the notion of regular ternary semigroups. Furthermore, they consider a special class of n-ary functions, the so-called left translations, and prove that the set of left translations can be endowed with the structure of a ternary Menger algebra.

The authors of [9] consider certain physically relevant finite geometries of binary symplectic polar spaces of small rank, when the points of these spaces canonically encode multi-qubit observables. In particular, they present a complete taxonomy of polar subspaces of $W(2N - 1, 2)$ for $2 \leq N \leq 4$, whose rank is $N - 1$. The results required extensive computer-aided proofs.

The aim of [10] is to extend the concept of reducibility to hyperrings, which is a well-known concept in hypergroup theory. The authors define this extension in general hyperrings, where addition and multiplication are both multivalued operations, and then apply this novel definition to particular classes of hyperrings, e.g., hyperrings of formal series, hyperrings with P-hyperoperations, and complete hyperrings. The main results provide conditions under which these hyperrings are or are not reduced.

We are grateful to all authors who contributed their manuscripts. We would also like to thank all the reviewers for their valuable comments, which improved the quality of the submissions. Finally, we express our warmest gratitude to the MDPI Editor, Dr Syna Mu, who assisted us at every stage of the editorial process with excellent timeliness and professionalism.

The aim of this Special Issue was to attract high-quality, novel papers in the field of hypercompositional algebra. The response from the international scientific community and the number of manuscripts submitted for consideration exceeded our expectations. We would like to mention that the MDPI publishing house has already issued two books from Special Issues that have dealt explicitly with algebraic hyperstructures, namely that edited by C. Massouros in 2021, *Hypercompositional Algebra and Applications*, and *Symmetry in Classical and Fuzzy Algebraic Hypercompositional Structures* edited by I. Cristea in 2020. We hope that the papers included in all these collections will be influential for the scientific community and will motivate further research in this exciting, active, and engaging research area.

Author Contributions: Conceptualization, investigation, project administration, writing—original draft, D.F. (Dario Fasino) and D.F. (Domenico Freni); Writing—review and editing, D.F. (Dario Fasino). All authors have read and agreed to the published version of the manuscript.

Conflicts of Interest: The authors declare no conflict of interest.

References

1. Massouros, C.; Massouros, G. On the Borderline of Fields and Hyperfields. *Mathematics* **2023**, *11*, 1289. [CrossRef]
2. Kazancı, O.; Hoskova-Mayerova, S.; Davvaz, B. Algebraic Hyperstructure of Multi-Fuzzy Soft Sets Related to Polygroups. *Mathematics* **2022**, *10*, 2178. [CrossRef]
3. Innamorati, S.; Zuanni, F. A Combinatorial Characterization of $H(4, q^2)$. *Mathematics* **2022**, *10*, 1707. [CrossRef]
4. Chvalina, J.; Smetana, B.; Vyroubalová, J. Construction of an Infinite Cyclic Group Formed by Artificial Differential Neurons. *Mathematics* **2022**, *10*, 1571. [CrossRef]
5. De Salvo, M.; Fasino, D.; Freni, D.; Lo Faro, G. Commutativity and Completeness Degrees of Weakly Complete Hypergroups. *Mathematics* **2022**, *10*, 981. [CrossRef]
6. Lo Faro, G.; Milici, S.; Tripodi, A. Uniform (C_k, P_{k+1})-Factorizations of $K_n - I$ When k Is Even. *Mathematics* **2022**, *10*, 936. [CrossRef]
7. De Salvo, M.; Fasino, D.; Freni, D.; Lo Faro, G. *G*-Hypergroups: Hypergroups with a Group-Isomorphic Heart. *Mathematics* **2022**, *10*, 240. [CrossRef]
8. Nongmanee, A.; Leeratanavalee, S. *v*-Regular Ternary Menger Algebras and Left Translations of Ternary Menger Algebras. *Mathematics* **2021**, *9*, 2691. [CrossRef]
9. Saniga, M.; de Boutray, H.; Holweck, F.; Giorgetti, A. Taxonomy of Polar Subspaces of Multi-Qubit Symplectic Polar Spaces of Small Rank. *Mathematics* **2021**, *9*, 2272. [CrossRef]
10. Cristea, I.; Kankaraš, M. The Reducibility Concept in General Hyperrings. *Mathematics* **2021**, *9*, 2037. [CrossRef]

Disclaimer/Publisher's Note: The statements, opinions and data contained in all publications are solely those of the individual author(s) and contributor(s) and not of MDPI and/or the editor(s). MDPI and/or the editor(s) disclaim responsibility for any injury to people or property resulting from any ideas, methods, instructions or products referred to in the content.

Article

On the Borderline of Fields and Hyperfields

Christos G. Massouros [1,*] and Gerasimos G. Massouros [2,*]

[1] Core Department, Euripus Campus, National and Kapodistrian University of Athens, GR 34400 Euboia, Greece
[2] School of Social Sciences, Hellenic Open University, GR 26335 Patra, Greece
* Correspondence: chrmas@uoa.gr or ch.massouros@gmail.com (C.G.M.); germasouros@gmail.com (G.G.M.)

Abstract: The hyperfield came into being due to a mathematical necessity that appeared during the study of the valuation theory of the fields by M. Krasner, who also defined the hyperring, which is related to the hyperfield in the same way as the ring is related to the field. The fields and the hyperfields, as well as the rings and the hyperrings, border on each other, and it is natural that problems and open questions arise in their boundary areas. This paper presents such occasions, and more specifically, it introduces a new class of non-finite hyperfields and hyperrings that is not isomorphic to the existing ones; it also classifies finite hyperfields as quotient hyperfields or non-quotient hyperfields, and it gives answers to the question that was raised from the isomorphic problems of the hyperfields: when can the subtraction of a field F's multiplicative subgroup G from itself generate F? Furthermore, it presents a construction of a new class of hyperfields, and with regard to the problem of the isomorphism of its members to the quotient hyperfields, it raises a new question in field theory: when can the subtraction of a field F's multiplicative subgroup G from itself give all the elements of the field F, except the ones of its multiplicative subgroup G?

Keywords: fields; hyperfields; rings; hyperrings; multiplicative subgroups; hypergroups; canonical hypergroups

MSC: 12-11; 12K99; 12E20; 16Y20; 20N20

1. Introduction

The hypergroup is the very first hypercompositional structure that appeared in Algebra. It was introduced in 1934 by F. Marty while he was studying problems in non-commutative algebra, such as cosets determined by non-invariant subgroups. Unfortunately, Marty was killed in 1940, at the age of 29, during World War II, while he was serving in the French Air Force as a lieutenant and hence his mathematical heritage on hypergroups was only three papers [1–3]. Nevertheless, his ideas did not remain in France only. They spread quickly throughout Europe and across the pond. Already, by the end of the 1930s and in the 1940s, both in Europe and in the USA, important mathematicians such as M. Krasner [4–8], J. Kuntzmann [8–10], H. Wall [11], O. Ore [12–14], M. Dresher [13], E. J. Eaton [14,15], L. W. Griffiths [16], W. Prenowitz [17–19], and A.P. Dietzman [20], studied the general form of the hypergroup as well as other, special forms of this algebraic structure, resulting to its enrichment with additional axioms. The basic concept behind the hypergroup is the hypercomposition. A *hypercomposition* or *hyperoperation* over a non-empty set E is a mapping from the cartesian product $E \times E$ into the power set $P(E)$ of E. A *hypergroup* is a non-empty set E enriched with a hypercomposition "·", which satisfies the following two axioms:

(i) The axiom of *associativity*:

$$a \cdot (b \cdot c) = (a \cdot b) \cdot c, \quad \text{for all} \quad a, b, c \in E$$

(ii) The axiom of *reproductivity*:

$$a \cdot E = E \cdot a = E, \quad \text{for all} \quad a \in E$$

Papers [21,22] present in detail that the group is defined with exactly the same axioms. Namely, a *group* is a non-empty set E that is enriched with a composition (i.e., a mapping from the cartesian product $E \times E$ into the set E) that satisfies the axioms (i) and (ii).

If "·" is an internal composition on a set E and A, B are subsets of H, then $A \cdot B$ signifies the set $\{a \cdot b | (a,b) \in A \times B\}$, while if "·" is a hypercomposition then $A \cdot B$ is the union $\bigcup_{(a,b) \in A \times B} a \cdot b$. Ab and aB have the same meaning as $A\{b\}$ and $\{a\}B$ respectively. In general, the singleton $\{a\}$ is identified with its member a.

Theorem 1. *If either $A = \emptyset$ or $B = \emptyset$, then $AB = \emptyset$ and vice versa.*

Proof. The proof will be given with the use of symbolic logic. So, it must be proved that:

$$A \times B = \emptyset \Leftrightarrow (A = \emptyset) \vee (B = \emptyset)$$

or equivalently that:

$$A \times \emptyset = \emptyset = \emptyset \times B$$

To this end, we have the following equivalent statements:

$A \times B \neq \emptyset \Leftrightarrow$
$\Leftrightarrow \exists\, (s,t) \in A \times B$ (definition of the Empty Set)
$\Leftrightarrow \exists\, s \in A \wedge \exists\, t \in B$ (definition of the Cartesian Product)
$\Leftrightarrow A \neq \emptyset \wedge B \neq \emptyset$ (definition of the Empty Set)
$\Leftrightarrow \neg\, (A = \emptyset \vee B = \emptyset)$ (De Morgan's Laws)

Hence, by the law of contraposition:

$$(A = \emptyset) \vee (B = \emptyset) \Leftrightarrow A \times B = \emptyset \quad \square$$

Theorem 2. *Refs. [21,22] The result of the hypercomposition of any two elements in a hypergroup H is always non-void.*

Proof. Suppose that $ab = \emptyset$, for some $a, b \in H$. By the reproductive axiom, $aH = H$ and $bH = H$. Hence:

$$H = aH = a(bH) = (ab)H = \emptyset H = \emptyset$$

which is absurd. $\square$

The second hypercompositional structure that appeared in Algebra was the *hyper-field*. It was introduced by M. Krasner in 1956 for the purpose of defining a certain approximation of a complete valued field by a sequence of such fields [23]. Its construction is as follows:

Let K be a valued field and let $|\cdot|$ be its valuation. Let ρ be a real number such that $0 \leq \rho < 1$ and let π_ρ be the equivalence relation in K, which is defined as follows:

$$a \equiv 0 \Leftrightarrow 0 \equiv a, \text{ if } a = 0$$
$$b \equiv a \Leftrightarrow \left| \frac{b}{a} - 1 \right| \leq \rho \Leftrightarrow |b - a| \leq \rho |a|, \text{ if } a \neq 0$$

The classes mod π_ρ are circles $C_\xi = C(\xi, \rho|\xi|)$ of center $\xi \in K$ and radius $\rho|\xi|$. It turns out that the element-wise (pointwise) multiplication of any two classes (i.e., each element of one class with all elements of the other) is a class, while their element-wise sum is a union of classes. Certain properties apply in the set K/π_ρ of these equivalence classes. These properties were the defining axioms of the hyperfield. So, a *hyperfield* is an algebraic structure $(H,+,\cdot)$ where H is a non-empty set, "$\cdot$" is an internal composition on H, and "+" is a hypercomposition on H, which satisfies the axioms:

I.   *Multiplicative axiom*
     $H = H^*\cup\{0\}$, where $(H^*,\cdot)$ is a multiplicative group and 0 is a bilaterally absorbing element of H, i.e., $0x = x0 = 0$, for all $x \in H$

II. *Additive axioms*
     i.   associativity:
          $a \cdot (b \cdot c) = (a \cdot b) \cdot c$, for all $a,b,c \in H$
     ii. commutativity:
          $a \cdot b = b \cdot a$, for all $a,b \in H$
     iii. for every $a \in H$ there exists one and only one $a' \in H$ such that $0 \in a+a'$. a' is written $-a$ and called the opposite of a; moreover, instead of $a+(-b)$ we write $a-b$.
     iv. reversibility:
          if $a \in b+c$, then $c \in a-b$

III. *Distributive axiom*
     $a \cdot (b+c) = a \cdot b + a \cdot c$, $(b+c) \cdot a = b \cdot a + c \cdot a$, for all $a,b,c \in H$

By virtue of axioms II.iii and II.iv it holds that $a+0=a$ for all $a \in H$. Indeed, $0 \in a-a$; therefore, $a \in a+0$. Next, if for any $x \in H$, it is true that $x \in a+0$, then $0 \in a-x$, consequently, $x=a$. If the multiplicative axiom *I* is replaced by the axiom:

I'. H^* is a multiplicative semigroup having a bilaterally absorbing element 0,

then, a more general structure is obtained which is called *hyperring* [24].

It is easy to see that a non-empty set H enriched with the additive axioms II is a hypergroup. This special hypergroup was named *canonical hypergroup* by Jean Mittas, who studied it in depth and presented his research results through a multitude of papers, e.g., [25–28].

Apparently, fields and rings satisfy the above axioms, and hence, they are also called *trivial hyperfields* and *trivial hyperrings*, respectively. It is worth mentioning, though, that several algebraic properties which are valid for the rings and the fields are not transferred in the hyperrings and hyperfields. The following proposition is such an example.

Proposition 1. *Let P be a hyperring. Then,*

$$(a+b)(c+d) \subseteq ac + ac + ad + bd$$

for any $a,b,c,d \in P$.

Proof.

$$(a+b)(c+d) = \bigcup_{x \in a+b} x(c+d) = \bigcup_{x \in a+b}(xc+xd) \subseteq \bigcup_{z \in a+b} zc + \bigcup_{w \in a+b} wc =$$
$$= (a+b)c + (a+b)d = ac + ac + ad + bd$$

(see also [29]) □

Another example is the polynomials over a hyperring P. As in the case of rings, a polynomial p over a hyperring P is defined as an ordered set $(a_0, a_1, \ldots)$ where all the a_i's after a certain one (say after a_n) are zero. The elements a_i are the coefficients of p and n is the degree of p. If $p = (a_i)$ and $q = (b_j)$ then

$$p+q = \{(c_i) | c_i \in a_i + b_i\} \quad \text{and} \quad pq = \left\{(c_i) \Big| c_i \in \sum_{j+k=i} a_j b_k\right\}$$

The set of the polynomials over P is not a hyperring since its multiplicative part is not a semigroup, but it is a semihypergroup. This algebraic structure was named *superring* by J. Mittas [30,31]. In [32], R. Ameri, M. Eyvazi, and S. Hoskova-Mayerova proved that the distributive axiom is not valid for the multiplication of the polynomials over a hyperring. More precisely, it is indicated that the weak distributive axiom holds, i.e.,

$$r \cdot (p+q) \subseteq r \cdot p + r \cdot q, \ (p+q) \cdot r \subseteq p \cdot r + q \cdot r$$

Moreover, as it is proved in [33] (Theorem 16), the direct sum of hypermodules is not a hypermodule but a weak hypermodule in the sense that it satisfies the weak distributive axiom. Unfortunately, there are numerous published papers that contain incorrect results as they are based on the erroneous assumption that the direct sum of hypermodules is a hypermodule or that the distributivity holds for the multiplication of polynomials over a hyperring.

Krasner named the hyperfields, which he used for the approximation of the complete valued field, *residual hyperfields*. Next, while working on the question of how rich the class of the hyperrings and hyperfields is, he was led to the construction of a more general class of hyperrings and hyperfields, i.e., the class of the *quotient hyperfields* and the *quotient hyperrings* [24].

Note on the notation: In the following pages, in addition to the typical algebraic notations, we are using Krasner's notation for the complement and the difference [34]. So, we denote by $A \cdot\cdot B$ the set of elements that are in the set A but not in the set B. If K is a field or a hyperfield, then K^* denotes the set $K \cdot\cdot \{0\}$.

2. The Quotient Hyperfield/Hyperring

The construction of the quotient hyperfield or hyperring is based on a field or ring, respectively. Let F be a field and G a subgroup of F's multiplicative group F^*. Then, the multiplicative classes modulo G in F form a partition of F. Krasner observed that the product of two such classes, considered as subsets of F, is also a class modulo G, while their sum is a union of such classes. Next, he proved that the set F/G of the classes of this partition becomes a hyperfield if the multiplication and the addition are defined as follows:

$$xG \cdot yG = xyG$$
$$xG \dagger yG = \{(xp+yq)G \mid p,q \in G\}$$

for all $xG, yG \in F/G$.

Moreover, Krasner proved that if R is a ring and G is a normal subgroup of its multiplicative group, then the above construction gives a hyperring [24].

From the proof that R/G is a hyperring, it derives that the definition of the addition in R/G as well as the proof of the additive axioms do not require the normality of G. On the other hand, the definition of the multiplication and the proof of the multiplicative and distributive axioms require only that the equality:

$$xG \cdot yG = \{xg_1 yg_2 \mid g_1, g_2 \in G\} = \{xyg \mid g \in G\} = xyG$$

holds. But the validity of this equality is equivalent to the normality of G only when G is a subgroup of a group and not when G is a subgroup of a semigroup, which is the case when R is a ring. This was proved by Ch. Massouros [35] via an example, which is generalized below.

Example 1. *Let R_o be a unitary ring such that $2 \neq 0$. Let us consider the cartesian product $R = R_o{}^n$. R is enriched with the following addition and multiplication:*

$$(a_1, \ldots, a_n) + (b_1, \ldots, b_n) = (a_1+b_1, \ldots, a_n+b_n)$$
$$(a_1, \ldots, a_n)(b_1, \ldots, b_n) = (a_1(b_1+\ldots+b_n), \ldots, a_n(b_1+\ldots+b_n))$$

It is well known that $(R, +)$ is a group. Next, observe that the multiplication is not commutative. Indeed:

$$(a_1, \ldots, a_n)(b_1, \ldots, b_n) = (a_1(b_1 + \ldots + b_n), \ldots, a_n(b_1 + \ldots + b_n))$$

while:

$$(b_1, \ldots, b_n)(a_1, \ldots, a_n) = (b_1(a_1 + \ldots + a_n), \ldots, b_n(a_1 + \ldots + a_n))$$

On the contrary, the multiplication is associative:

$$\begin{aligned}
&(a_1, \ldots, a_n)[(b_1, \ldots, b_n)(c_1, \ldots, c_n)] = \\
&= (a_1, \ldots, a_n)(b_1(c_1 + \ldots + c_n), \ldots, b_n(c_1 + \ldots + c_n)) = \\
&= \begin{pmatrix} a_1(b_1(c_1 + \ldots + c_n) + \ldots + b_n(c_1 + \ldots + c_n)), \ldots \\ \ldots, a_n(b_1(c_1 + \ldots + c_n) + \ldots + b_n(c_1 + \ldots + c_n)) \end{pmatrix} = \\
&= (a_1(b_1 + \ldots + b_n)(c_1 + \ldots + c_n), \ldots, a_n(b_1 + \ldots + b_n)(c_1 + \ldots + c_n)) = \\
&= (a_1(b_1 + \ldots + b_n), \ldots, a_n(b_1 + \ldots + b_n))(c_1, \ldots, c_n) = \\
&= [(a_1, \ldots, a_n)(b_1, \ldots, b_n)](c_1, \ldots, c_n)
\end{aligned}$$

and distributive:

$$\begin{aligned}
&(a_1, \ldots, a_n)[(b_1, \ldots, b_n) + (c_1, \ldots, c_n)] = \\
&= (a_1, \ldots, a_n)(b_1 + c_1, \ldots, b_n + c_n) = \\
&= (a_1(b_1 + c_1 + \ldots b_n + c_n), \ldots, a_n(b_1 + c_1 + \ldots b_n + c_n)) = \\
&= (a_1(b_1 + \ldots + b_n) + a_1(c_1 + \ldots + c_n), \ldots, a_n(b_1 + \ldots + b_n) + a_n(c_1 + \ldots + c_n)) = \\
&= (a_1(b_1 + \ldots + b_n), \ldots, a_n(b_1 + \ldots + b_n)) + (a_1(c_1 + \ldots + c_n), \ldots, a_n(c_1 + \ldots + c_n)) = \\
&= (a_1, \ldots, a_n)(b_1, \ldots, b_n) + (a_1, \ldots, a_n)(c_1, \ldots, c_n)
\end{aligned}$$

Thus $(R, +, \cdot)$ is a ring. A non-zero element $(a_1, \ldots, a_n)$ of R is idempotent if $a_1 + \ldots + a_n = 1$. Indeed:

$$(a_1, \ldots, a_n)^2 = (a_1(a_1 + \ldots + a_n), \ldots, a_n(a_1 + \ldots + a_n)) = (a_1 \cdot 1, \ldots, a_n \cdot 1) = (a_1, \ldots, a_n)$$

Thus, the elements $e_1 = (1, \ldots, 0), \ldots, e_n = (0, \ldots, 1)$ are idempotent. Moreover, the opposite of the $e_i = (0, \ldots, 1, \ldots, 0)$, $i = 1, \ldots, n$ is $-e_i = (0, \ldots, -1, \ldots, 0)$, which is different from the e_i because $2e_i = (0, \ldots, 2, \ldots, 0) \neq (0, \ldots, 0) = 0$. Since $(-e_i)^2 = e_i^2 = e_i$, the 2-element sets $G_i = \{-e_i, e_i\}$, $i = 1, \ldots, n$ are multiplicative subgroups of R. Next, if $a = (a_1, \ldots, a_n)$ is an element in R, then:

$$\begin{aligned}
aG_i &= (a_1, \ldots, a_n)\{-e_i, e_i\} = \\
&= \{(a_1, \ldots, a_n)(0, \ldots, -1, \ldots, 0), (a_1, \ldots, a_n)(0, \ldots, -1, \ldots, 0)\} = \\
&= \{(-a_1, \ldots, -a_n), (a_1, \ldots, a_n)\} = \{-a, a\}
\end{aligned}$$

while

$$\begin{aligned}
G_i a &= \{-e_i, e_i\}(a_1, \ldots, a_n) = \\
&= \{(0, \ldots, -1, \ldots, 0)(a_1, \ldots, a_n), (0, \ldots, -1, \ldots, 0)(a_1, \ldots, a_n)\} = \\
&= \{(0, \ldots, -a_1 - \ldots - a_n, \ldots, 0), (0, \ldots, a_1 + \ldots + a_n, \ldots, 0)\}
\end{aligned}$$

Consequently, the multiplicative subgroups G_i, $i = 1, \ldots, n$ are not normal. Nevertheless, they satisfy the condition:

$$(aG_i)(bG_i) = abG_i$$

Indeed,

$$(aG_i)(bG_i) = \{-a, a\}\{-b, b\} = \{(-a)(-b), (-a)b, a(-b), ab\} = \{-ab, ab\} = abG_i$$

Therefore, the quotients R/G_i, $i = 1, \ldots, n$ are hyperrings. Observe that G_i is a right neutral element for multiplication in R/G_i, but it is not a left one as well. In contrast, when the quotient hyperring is constructed via a normal subgroup G of the ring's multiplicative semigroup, then G is a bilateral neutral element for the multiplication in the quotient hyperring.

The aforementioned hyperrings, although they are not quotient hyperrings of a ring by a normal subgroup of its multiplicative semigroup, they are still embeddable in such quotient hyperrings [35,36].

A large number of papers has been published on the hyperfields and hyperrings, starting from the pioneer work of J. Mittas [37–44] and continuing with a plenitude of researchers, such as Ch. Massouros [29,35,45–51], A. Nakassis [36], G. Massouros [50–54], R. Rota [55,56], S. Jančic-Rašović [57–59], I. Cristea [58–64], H. Bordbar [59–61], M. Kankaraš [62], V. Vahedi et al. [63–65], M. Jafarpour et al. [63–66], A. Connes and C. Consani [67,68], O. Viro [69,70], R. Ameri, M. Eyvazi and S. Hoskova-Mayerova [32,71], M. Baker et al. [72–74], J. Jun [75], O. Lorscheid [76], Z. Liu [77], H. Shojaei and F. Dasino [78], K. Das et al. [79], K. Roberto et al. [80–82], P. Corsini [83], B. Davvaz, V. Leoreanu-Fotea [84], C. Yatras [85–87], S. Atamewoue Tsafack, S. Wen, B.O. Onasanya, et al. [88], A. Linz, and P. Touchard [89], S. Creech [90], T. Gunn [91], etc. In the recent years, several hyperfields which belong to the class of quotient hyperfields have appeared, a fact that is not mentioned or even noticed, while, sometimes, an unsuccessful terminology is used for them. More specifically:

(a) In the papers [67,68] by A. Connes and C. Consani and afterward in many subsequent papers (e.g. [69,72,75,76]), the name «*Krasner's hyperfield*» is used for the hyperfield, which is constructed over the set {0, 1} via the hypercomposition:

$$0 + 0 = 0, \ 0 + 1 = 1 + 0 = 1, \ 1 + 1 = \{0, 1\}$$

Oleg Viro, in his paper [69], justifiably states about this hyperfield: «*To the best of my knowledge, K did not appear in Krasner's papers*». His remark is absolutely correct. Actually, the above is a special case of a quotient hyperfield, and in this sense, it belongs to a special class of Krasner hyperfields. Indeed, for a field F and its multiplicative subgroup F*, the quotient hyperfield F/F* = {0,F*} is isomorphic to the hyperfield considered by A. Connes and C. Consani. More precisely, in the case of hyperfields with cardinality 2, the following theorem holds:

Theorem 3. *The two-element non-trivial hyperfield is isomorphic to a quotient hyperfield.*

(b) In the papers [67,68] by A. Connes and C. Consani, a hyperfield is considered over the set {−1, 0, 1} with the following hypercomposition:

$$0 + 0 = 0, \ 0 + 1 = 1 + 0 = 1, \ 1 + 1 = 1, \ -1 - 1 = -1, \ 1 - 1 = -1 + 1 = \{-1, 0, 1\}$$

This hyperfield is now called «*sign hyperfield*» by some authors. Nevertheless, this hyperfield is a quotient hyperfield as well. Indeed, let F be an ordered field and let F^+ be its positive cone. Then the quotient hyperfield F/F^+ = {$-F^+$,0,F^+} is isomorphic to the sign hyperfield.

(c) The «*phase hyperfield*» that appeared recently in the bibliography (see, e.g., [69,72]) is just the quotient hyperfield $\mathbb{C}/\mathbb{R}^+$, where $\mathbb{C}$ is the field of complex numbers and $\mathbb{R}^+$ is the set of positive real numbers. The elements of this hyperfield are the rays of the complex field with origin at the point (0,0). The sum of two elements $z\mathbb{R}^+$, $w\mathbb{R}^+$ of $\mathbb{C}/\mathbb{R}^+$ is the set $\{(zp + wq)\mathbb{R}^+ \mid p, q \in \mathbb{R}^+\}$. When $z\mathbb{R}^+ \neq w\mathbb{R}^+$, this sum consists of all the interior rays $x\mathbb{R}^+$ of the convex angle which is created from $z\mathbb{R}^+$ and $w\mathbb{R}^+$, while if $w\mathbb{R}^+ = -z\mathbb{R}^+$ then, the sum of the two opposite rays $z\mathbb{R}^+$, $-z\mathbb{R}^+$ is the set $\{0, -z\mathbb{R}^+, z\mathbb{R}^+\}$. This hyperfield is presented in detail in [46].

Note on the notation: In the following theorems, new hyperfields are constructed via other hyperfields or fields. To avoid any confusion between the new and the old hypercomposition we use + as the sign for the initial addition and symbols such as $\dotplus$, $\hat{+}$, $\tilde{+}$, etc., to denote the new one.

Theorem 4. Let $(F, +, \cdot)$ be a field. If we define the hypercomposition $\dotplus$ on F as follows:

$$\begin{aligned}
x \dotplus y &= \{\, x, y, x \dotplus y \,\}, & \text{if } y \neq -x \text{ and } x, y \neq 0, \\
x \dotplus (-x) &= F, & \text{for all } x \in F_*, \\
x \dotplus 0 &= 0 \dotplus x = x, & \text{for all } x \in F,
\end{aligned}$$

then $(F, \dotplus, \cdot)$ is a hyperfield isomorphic to a quotient hyperfield.

Proof. From the verification of the axioms, it follows that $(F, \dotplus, \cdot)$ is a hyperfield (see also [46]). Next, since $(F, +, \cdot)$ is a field, the polynomial ring $F[x]$ is an integral domain, and so the field $F(x)$ of the rational functions over F can be defined. We can then assume that in all rational functions, the coefficient of the highest power of the denominator's polynomial is 1 since, if this is not the case, we can make it via the appropriate division. Now, let G be the set

$$G = \{\, \pi(x) \in F(x) \mid a_m = 1 \,\}$$

where a_m is the coefficient of the numerator's highest power. G is a multiplicative subgroup of the multiplicative group of $F(x)$. Therefore, we can consider the quotient hyperfield $(F(x)/G, \dotplus, \cdot)$. The function $\varphi : F \to F(x)/G$, with $\varphi(a) = aG$, for each $a \in F$, is one-to-one, since if a, b are distinct elements in F, then

$$aG = \{\pi(x) \in F(x) \text{ with } a_m = a\} \quad \text{and} \quad bG = \{\pi(x) \in F(x) \text{ with } a_m = b\}$$

are distinct elements of $F(x)/G$. Moreover, φ is a surjection since every element aG of $F(x)/G$ is the image of the corresponding element a of F. Next, let

$$\pi_1(x) = \dfrac{\sum_{i=1}^{k} a_i t^{a_i}}{\sum_{j=1}^{l} b_j t^{b_j}},\ a_k = 1,\ b_l = 1 \quad \text{and} \quad \pi_2(x) = \dfrac{\sum_{i=1}^{n} a_i' t^{a_i}}{\sum_{j=1}^{l} b_j t^{b_j}},\ a_n' = 1,\ b_l = 1$$

be two elements in G. We assume that $\pi_1(x)$ and $\pi_2(x)$ have the same denominator because if they are rational expressions with unlike denominators, we can convert them into rational expressions with common denominators. Let us consider the sum:

$$aG \dotplus bG = \{\, [a\pi_1(x) + b\pi_2(x)]G \mid \pi_1(x), \pi_2(x) \in G \,\} \text{ with } bG \neq -aG$$

Then:

(i) If $k > n$, then the coefficient of the highest power of the polynomial $a\pi_1(x) + b\pi_2(x)$ is a, thus $a\pi_1(x) + b\pi_2(x) \in aG$, and therefore $aG \in aG \dotplus bG$. On the other hand, the coefficient of the highest power of the polynomial $b\pi_1(x) + a\pi_2(x)$ is b, thus $b\pi_1(x) + a\pi_2(x) \in bG$ and therefore $bG \in aG \dotplus bG$.

(ii) If $k = n$, then the coefficient of the highest power of the polynomial $a\pi_1(x) + b\pi_2(x)$ is $a + b$, thus $a\pi_1(x) + b\pi_2(x) \in (a+b)G$, and therefore $(a+b)G \in aG \dotplus bG$.

Consequently, φ is an isomorphism, and thus the Theorem. □

It needs to be clarified here that the definition of the hypercomposition for the non-opposite elements, in combination with the axioms of the hyperfield, allows no different way for the definition of the hypercomposition of two opposite elements. More precisely, we have the following two Propositions (for their proofs see [46]):

Proposition 2. In a hyperfield K, with $\operatorname{card} K > 3$, the sum $x + y$ of any two elements $x, y \neq 0$ contains these two elements if and only if the difference $x - x$ equals K for all $x \neq 0$.

Proposition 3. *In a hyperfield K, with cardK > 3, the sum x+y of any two non-opposite elements x,y≠0 does not contain the participating elements if and only if the difference x−x equals to {−x,0,x}, for all x≠0.*

The hypercomposition that appears in Proposition 2 is called *closed* (or *containing*; sometimes it is also called *extensive* [92]), while the hypercomposition that appears in Proposition 3 is called *open* [93]. In particular, a hypercomposition in a hypergroupoid (E,+) is called *right closed* if $a \in b+a$ for all $a,b \in E$, *left closed* if $a \in a+b$ for all $a,b \in E$, and *closed* if $\{a,b\} \subseteq a+b$ for all $a,b \in E$. A hypercomposition is called *right open* if $a \notin b+a$ for all $a,b \in E$ with $b \neq a$ while it is called *left open* if $a \notin a+b$ for all $a,b \in E$ with $b \neq a$. A hypercomposition is called *open* if it is both right and left open. Right closed hypercompositions are left open, and left closed compositions are right open. If the commutativity is valid, then the right/left closed and the closed (resp. the right/left open and the open) hypercompositions coincide.

The following Theorem presents the construction of a hyperfield that is equipped with a closed hypercomposition, and therefore, the definition of the sum of two opposite elements in it is restricted by the provisions of Proposition 2.

Theorem 5. *Ref. [46] Let* $(H, +, \cdot)$ *be a hyperfield. If we define a new hypercomposition «$\dotplus$» on H as follows:*

$$\begin{array}{ll}
x \dotplus y = \{x,y\} \cup x + y, & \text{for all } x,y \in H^*, \text{ with } y \neq -x, \\
x \dotplus (-x) = H, & \text{for all } x \in H^*, \\
x \dotplus 0 = 0 \dotplus x = x, & \text{for all } x \in H,
\end{array}$$

then, $(H, \dotplus, \cdot)$ *is a hyperfield and when* $(H, +, \cdot)$ *is a quotient hyperfield, then* $(H, \dotplus, \cdot)$ *is also a quotient hyperfield.*

The proof of this theorem can be found in [46].

The hyperfield, which is constructed by the above Theorems 4 and 5, will be termed *augmented hyperfield* because the composition or the hypercomposition is augmented to contain the two addends. The augmented hyperfield of a field or a hyperfield F is denoted by [F]. The augmented hyperfield's distinctive feature is that it always provides the information (the elements) that produced the result. As shown in the following sections, different hyperfields can have the same augmented hyperfield.

Theorems 4 and 5 ensure that the augmented hyperfield of a field or a quotient hyperfield is a quotient hyperfield, but it is not known yet whether all the members of a family of hyperfields whose augmented hyperfield is a quotient hyperfield are quotient hyperfields.

In the following construction Theorems, Proposition 2 is used to define the sum of two opposite elements:

Theorem 6. *Ref [46] Let G be a non-unitary multiplicative group and let* $(H^*, \cdot)$ *be its direct product with the multiplicative group* $\{-1,1\}$. *Consider the set* $H = H^* \cup \{0\}$, *where 0 is a bilaterally absorbing element in H, i.e.,* $0w = w0 = 0$, *for all* $w \in H$. *The following hypercomposition is introduced on H:*

$$\begin{array}{ll}
(x,i) \dotplus (y,j) = \{(x,i), (y,j)\}, & \text{if } (y,j) \neq (x,-i), \\
(x,i) \dotplus (x,-i) = H, & \text{for all } (x,i) \in H^*, \\
(x,i) \dotplus 0 = 0 \dotplus (x,i) = (x,i) \text{ and } 0 \dotplus 0 = 0 & \text{for all } (x,i) \in H^*.
\end{array}$$

Then, $(H, \dotplus, \cdot)$ *is a hyperfield.*

Theorem 7. Ref. [46] Let $(G, \cdot)$ be a non-unitary multiplicative group and 0 a bilaterally absorbing element. If we define a hypercomposition $\hat{+}$ on $H = G \cup \{0\}$ as follows:

$$\begin{aligned} x \hat{+} y &= \{x, y\}, & \text{for all } x, y \in G, \text{ with } y \neq x, \\ x \hat{+} x &= H, & \text{for all } x, y \in G, \\ x \hat{+} 0 &= 0 \hat{+} x = x, & \text{for all } x \in H, \end{aligned}$$

then, the triplet $(H, \hat{+}, \cdot)$ becomes a hyperfield.

In [46], it is proved that the above Theorem constructs a family of hyperfields, which contains quotient hyperfields, but it is not known yet whether this family contains non-quotient hyperfields as well.

Theorem 8. Let Q be a multiplicative group that has more than two elements and let 0 be a multiplicatively bilaterally absorbing element. If we define a hypercomposition $\hat{+}$ on $H = Q \cup \{0\}$ as follows:

$$\begin{aligned} x \hat{+} y &= Q, & \text{for all } x, y \in Q, \text{ with } y \neq x, \\ x \hat{+} x &= H \cdot \cdot \{x\}, & \text{for all } x \in Q, \\ x \hat{+} 0 &= 0 \hat{+} x = x, & \text{for all } x \in H, \end{aligned}$$

then, the triplet $H(Q) = (Q \cup \{0\}, \hat{+}, \cdot)$ is a hyperfield.

The following example proves the existence of quotient hyperfields which are constructed according to the above Theorem.

Example 2. (i) Consider the field $\mathbb{Z}_{41}$. This field's multiplicative subgroup of order 4

$$G = \{1, 4, 10, 16, 18, 23, 25, 31, 37, 40\}$$

has the property $G - G = G + G = \mathbb{Z}_{41} \cdot \cdot G$ and $xG + yG = \mathbb{Z}_{41} \cdot \cdot \{0\}$ when $x \neq y$ with $x, y \in \left\{ 3^k \mid k = 0, 1, 2, 3 \right\}$. Therefore, the quotient hyperfield

$$\mathbb{Z}_{41}/G = \left\{0,\ G,\ 3G,\ 3^2 G,\ 3^3 G\right\}$$

is of the type of hyperfields of Theorem 8.

(ii) Consider the field $\mathbb{Z}_{71}$. Its multiplicative subgroup of order 5 is

$$G = \{1, 20, 23, 26, 30, 32, 34, 37, 39, 41, 45, 48, 51, 70\}$$

and it has the property $G - G = G + G = \mathbb{Z}_{71} \cdot \cdot G$ and $xG + yG = \mathbb{Z}_{71} \cdot \cdot \{0\}$ when $x \neq y$ with $x, y \in \left\{ 2^k \mid k = 0, 1, 2, 3, 4 \right\}$. Therefore, the quotient hyperfield

$$\mathbb{Z}_{71}/G = \left\{0,\ G,\ 2G,\ 2^2 G,\ 2^3 G,\ 2^4 G\right\}$$

is of the type of hyperfields of Theorem 8.

(iii) Consider the field $\mathbb{Z}_{101}$. This field's multiplicative subgroup of order 5

$$G = \{1, 6, 10, 14, 17, 32, 36, 39, 41, 44, 57, 60, 62, 65, 69, 84, 87, 91, 95, 100\}$$

has the property $G - G = G + G = \mathbb{Z}_{101} \cdot \cdot G$ and $xG + yG = \mathbb{Z}_{101} \cdot \cdot \{0\}$ when $x \neq y$ with $x, y \in \left\{ 2^k \mid k = 0, 1, 2, 3, 4 \right\}$. Therefore, the quotient hyperfield

$$\mathbb{Z}_{101}/G = \left\{0,\ G,\ 2G,\ 2^2 G,\ 2^3 G,\ 2^4 G\right\}$$

is of the type of hyperfields of Theorem 8.

The hyperfields of Theorems 6 and 7 are called *b-hyperfields* due to the binary result of the hypercomposition, which consists of the two addends when they are different elements. Moreover, the hyperfields of Theorems 4, 5, 6, and 7 were termed *monogenic (monogène)* because they are generated by just a single element of the hyperfield [46]. Additionally, the hyperfield which is constructed by Theorem 8 is monogenic (monogène) because $H = x \mp x \mp x \mp x$. The *monogenic (monogène)* canonical hypergroup was introduced and studied in depth by J. Mittas [26]. The set of the canonical subhypergroups of a canonical hypergroup H is a complete lattice, thus for a given subset X of H there always exists the least (in the sense of inclusion) canonical subhypergroup $\overline{X}$ of H which contains X. Now, if X is the singleton $\{x\}$, then the canonical subhypergroup that is generated from it, is called *monogenic (monogène)*. If $H = \overline{\{x\}}$, then H itself is called *monogenic (monogène)*. The study of the monogenic (monogène) hypergroups led to the definition of the *order* of a canonical hypergroup's elements [26] and sequentially to the order of the elements of a hyperfield [41]. Since:

$$mx + nx = \begin{cases} (m+n)x, & \text{if } mn > 0 \\ (m+n)x + \min\{|m|, |n|\}(x-x), & \text{if } mn < 0 \end{cases}$$

for the monogenic (monogène) hypergroup it holds:

$$\overline{\{x\}} = mx + n(x - x), \quad m, n \in \mathbb{Z}$$

and as it is true that $-(x-x)=x-x$, we can assume that $(m,n) \in \mathbb{Z} \times \mathbb{N}$ instead of $\mathbb{Z} \times \mathbb{Z}$. Thus, two mutually exclusive cases can appear:

(I) For every $(m,n) \in \mathbb{Z} \times \mathbb{N}$, with $m \neq 0$, $0 \notin mx+n(x-x)$, in which case x, as well as $\overline{\{x\}}$ are said to be of *infinite order* denoted by $\omega(x)=+\infty$.

Proposition 4. *Ref. [26]* $\omega(x)=+\infty$ *if and only if* $m'x \cap m''x = \emptyset$, *for every* $m', m'' \in \mathbb{Z}$ *with* $m' \neq m''$.

(II) There exists $(m,n) \in \mathbb{Z} \times \mathbb{N}$, with $m \neq 0$, such that $0 \in mx+n(x-x)$. In the following, p will denote the minimum positive integer for which there exists $n \in \mathbb{N}$, such that $0 \in px+n(x-x)$.

Proposition 5. *Ref. [26] For a given* $m \in \mathbb{Z}$ *there exists* $n \in \mathbb{N}$ *such that* $0 \in mx+n(x-x)$, *if and only if m is divided by p.*

For $m=kp$, $k \in \mathbb{Z}^*$, let $q(k)$ be the minimum nonnegative integer such that $0 \in kpx+q(k)(x-x)$. Then q is a function from $\mathbb{Z}$ to $\mathbb{N}$. Mittas called the pair $\omega(x)=(p,q)$ order of both x and $\overline{\{x\}}$. Also, he named p the *principal order* of x and q the *associative order* of x [26,41]. Therefore, the order of all the elements of the hyperfields which are constructed by the Theorems 4, 5 and 6 is (1,1) because $0 \in x+(x-x)$, while the order of the elements of the monogenic (monogène) hyperfield of Theorem 7 is (2,0), since $0 \in x+x=2x+0(x-x)$ and of the hyperfield of Theorem 8 is (4,0), since $0 \in x+x+x+x=4x+0(x-x)$.

These definitions were later used in other hypercompositional structures, such as the fortified transposition hypergroups [22], the hyperringoids [52], the M-polysymmetrical hyperrings [86] etc.

3. The Non-Quotient Hyperfields/Hyperrings

M. Krasner realized that the existence of non-quotient hyperfields and hyperrings was an essential question for the self-sufficiency of the theory of hyperfields and hyperrings vis-à-vis that of fields and rings, since if all hyperrings and hyperfields could be isomorphically embedded into the quotient hyperrings, then several conclusions of their theory could have been obtained in a direct and straightforward way, through the use of the ring, field and modules theories, instead of developing new techniques and proof methodologies. Therefore, in his paper [24], he raised the relevant question. The answer to this question led to the construction of two classes of hyperfields and hyperrings, which contain elements that are not isomorphic to the quotient ones. The following Theorems 9 and 10 which were proved by Ch. Massouros, refer to hyperfields with closed hypercompositions and they prove the existence of finite and infinite non-quotient hyperfields. The subsequent Theorem 11 was proved by A. Nakassis, it is on hyperfields with open hypercompositions and it reveals another class of finite non-quotient hyperfields. Moreover, Theorem 12 gives a new class of infinite non-quotient hyperfields which do not belong to the previous two classes, and Theorem 13 uncovers a new class of infinite non-quotient hyperrings.

Theorem 9. *Ref. [35] Let Θ be a multiplicative group that has more than two elements and let $(K^*, \cdot)$ be its direct product with the multiplicative group $\{-1, 1\}$. Consider the set $K = K^* \cup \{0\}$, where 0 is a bilaterally absorbing element in K, i.e., $0w = w0 = 0$, for all $w \in K$. The following hypercomposition is introduced on K:*

$$
\begin{aligned}
(x,i) + (y,j) &= \{(x,i), (y,j), (x,-i), (y,-j)\}, \quad \text{if } (y,j) \neq (x,i), (x,-i) \\
(x,i) + (x,i) &= K \cdot \cdot \{(x,i), (x,-i), 0\} \\
(x,i) + (x,-i) &= K \cdot \cdot \{(x,i), (x,-i)\} \\
(x,i) + 0 &= 0 + (x,i) = (x,i) \text{ and } 0 + 0 = 0
\end{aligned}
$$

Then, the triplet $K(\Theta) = (K, +, \cdot)$ is a hyperfield that does not belong to the class of quotient hyperfields when Θ is a periodic group.

For the proof of the above Theorem, see [35].

Theorem 10. *Refs. [29,47] Let Θ be a multiplicative group which has more than two elements and let 0 be a multiplicatively bilaterally absorbing element. If we define a hypercomposition $+$ on $H = \Theta \cup \{0\}$ as follows:*

$$
\begin{aligned}
x + y &= \{x, y\}, &&\text{for all } x, y \in \Theta, \text{ with } y \neq x, \\
x + x &= H \cdot \cdot \{x\}, &&\text{for all } x \in \Theta, \\
x + 0 &= 0 + x = x, &&\text{for all } x \in H,
\end{aligned}
$$

then, the triplet $H(\Theta) = (\Theta \cup \{0\}, +, \cdot)$ is a hyperfield which is not isomorphic to a quotient hyperfield when Θ is a periodic group.

For the proof of the above Theorem, see [29,47].

Proposition 6. *Ref. [36] Let $(T, \cdot)$ be a multiplicative group of order m, with $m > 3$. Additionally, let $H = T \cup \{0\}$ where 0 is a multiplicatively absorbing element. If H is equipped with the hypercomposition:*

$$
\begin{aligned}
x + y &= H \cdot \cdot \{0, x, y\} &&\text{for all } x, y \in T, \text{ with } y \neq x, \\
x + x &= \{0, x\}, &&\text{for all } x \in T, \\
x + 0 &= 0 + x = x, &&\text{for all } x \in H,
\end{aligned}
$$

then, $H(T) = (T \cup \{0\}, +, \cdot)$ is a hyperfield.

It is worth noting here that the elements of the above hyperfield are self-opposite, and since the hypercomposition is open, Proposition 3 imposes the definition of the sum of the self-opposite elements so that $H(T)$ fulfills the axioms of the hyperfield.

Theorem 11. *Ref. [36] If T is a finite multiplicative group of m, m>3 elements and if the hyperfield H(T) is isomorphic to a quotient hyperfield F/Q, then Q∪{0} is a field of m−1 elements while F is a field of (m−1)² elements.*

Obviously, the cardinality of *T* can be chosen in such a way that *H(T)* cannot be isomorphic to a quotient hyperfield.

For the proof of Theorem 11, the following important counting lemma was introduced and used by A. Nakassis.

Lemma 1. *Ref. [36] Let H be a hyperfield equipped with a hypercomposition such that the differences x−x, x∈H have only 0 in common. If H is isomorphic to a quotient hyperfield F/Q, then the cardinality of the sum of any two non-opposite elements is equal to the cardinality of Q.*

Proof. Suppose that *H* is a hyperfield equipped with a hypercomposition such that $(x-x) \cap (y-y) = \{0\}$ for all $x,y \in H$ with $x \neq y$. Assume that *H* is isomorphic to a quotient hyperfield *F/Q*. Let a',b' with $a' \neq b'$ be two elements in *H* and let *aQ, bQ* be their homomorphic images in *F/Q*. Then $a'+b'$ has the same cardinality with $aQ + bQ = \{(a+bq)Q \mid q \in Q\}$. Next, if $(a+bq)Q = (a+bp)Q$, then

$$a + bq = (a + bp)r \Leftrightarrow a - ar = bq - bpr \Rightarrow (aQ - aQ) \cap (bQ - bQ) \neq \varnothing$$

However, since the equality $(aQ - aQ) \cap (bQ - bQ) = \{0\}$ is valid, it follows that $a-ar=0$. Therefore $r=1$ and consequently $bq-bp=0$ or equivalently $q=p$. Hence $card(aQ+bQ)=cardQ$ and so the lemma. □

A direct consequence of Nakassis' lemma is that if a hyperfield *H* is isomorphic to a quotient hyperfield and the differences $x-x$, $x \in H$ have only 0 in common, then the sums of the non-opposite elements have the same cardinality. This result is very useful to the classification of hyperfields which is presented in Section 5.

In the following, the class of non-quotient hyperrings and hyperfields will be enriched with another family of such structures.

J. Mittas in the first section of [41], constructed the following hyperfield, which is called *tropical hyperfield* nowadays (see, e.g., [69,70,72,75,76]) because it is proved to be a suitable and effective algebraic tool for the study of tropical geometry:

Example 3. *Ref. [41] Let (E,·) be a totally ordered multiplicative semigroup, having a minimum element 0, which is bilaterally absorbing with regard to the multiplication. The following hypercomposition is defined on E:*

$$x \hat{+} y = \begin{cases} \max\{x,y\} & \text{if } x \neq y \\ \{z \in E \mid z \leq x\} & \text{if } x = y \end{cases}$$

Then $(E, \hat{+}, \cdot)$ is a hyperring. If $E \cdot \cdot \{0\}$ is a multiplicative group, then $(E, \hat{+}, \cdot)$ is a hyperfield.

A slight modification of the definition of the above hypercomposition, when *x* is equal to *y*, gives the following Theorem:

Theorem 12. *Let (E,·) be a totally ordered multiplicative semigroup, having a minimum element 0, which is bilaterally absorbing with regard to the multiplication. The following hypercomposition is defined on E:*

$$x \check{+} y = \begin{cases} \max\{x,y\} & \text{if } x \neq y \\ \{z \in E \mid z < x\} & \text{if } x = y \end{cases}$$

Then $(E, \check{+}, \cdot)$ is a non-quotient hyperring. If $E \cdot \cdot \{0\}$ is a multiplicative group, then $(E, \check{+}, \cdot)$ is a non-quotient hyperfield.

Proof. The verification of the axioms of the hyperring and the hyperfield proves that $(E, +, \cdot)$ is such a structure. Next suppose that $(E, +, \cdot)$ is isomorphic to a quotient hyperring $(R/Q, +, \cdot)$. As $x \notin x+x$, for all $x \in E$ and because $2 = 1+1 \in Q+Q$, it follows that $2 \notin Q$. Hence $2Q$ is a class different from Q which belongs to $Q + Q$, therefore $2Q < Q$ and so $2Q + Q = Q$. Next:

$3 = 2+1 \in 2Q+Q = Q$
$4 = 3+1 \in Q+Q$, thus $4 \notin Q$
$4 = 2+2 \in 2Q+2Q$, thus $4 \notin 2Q$

Consequently $4Q$ is a new class different from Q and $2Q$ and furthermore, since it belongs to $Q + Q$, it holds that $4Q < Q$. Therefore:

$4Q + Q = Q$
$5 = 4+1 \in 4Q+Q = Q$
$6 = 2 \cdot 3 \in 2Q \cdot Q = 2Q$

Hence, for 7, we have:

on the one hand $7 = 6+1 \in 2Q+Q = Q$
while, on the other hand, $7 = 4+3 \in Q+Q$, subsequently $7 \notin Q$.

This is a contradiction and therefore $(E, +, \cdot)$ does not belong to the class of quotient hyperrings or hyperfields. □

Note that Theorem 12's hypercomposition is neither open nor closed. Also, note that the above Theorem enriches the class of non-quotient hyperrings with many new members in addition to the ones it is constructing. Indeed, [35] gives a method of constructing non-quotient hyperrings when at least one non-quotient hyperfield is known. In particular, the following Theorem is valid:

Theorem 13. *Ref. [35] The direct sum of the hyperrings S_i, $i \in I$ is not isomorphic to a sub-hyperring of a quotient hyperring if at least one of the S_i is not a quotient hyperfield.*

Thus, for example, if $\mathbb{R}$ is the field of the real numbers and $\check{\mathbb{R}}_+$ the hyperfield of Theorem 12 which is constructed over the set of the non-negative real numbers, then $\mathbb{R} \oplus \check{\mathbb{R}}_+$ is a non-quotient hyperring.

Another class of non-quotient hyperrings was constructed by Nakassis in [36]. Nakassis' hyperrings are endowed with open hypercompositions.

4. Problems in the Theory of Fields that arose from a Question in the Theory of Hyperfields

The constructions of specific monogenic (monogène) hyperfields in the early 1980s, led directly to the hitherto open question of whether these constructions can produce non-quotient hyperfields as well [35,49,94]. It should be noted that to date they have given several hyperfields all of which are quotient [46,47,49]. Theorem 4 gives a family of such monogenic quotient hyperfields. If $x-x=H$, $x \in H^*$ is valid in a monogenic (monogène) hyperfield H which is isomorphic to a quotient hyperfield F/G, then $G-G = F$. Hence, the problem of the isomorphism of monogenic hyperfields to quotient hyperfields, simultaneously brought into being the following problem in the theory of fields:

When can a subgroup G of the multiplicative group of a field F generate F via the subtraction of G from itself?

The answer to this question for subgroups of finite fields of index 2 and 3 was given in [49]. The following Theorem presents the results of papers [47,49,95,96] collectively:

Theorem 14. Refs. [33,48] *Let F be a finite field and G be a subgroup of its multiplicative group of index n and order m. Then, G−G=F, if and only if:*

$$n = 2 \text{ and } m > 2,$$
$$n = 3 \text{ and } m > 5,$$
$$n = 4, -1 \in G \text{ and } m > 11,$$
$$n = 4, -1 \notin G \text{ and } m > 3,$$
$$n = 5, charF = 2 \text{ and } m > 8,$$
$$n = 5, charF = 3 \text{ and } m > 9,$$
$$n = 5, charF \neq 2,3 \text{ and } m > 23$$

Remark 1. From the above Theorem, it becomes apparent that the validity of the equality G−G=F depends on the cardinality of G. However, this does not mean that any subset S of the field F with the same cardinality as G has the property S−S=F. For example, if $F=\mathbb{Z}_{19}$, then its multiplicative subgroup of index 3, G={1,7,8,11,13,17} satisfies the equality G−G=F, while its subset S={1,6,8,11,13,17}, which has the same cardinality as G, does not. It must also be noted that G's cosets have the same property as G.

Working with the subgroups of index 6, in light of the above Theorem, we have the following Proposition:

Proposition 7. *If G is a subgroup of index 6 of the multiplicative group of a finite field F such that G−G=F and $-1 \notin G$, then G has more than 10 elements.*

Proof. −G and G have the same number of elements and −G∩G=∅. Moreover, (−G)(−G)=G. Consequently W=−G∪G is a subgroup of index 3 of the multiplicative group of F. Thus, by Theorem 14, *cardW*>5 and therefore *cardG*>10. □

Proposition 7 provides a very accurate result. Indeed, the field with the fewest elements which has a multiplicative subgroup of index 6 that satisfies the assumptions of the above Proposition is $\mathbb{Z}_{67}$ and this field's multiplicative subgroup of index 6 is G = {1,9,14,15,22,24,25,40,59,62,64}. As shown in Cayley Table 1, G−G=$\mathbb{Z}_{67}$ is valid.

Table 1. The Cayley table of the subtraction G−G.

	1	9	14	15	22	24	25	40	59	62	64
1	0	8	13	14	21	23	24	39	58	61	63
9	59	0	5	6	13	15	16	31	50	53	55
14	54	62	0	1	8	10	11	26	45	48	50
15	53	61	66	0	7	9	10	25	44	47	49
22	46	54	59	60	0	2	3	18	37	40	42
24	44	52	57	58	65	0	1	16	35	38	40
25	43	51	56	57	64	66	0	15	34	37	39
40	28	36	41	42	49	51	52	0	19	22	24
59	9	17	22	23	30	32	33	48	0	3	5
62	6	14	19	20	27	29	30	45	64	0	2
64	4	12	17	18	25	27	28	43	62	65	0

Lemma 1. *Fields of characteristic 2 have no multiplicative subgroups of index 6.*

Proof. The multiplicative subgroup of a field of characteristic 2 has $2^k - 1$ elements. Therefore, it is not divisible by 6, because it has an odd number of elements. □

Lemma 2. *Fields of characteristic 3 have no multiplicative subgroups of index 6.*

Proof. The multiplicative subgroup of a field of characteristic 3 has $3^k - 1$ elements, which is a non-multiple of number 3 and hence non-divisible by 6. □

Taking into consideration Proposition 7, Lemmas 1, 2 and applying techniques that are similar to the ones developed in [47,49,95,96], we have the Theorem:

Theorem 15. *Let F be a finite field and G be a subgroup of its multiplicative group of index 6 and order m. Then, G–G=F, if and only if:*

$$-1 \notin G, \text{ and } m \geq 11,$$
$$-1 \in G, \text{ charF} = 11 \text{ and } m \geq 20,$$
$$-1 \in G, \text{ charF} = 13 \text{ and } m \geq 28,$$
$$-1 \in G, \text{ charF} \neq 11, 13 \text{ and } m \geq 30.$$

The conclusions of the above Theorem are sharp. The examples that follow are indicative of this fact.

Example 4. *The field $GF[11^2]$ consists of all the linear polynomials with coefficients in the field of residues modulo 11. In $GF[11^2]$, the polynomial $x^2 + 1$ is irreducible. Thus, in the multiplication the polynomials are combined according to the ordinary rules, setting $x^2 = -1 = 10$, and working modulo 11. $GF[11^2]$ has the following multiplicative subgroup of index 6,*

$$G = \{1, 2, 3, 4, 5, 6, 7, 8, 9, 10, x, 2x, 3x, 4x, 5x, 6x, 7x, 8x, 9x, 10x\}$$

which has 20 elements. It can be verified that $G–G=GF[11^2]$.

Example 5. *The field $GF[13^2]$ consists of all the linear polynomials with coefficients in the field of residues modulo 13. The addition and the multiplication are defined in the usual way, replacing x^2 by 11, since the polynomial $x^2 + 2$ is irreducible. $GF[13^2]$ has the following multiplicative subgroup of index 6,*

$$G = \left\{ \begin{array}{l} 1, 5, 8, 12, \\ 5x+1, 8x+1, 2x+2, 11x+2, 3x+3, 10x+3, 5x+4, 8x+4, x+5, 12x+5, \\ x+6, 12x+6, x+7, 12x+7, x+8, 12x+8, 5x+9, 8x+9, \\ 3x+10, 10x+10, 2x+11, 11x+11, 5x+12, 8x+12 \end{array} \right\}$$

which has 28 elements. It can be verified that $G–G=GF[13^2]$.

Example 6. *The field $\mathbb{Z}_{181}$ of residues modulo 181 has the following multiplicative subgroup of index 6,*

$$G = \left\{ \begin{array}{l} 1, 5, 25, 27, 29, 36, 42, 46, 48, 49, 56, 59, 64, 67, 82, 99, \\ 114, 117, 122, 125, 132, 133, 135, 139, 145, 152, 154, 156, 176, 180 \end{array} \right\}$$

which has 30 elements. It can be verified that $G - G = \mathbb{Z}_{181}$.

Similar conclusions to those of Theorem 14 for the multiplicative subgroups of index 3 have been published in [97] without however mentioning the mathematical necessity that led to this problem. The papers [98–100] also deal with this problem without proving though the clear and accurate results that are given by Theorems 14 and 15. On the other hand, in [98–100], the following Theorem is proved:

Theorem 16. *Refs. [98–100] If G is a subgroup of finite index in the multiplicative group of an infinite field F, then G−G=F.*

The above Theorem leads to an extension of Theorems 9 and 10. Indeed, since all finite groups are periodic, while there also exists infinite periodic groups, Theorems 9 and 10 generate finite and infinite non-quotient hyperfields. However, according to Theorem 16, if a hyperfield H is the quotient of an infinite field with a multiplicative subgroup of finite index, then $x-x=H$ for all $x \in H$. Thus, the following Theorem holds:

Theorem 17. *There do not exist finite quotient hyperfields with the hypercompositions which are defined in Theorems 9 and 10.*

Furthermore, Theorem 8 sets a new question in the theory of fields:

When can a subgroup G of the multiplicative group of a field F generate F··G via its subtraction from itself?

Example 2 presents three finite fields which have a multiplicative subgroup G possessing the above property, while the sum of any two of its cosets gives all the non-zero elements of the field F. It is worth mentioning that the rather old paper [101] investigates conditions under which the sum of two cosets of a multiplicative subgroup G of a finite field has a nonempty intersection with at least 3 cosets of G.

5. Classification of Finite Hyperfields into Quotient and Non-Quotient Hyperfields

The enumeration of certain finite hyperfields has been conducted in several papers [66,71,73,77]. Paper [66] deals with hyperfields of order less than or equal to 4, [73,77] deals with hyperfields of order less than or equal to 5, and [71] deals with hyperfields of order less than or equal to 6. In [71], R. Ameri, M. Eyvazi, and S. Hoskova-Mayerova make a thorough check of the isomorphism of these hyperfields to the quotient hyperfields using conclusions from the papers [46–48,95–97]. This section addresses the isomorphism problems with the use of the techniques which were developed from the above study, while it covers some of the gaps that appear in [71].

5.1. Hyperfields of Order 2

According to Theorem 3 there is one two-element non-trivial hyperfield, which is isomorphic to the quotient hyperfield F/F^*, where F is any field with $cardF>2$ and F^* is its multiplicative group. Hence, there exist two hyperfields of order 2, the above and $\mathbb{Z}_2$.

5.2. Hyperfields of Order 3

Hyperfields of order 3 have two non-zero elements. There are five isomorphism classes of these hyperfields [66,71,73,77]. The trivial hyperfield $\mathbb{Z}_3$ is the first of them. Next, there are three hyperfields of order 3, which derive as quotients of a finite field F by an index 2 multiplicative subgroup G of its multiplicative group. According to Theorem 14, the following three cases can be valid for the subgroup G:

i. $G-G \neq F$, which applies only when $F = \mathbb{Z}_5$ and $G = \{1,4\}$
ii. $-1 \notin G$ (i.e., $\{-1, 1\} \nsubseteq G$) and $G-G = F$, which applies when

$$cardF = 2(cardG) + 1 = 2(2k+1) + 1 = 4k + 3$$

iii. $-1 \in G$ (i.e., $\{-1, 1\} \subseteq G$) and $G-G = G + G = F$, which applies when

$$cardF = 2(cardG) + 1 = 2(2k) + 1 = 4k + 1, \ k > 2$$

Therefore, there exist the corresponding three isomorphism classes of quotient hyperfields of order 3 constructed from finite fields:

i. $\mathbb{Z}_5/\{1,4\}$
ii. $GF[p^q]/G$, $p^q=3(mod 4)$
iii. $GF[p^q]/G$, $p^q=1(mod 4)$

The above classification can also derive as follows:

The first two classes are the field $\mathbb{Z}_3$ and its augmented hyperfield. The Cayley tables of their additive parts are shown in the following Table 2:

Table 2. The Cayley tables of the additive group of $\mathbb{Z}_3$ and of the additive canonical hypergroup of its augmented hyperfield $[\mathbb{Z}_3]$.

$\mathbb{Z}_3$	0	1	2
0	0	1	2
1	1	2	0
2	2	0	1

$[\mathbb{Z}_3]$	0	1	2
0	0	1	2
1	1	{1,2}	{0,1,2}
2	2	{0,1,2}	{1,2}

By Theorem 4, the augmented hyperfield of $\mathbb{Z}_3$ is a quotient hyperfield. Observe that $[\mathbb{Z}_3]$ is isomorphic to the quotient hyperfield $\mathbb{Z}_7/\{1,2,4\}$. More generally, the augmented hyperfield of $\mathbb{Z}_3$ is isomorphic to the quotient hyperfield $GF[p^q]/G$, $p^q = 3(mod 4)$, G being an index 2 multiplicative subgroup of the field's multiplicative group.

The next two classes are the quotient hyperfield $\mathbb{Z}_5/\{1,4\}$ and its augmented hyperfield $[\mathbb{Z}_5/\{1,4\}]$. Denoting by 1 the group $G=\{1,4\}$ and by a its coset $2G=\{2,3\}$, we have the following Cayley tables (Table 3) for the additive canonical hypergroups of $\mathbb{Z}_5/\{1,4\}$ and of its augmented hyperfield:

Table 3. The Cayley tables of the additive canonical hypergroups of the hyperfield $\mathbb{Z}_5/\{1,4\}$ and its augmented hyperfield $[\mathbb{Z}_5/\{1,4\}]$.

$\mathbb{Z}_5/\{1,4\}$	0	1	a
0	0	1	a
1	1	$\{0,a\}$	$\{1,a\}$
a	a	$\{1,a\}$	$\{0,1\}$

$[\mathbb{Z}_5/\{1,4\}]$	0	1	a
0	0	1	a
1	1	$\{0,1,a\}$	$\{1,a\}$
a	a	$\{1,a\}$	$\{0,1,a\}$

According to Theorem 5, the augmented hyperfield of a quotient hyperfield is a quotient hyperfield. Therefore, $[\mathbb{Z}_5/\{1,4\}]$ is a quotient hyperfield which is isomorphic to $\mathbb{Z}_{13}/\{1,3,4,9,10,12\}$. More generally, $[\mathbb{Z}_5/\{1,4\}]$ is isomorphic to the quotient hyperfield $GF[p^q]/G$, $p^q = 1(mod 4)$, G being an index 2 multiplicative subgroup of the field's multiplicative group.

The fifth and final class of the order 3 hyperfields is the quotient of an infinite field, and in particular, it is the quotient of an ordered field F by its positive cone F^+. This is the so-called «*sign hyperfield*» and the Cayley table of its canonical hypergroup is shown in Table 4:

Table 4. The Cayley table of the canonical hypergroup of the hyperfield F/F^+.

	0	−1	1
0	0	−1	1
−1	−1	−1	{−1,0,1}
1	1	{−1,0,1}	1

The above conclusions are summed up in the following Theorem:

Theorem 18. *All the hyperfields of order 3 are quotient hyperfields which are classified into 5 isomorphism classes having the following representatives:*

i. $\mathbb{Z}_3$ *and its augmented hyperfield* $[\mathbb{Z}_3]$.
ii. $\mathbb{Z}_5/\{1,4\}$ *and its augmented hyperfield* $[\mathbb{Z}_5/\{1,4\}]$.
iii. *The quotient hyperfield of an ordered field F by its positive cone* F^+.

Hence, the next Theorem holds:

Theorem 19. *All the hyperfields of order 2 and 3 are quotient hyperfields.*

5.3. Hyperfields of Order 4

There are 7 isomorphism classes of hyperfields of order 4, as they have been enumerated in [66,71,73,77]. These consist of the Galois field $GF[2^2]$, 4 classes of quotient hyperfields, and 2 classes of non-quotient hyperfields.

Note on the notation: In the subsequent paragraphs, we denote the quotient hyperfields by QHF_i^j and the non-quotient hyperfields by $NQHF_i^j$. The subscript i denotes the order of the hyperfield, while the superscript j lists the classes.

5.3.i. Quotient Hyperfields of Order 4

The first two classes are the field $GF[2^2]$ and its augmented hyperfield. Recall that, according to Theorem 4, the augmented hyperfield of $GF[2^2]$ is a quotient hyperfield. The Cayley tables of their additive parts are presented in Table 5:

Table 5. The Cayley tables of the additive group of $GF[2^2]$ and of the additive canonical hypergroup of its augmented hyperfield $[GF[2^2]]$, which is also denoted by QFH_4^1.

$GF[2^2]$	0	1	x	$x+1$
0	0	1	x	$x+1$
1	1	0	$x+1$	x
x	x	$x+1$	0	1
$x+1$	$x+1$	x	1	0

Table 5. Cont.

QFH_4^1	0	1	x	$x+1$
0	0	1	x	$x+1$
1	1	$\{0,1,x,x+1\}$	$\{1,x,x+1\}$	$\{1,x,x+1\}$
x	x	$\{1,x,x+1\}$	$\{0,1,x,x+1\}$	$\{1,x,x+1\}$
$x+1$	$x+1$	$\{1,x,x+1\}$	$\{1,x,x+1\}$	$\{0,1,x,x+1\}$

Regarding their multiplicative part, the four elements are combined according to the usual rules, working modulo 2 and writing x^2 as $x+1$ since x^2+x+1 is the irreducible polynomial of degree 2. Therefore, Table 6 is the Cayley table of the multiplicative group of the field $GF[2^2]$ and its augmented hyperfield:

Table 6. The Cayley table of the multiplicative group of the field $GF[2^2]$ and of its augmented hyperfield $[GF[2^2]]$.

	1	x	$x+1$
1	1	x	$x+1$
x	x	$x+1$	1
$x+1$	$x+1$	1	x

We keep using Theorem 14 to examine the next two classes. So, according to Theorem 14, for the fields F with cardinality less than or equal to 16, it holds G—$G \neq F$, when G is a multiplicative subgroup of index 3. These fields are $\mathbb{Z}_7$, $\mathbb{Z}_{13}$, and $GF[2^4]$. $GF[2^4]$ is the field of all the polynomials of degree ≤ 3, with coefficients in $\mathbb{Z}_2$.

The multiplicative subgroup of index 3 in the field $\mathbb{Z}_7$ is $G=\{1,6\}$, and $2G$, 2^2G are its cosets. Denoting by $1, a, a^2$ the group G and its two cosets, respectively, we have the following Cayley table (Table 7) for the additive canonical hypergroup of the quotient hyperfield $\mathbb{Z}_7/\{1,6\}$.

Table 7. The Cayley table of the additive canonical hypergroup of the quotient hyperfield $\mathbb{Z}_7/\{1,6\}$.

QHF_4^2	0	1	a	a^2
0	0	1	a	a^2
1	1	$\{0,a\}$	$\{1,a^2\}$	$\{a,a^2\}$
a	a	$\{1,a^2\}$	$\{0,a^2\}$	$\{1,a\}$
a^2	a^2	$\{a,a^2\}$	$\{1,a\}$	$\{0,1\}$

The multiplicative subgroup of index 3 in the field $\mathbb{Z}_{13}$ is $G=\{1,5,8,12\}$ and $2G$, 2^2G are its cosets. Denoting by $1, a, a^2$ the group G and its two cosets, respectively, we have the following Cayley Table 8 for the additive canonical hypergroups of the quotient hyperfield $\mathbb{Z}_{13}/\{1,5,8,12\}$.

Table 8. The Cayley table of the additive canonical hypergroup of the quotient hyperfield $\mathbb{Z}_{13}/\{1,5,8,12\}$.

QHF_4^3	0	1	a	a^2
0	0	1	a	a^2
1	1	$\{0,a,a^2\}$	$\{1,a,a^2\}$	$\{1,a,a^2\}$
a	a	$\{1,a,a^2\}$	$\{0,1,a^2\}$	$\{1,a,a^2\}$
a^2	a^2	$\{1,a,a^2\}$	$\{1,a,a^2\}$	$\{0,1,a\}$

In the field $GF[2^4]$ of all polynomials of degree ≤ 3 with coefficients in $\mathbb{Z}_2$, the addition and the multiplication of the polynomials are defined in the usual way, by replacing x^4 with $x+1$, since x^4+x+1 is the irreducible polynomial of degree 4. The multiplicative subgroup of index 3 in the field $GF[2^4]$ is

$$G = \left\{1,\ x^3+x^2,\ x^3+x^2+x+1,\ x^3,\ x^3+x\right\}$$

and xG, x^2G are its cosets. Observe that the quotient hyperfield

$$GF[2^4] \Big/ \left\{1,\ x^3+x^2,\ x^3+x^2+x+1,\ x^3,\ x^3+x\right\}$$

is isomorphic to $\mathbb{Z}_{13}/\{1,5,8,12\}$.

Notice that QHF_4^1 is the augmented hyperfield of both QHF_4^2 and QHF_4^3. Moreover, according to Theorem 14, the hyperfield QHF_4^1 is isomorphic to the quotient hyperfield of a finite field F by a subgroup of its multiplicative group of index 3, when $card\ F > 3\cdot 5 + 1$. The hyperfield $\mathbb{Z}_{19}/\{1,7,8,11,12,18\}$ is a representative of this class of quotient hyperfields.

All the above classes of quotient hyperfields derive from the quotient of finite fields with their multiplicative subgroups, but the last one derives from an infinite field. The Cayley table of the canonical hypergroup of this hyperfield appears in Table 9:

Table 9. The Cayley table of the canonical hypergroup of the quotient hyperfield of an infinite field by a multiplicative subgroup of index 3.

QHF_4^4	0	1	a	a^2
0	0	1	a	a^2
1	1	$\{0,1,a,a^2\}$	$\{1,a\}$	$\{1,a^2\}$
a	a	$\{1,a\}$	$\{0,1,a,a^2\}$	$\{a,a^2\}$
a^2	a^2	$\{1,a^2\}$	$\{a,a^2\}$	$\{0,1,a,a^2\}$

Observe that the hyperfield QHF_4^4 is a monogenic b-hyperfield. In [46], it is proved that there exist monogenic b-hyperfields, which are quotient hyperfields. The above monogenic b-hyperfield is such a hyperfield. Indeed, as it is shown in [97], the multiplicative subgroup $G=v^{-1}(3\mathbb{Z})=\{p^{3k}v \mid k\in\mathbb{Z}$ and v is a p-adic unit$\}$ of the field $\mathbb{Q}_p$ of the p-adic numbers with p-adic valuation v, is of index 3 and $G\subseteq G+aG$, while $a^2G\not\subset G+aG$. Therefore, because of Proposition 2, for the quotient hyperfield $\mathbb{Q}_p/G$ it holds that $xG-xG=\mathbb{Q}_p$, $x=1,a,a^2$, and so QHF_4^4 is a quotient hyperfield.

Remark 2. In [71], it is inaccurately stated that the hyperfield QHF_4^4 is isomorphic to $GF[2^4] \big/ \{1, x^3, x^3+x, x^3+x^2, x^3+x^2+x+1\}$. This is not true because, as it is shown above, this is isomorphic to QHF_4^3.

5.3.ii. Non-Quotient Hyperfields of Order 4

The non-quotient hyperfields of order 4 are presented next. Since the multiplicative group of the hyperfields of order 4 has 3 elements, Theorem 10 can be applied to construct a non-quotient hyperfield. The Cayley table of the canonical hypergroup of this hyperfield is presented in Table 10:

Table 10. The Cayley table of the additive canonical hypergroup of the non-quotient b-hyperfield with 4 elements.

$NQHF_4^1$	0	1	a	a^2
0	0	1	a	a^2
1	1	$\{0,a,a^2\}$	$\{1,a\}$	$\{1,a^2\}$
a	a	$\{1,a\}$	$\{0,1,a^2\}$	$\{a,a^2\}$
a^2	a^2	$\{1,a^2\}$	$\{a,a^2\}$	$\{0,1,a\}$

Table 11 shows the additive canonical hypergroup of the seventh hyperfield of order 4:

Table 11. The Cayley table of the additive canonical hypergroup of the non-quotient hyperfield $NQHF_4^2$.

$NQHF_4^2$	0	1	a	a^2
0	0	1	a	a^2
1	1	$\{0,1,a\}$	$\{1,a^2\}$	$\{a,a^2\}$
a	a	$\{1,a^2\}$	$\{0,a,a^2\}$	$\{1,a\}$
a^2	a^2	$\{a,a^2\}$	$\{1,a\}$	$\{0,1,a^2\}$

$NQHF_4^2$ is a non-quotient hyperfield. Indeed, having analyzed above all the cases of quotient hyperfields that derive from finite fields, we conclude that if $NQHF_4^2$ belongs to the quotient hyperfields it must originate from a quotient of an infinite field F by some multiplicative subgroup G of index 3. But in this case, G is a subgroup of finite index in the multiplicative group of the infinite field F. Therefore, by Theorem 16, the equality $G - G = F$ must hold. However, this is not true in $NQHF_4^2$. Consequently, $NQHF_4^2$ is not a quotient hyperfield.

5.4. Hyperfields of Order 5

Since the multiplicative group of finite fields is cyclic, the multiplicative group of the quotient hyperfields resulting from finite fields is cyclic as well. Therefore,

Proposition 8. *Finite hyperfields whose multiplicative part is a non-cyclic group cannot be derived from quotients of finite fields.*

Thus, the finite hyperfields whose multiplicative part is a non-cyclic group derive only from quotients of infinite fields. On the other hand, because of Theorem 14, if G is a subgroup of finite index in the multiplicative group of an infinite field F, then $G-G = F$, and therefore, if H is a finite hyperfield isomorphic to a quotient hyperfield of an infinite field F by a subgroup G of its multiplicative group, then $x-x = H$ must hold for all $x \in H^*$. Consequently, the next Theorem holds:

Theorem 20. *If the multiplicative group of a finite hyperfield H is not cyclic and $x\text{-}x \neq H$, $x \in H^*$, then H is not isomorphic to a quotient hyperfield.*

There exist two groups of order 4, both of which are Abelian. One is the cyclic group C_4 ($\cong \mathbb{Z}/4\mathbb{Z}$), and the other is F. Klein's Vierergruppe V ($\cong C_2 \times C_2$), which is not cyclic. Moreover, it is known that the multiplicative group of the finite fields is cyclic. However, this is not valid for non-trivial hyperfields. Papers [29,35,46] show how to construct hyperfields from any abelian multiplicative group. Therefore, hyperfields can be constructed from the Vierergruppe as well, and thus, the smallest hyperfield with a non-cyclic multiplicative group has 5 elements.

5.4.1. Hyperfields with the Vierergruppe as Their Multiplicative Group

In [71], it has been shown that there exist 11 hyperfields whose multiplicative group is the Vierergruppe. Recall that the Cayley table of the Vierergruppe is the following Table 12:

Table 12. The Cayley table of the Vierergruppe.

	1	a	b	c
1	1	a	b	c
a	a	1	c	b
b	b	c	1	a
c	c	b	a	1

As the Vierergruppe is not a cyclic group, the next Corollary follows from the above Theorem 20:

Corollary 1. *If the multiplicative part of a hyperfield H is the Vierergruppe and if $x-x \neq H$, $x \in H^*$, then H is a non-quotient hyperfield.*

By Corollary 1, among the 11 hyperfields whose multiplicative part is the Vierergruppe, the following 4, which are shown in Table 13, are non-quotient hyperfields.

Table 13. The Cayley tables of the additive canonical hypergroups of the non-quotient hyperfields whose multiplicative group is the Vierergruppe.

$NQHF_5^1$	0	1	a	b	c
0	0	1	a	b	c
1	1	{0, 1}	{b, c}	{a, c}	{a, b}
a	a	{b, c}	{0, a}	{1, c}	{1, b}
b	b	{a, c}	{1, c}	{0, b}	{1, a}
c	c	{a, b}	{1, b}	{1, a}	{0, c}

$NQHF_5^2$	0	1	a	b	c
0	0	1	a	b	c
1	1	{a, b, c}	{1, a, b, c}	{1, a, b, c}	{0, a, b}
a	a	{1, a, b, c}	{1, b, c}	{0, 1, c}	{1, a, b, c}
b	b	{1, a, b, c}	{0, 1, c}	{1, a, c}	{1, a, b, c}
c	c	{0, a, b}	{1, a, b, c}	{1, a, b, c}	{1, a, b}

$NQHF_5^3$	0	1	a	b	c
0	0	1	a	b	c
1	1	{0, a, b, c}	{1, a, b, c}	{1, a, b, c}	{1, a, b, c}
a	a	{1, a, b, c}	{0, 1, b, c}	{1, a, b, c}	{1, a, b, c}
b	b	{1, a, b, c}	{1, a, b, c}	{0, 1, a, c}	{1, a, b, c}
c	c	{1, a, b, c}	{1, a, b, c}	{1, a, b, c}	{0, 1, a, b}

$NQHF_5^4$	0	1	a	b	c
0	0	1	a	b	c
1	1	{0, a, b, c}	{1, a}	{1, b}	{1, c}
a	a	{1, a}	{0, 1, b, c}	{a, b}	{a, c}
b	b	{1, b}	{a, b}	{0, 1, a, c}	{b, c}
c	c	{1, c}	{a, c}	{b, c}	{0, 1, a, b}

The following is an alternative proof that the hyperfields $NQHF_5^2$ and $NQHF_5^4$ are non-quotient hyperfields, which is not based on Corollary 1. Indeed:

(α) For $NQHF_5^2$ observe that the opposite of 1 is c, the opposite of a is b and moreover that:
$$(1+c) \cap (a+b) = \{0, a, b\} \cap \{0, 1, c\} = \{0\}$$

Therefore, according to Lemma 1, if $NQHF_5^2$ were isomorphic to a quotient hyperfield, then the sums of any two non-opposite elements should have the same cardinality. However, this is not the case because, for example:

$$\text{card}(1+a)=4 \quad \text{while} \quad \text{card}(1+1)=3.$$

(β) For $NQHF_5^4$ observe that it is the hyperfield constructed via Theorem 10, when the Vierergruppe is used. Since the Vierergruppe is periodic, Theorem 10 implies that the hyperfield $NQHF_5^4$ cannot be isomorphic to a quotient hyperfield.

The classification of the remaining 7 hyperfields that appear in [71] is a hitherto open problem, and it also raises the question of whether there exist quotient hyperfields that have the Vierergruppe as their multiplicative group. It is worth mentioning here that the hypercompositions in all 7 unclassified hyperfields are closed, and so $x - x$ contains all the elements of the hyperfield for each x in the Vierergruppe.

5.4.2. Hyperfields Having as Multiplicative Group the Cyclic Group C_4

In [71], it is shown that there exist 16 hyperfields whose multiplicative group is the cyclic group C_4. Some of them have been identified as quotient hyperfields. Their classification is completed in the following, starting with the quotient hyperfields.

5.4.2.i. Quotient Hyperfields with Multiplicative Group Being the Cyclic Group C_4

We begin with the field $\mathbb{Z}_5$ and then we continue with the quotient hyperfields of finite fields, along with their augmented hyperfields which, according to Theorems 4 and 5, are quotient hyperfields as well (Table 14).

Table 14. The Cayley tables of the additive group of $\mathbb{Z}_5$ and of the canonical hypergroup of its augmented hyperfield $[\mathbb{Z}_5]$, which is also denoted by QHF_5^1.

$\mathbb{Z}_5$	0	1	2	3	4
0	0	1	2	3	4
1	1	2	3	4	0
2	2	3	4	0	1
3	3	4	0	1	2
4	4	0	1	2	3

Table 14. Cont.

QHF_5^1	0	1	2	3	4
0	0	1	2	3	4
1	1	{2, 1}	{1, 2, 3}	{1, 3, 4}	{0, 1, 2, 3, 4}
2	2	{1, 2, 3}	{2, 4}	{0, 1, 2, 3, 4}	{2, 4, 1}
3	3	{1, 3, 4}	{0, 1, 2, 3, 4}	{1, 3}	{3, 4, 2}
4	4	{0, 1, 2, 3, 4}	{2, 4, 1}	{3, 4, 2}	{4, 3}

Next, $G = \{1,3,9\}$ is the multiplicative subgroup of order 4 of the field $\mathbb{Z}_{13}$ and $2G$, 2^2G, 2^3G, are its cosets. Denoting by $1, a, a^2, a^3$ the group G and its cosets, respectively, Table 15 gives the Cayley tables for the additive canonical hypergroups of $\mathbb{Z}_{13}/G$ and of its augmented hyperfield:

Table 15. The Cayley tables of the additive canonical hypergroup of the hyperfield $\mathbb{Z}_{13}/G$, which is also denoted by QHF_5^2 and of its augmented hyperfield $[\mathbb{Z}_{13}/G]$, which is also denoted by QHF_5^3.

QHF_5^2	0	1	a	a^2	a^3
0	0	1	a	a^2	a^3
1	1	$\{a,a^2\}$	$\{1,a,a^3\}$	$\{0,a,a^3\}$	$\{1,a^2,a^3\}$
a	a	$\{1,a,a^3\}$	$\{a^2,a^3\}$	$\{1,a,a^2\}$	$\{0,1,a^2\}$
a^2	a^2	$\{0,a,a^3\}$	$\{1,a,a^2\}$	$\{1,a^3\}$	$\{a,a^2,a^3\}$
a^3	a^3	$\{1,a^2,a^3\}$	$\{0,1,a^2\}$	$\{a,a^2,a^3\}$	$\{1,a\}$

QHF_5^3	0	1	a	a^2	a^3
0	0	1	a	a^2	a^3
1	1	$\{1,a,a^2\}$	$\{1,a,a^3\}$	$\{0,1,a,a^2,a^3\}$	$\{1,a^2,a^3\}$
a	a	$\{1,a,a^3\}$	$\{a,a^2,a^3\}$	$\{1,a,a^2\}$	$\{0,1,a,a^2,a^3\}$
a^2	a^2	$\{0,1,a,a^2,a^3\}$	$\{1,a,a^2\}$	$\{1,a^2,a^3\}$	$\{a,a^2,a^3\}$
a^3	a^3	$\{1,a^2,a^3\}$	$\{0,1,a,a^2,a^3\}$	$\{a,a^2,a^3\}$	$\{1,a,a^3\}$

Theorem 14 will continue to be used for the classification of the next classes of quotient hyperfields of order 5. Thus, in addition to the above, the fields with cardinality less than or equal to 4·11+1=45 are the following ones:

$$GF[3^2], GF[5^2], \mathbb{Z}_{17}, \mathbb{Z}_{29}, \mathbb{Z}_{37}, \text{ and } \mathbb{Z}_{41}.$$

$GF[3^2]$ consists of the 9 polynomials in x of degree 0 or 1 with coefficients in the field $\mathbb{Z}_3$ and writing x^2 as 2 whenever it occurs. $G = \{1,2\}$ is the multiplicative subgroup of index 4 in the field $GF[3^2]$. The hyperfield $GF[3^2]/G$ is the following one:

$$GF\left[3^2\right]\Big/G = \{G, xG, (x+1)G, (x+2)G\} = \left\{(x+1)^k G \,\middle|\, k=0,1,2,3\right\}$$

Denoting the coset $(x+1)G$ by a and G by 1, the additive canonical hypergroup of the $GF[3^2]/G$ is shown in Table 16:

Table 16. The Cayley table of the additive canonical hypergroup of the quotient hyperfield $GF[3^2]/G$, which is also denoted by QHF_5^4.

QHF_5^4	0	1	a	a^2	a^3
0	0	1	a	a^2	a^3
1	1	$\{0,1\}$	$\{a^2, a^3\}$	$\{a, a^3\}$	$\{a, a^2\}$
a	a	$\{a^2, a^3\}$	$\{0, a\}$	$\{1, a^3\}$	$\{1, a\}$
a^2	a^2	$\{a, a^3\}$	$\{1, a^3\}$	$\{0, a^2\}$	$\{1, a\}$
a^3	a^3	$\{a, a^2\}$	$\{1, a^2\}$	$\{1, a\}$	$\{0, a^3\}$

$GF[5^2]$ consists of the 25 polynomials in x of degree 0 or 1 with coefficients in the field $\mathbb{Z}_5$. Since x^2+3x+4 is the irreducible polynomial of degree 2 we are writing x^2 as $-3x-4 = 2x+1$ whenever it occurs. $G = \{1, 4, 2x, 3x+4, 3x, 2x+1\}$ is the multiplicative subgroup of index 4 in the field $GF[5^2]$. The hyperfield $GF[5^2]/G$ is the following:

$$GF\left[5^2\right]\Big/G = \{G, 2G, (x+1)G, (2x+2)G\} = \left\{(x+1)^k G \,\middle|\, k=0,1,2,3\right\}$$

Denoting the coset $(x+1)G$ by a and G by 1, the Cayley table for the additive canonical hypergroup of the $GF[5^2]/G$ is presented in Table 17:

Table 17. The Cayley table of the additive canonical hypergroup of the quotient hyperfield $GF[5^2]/G$, which is also denoted by QHF_5^5.

QHF_5^5	0	1	a	a^2	a^3
0	0	1	a	a^2	a^3
1	1	$\{0,1,a^2,a^3\}$	$\{a,a^2,a^3\}$	$\{1,a,a^2,a^3\}$	$\{1,a,a^2\}$
a	a	$\{a,a^2,a^3\}$	$\{0,1,a,a^3\}$	$\{1,a^2,a^3\}$	$\{1,a,a^2,a^3\}$
a^2	a^2	$\{1,a,a^2,a^3\}$	$\{1,a^2,a^3\}$	$\{0,1,a,a^2\}$	$\{1,a,a^3\}$
a^3	a^3	$\{1,a,a^2\}$	$\{1,a,a^2,a^3\}$	$\{1,a,a^3\}$	$\{0,a,a^2,a^3\}$

The multiplicative subgroup of index 4 in the field $\mathbb{Z}_{17}$ is $G = \{1,4,13,16\}$ and $5G$, 5^2G, 5^3G, are its cosets. Denoting by $1, a, a^2, a^3$ the group G and its three cosets, respectively, we have the following Cayley Table 18 for the additive canonical hypergroups of the quotient hyperfield $\mathbb{Z}_{17}/G$:

Table 18. The Cayley table of the additive canonical hypergroup of the quotient hyperfield $\mathbb{Z}_{17}/\{1,4,13,16\}$, which is also denoted by QHF_5^6.

QHF_5^6	0	1	a	a^2	a^3
0	0	1	a	a^2	a^3
1	1	$\{0,a,a^2\}$	$\{1,a^2,a^3\}$	$\{1,a,a^2,a^3\}$	$\{a,a^2,a^3\}$
a	a	$\{1,a^2,a^3\}$	$\{0,a^2,a^3\}$	$\{1,a,a^3\}$	$\{1,a,a^2,a^3\}$
a^2	a^2	$\{1,a,a^2,a^3\}$	$\{1,a,a^3\}$	$\{0,1,a^3\}$	$\{1,a,a^2\}$
a^3	a^3	$\{a,a^2,a^3\}$	$\{1,a,a^2,a^3\}$	$\{1,a,a^2\}$	$\{0,1,a\}$

Notice that the hyperfields QHF_5^4, QHF_5^5 and QHF_5^6 have the same augmented hyperfield QHF_5^7. Because of Theorem 4, this hyperfield is a quotient hyperfield. Furthermore, it can be verified that this hyperfield is isomorphic to the quotient hyperfield $\mathbb{Z}_{53}/\{1,10,13,15,16,24,28,36,42,44,46,47,49\}$. The Cayley table of the additive canonical hypergroup of this hyperfield appears in Table 19:

Table 19. The Cayley table of the additive canonical hypergroup of the augmented hyperfield of QHF_5^4, QHF_5^5 and QHF_5^6 which is simultaneously the additive hypergroup of the quotient hyperfield $\mathbb{Z}_{53}/\{1,10,13,15,16,24,28,36,42,44,46,47,49\}$.

QHF_5^7	0	1	a	a^2	a^3
0	0	1	a	a^2	a^3
1	1	$\{0,1,a,a^2,a^3\}$	$\{1,a,a^2,a^3\}$	$\{1,a,a^2,a^3\}$	$\{1,a,a^2,a^3\}$
a	a	$\{1,a,a^2,a^3\}$	$\{0,1,a,a^2,a^3\}$	$\{1,a,a^2,a^3\}$	$\{1,a,a^2,a^3\}$
a^2	a^2	$\{1,a,a^2,a^3\}$	$\{1,a,a^2,a^3\}$	$\{0,1,a,a^2,a^3\}$	$\{1,a,a^2,a^3\}$
a^3	a^3	$\{1,a,a^2,a^3\}$	$\{1,a,a^2,a^3\}$	$\{1,a,a^2,a^3\}$	$\{0,1,a,a^2,a^3\}$

The multiplicative subgroup of index 4 in the field $\mathbb{Z}_{29}$ is $G = \{1,7,16,20,23,24,25\}$ and $2G$, 2^2G, 2^3G, are its cosets. Denoting by 1, a, a^2, a^3 the group G and its three cosets, respectively, we have the following Cayley Table 20 for the additive canonical hypergroups of the quotient hyperfield $\mathbb{Z}_{29}/G$:

Table 20. The Cayley table of the additive canonical hypergroup of the quotient hyperfield $\mathbb{Z}_{29}/\{1,7,16,20,23,24,25\}$.

QHF_5^8	0	1	a	a^2	a^3
0	0	1	a	a^2	a^3
1	1	$\{1,a,a^3\}$	$\{1,a,a^2,a^3\}$	$\{0,1,a,a^2,a^3\}$	$\{1,a,a^2,a^3\}$
a	a	$\{1,a,a^2,a^3\}$	$\{1,a,a^2\}$	$\{1,a,a^2,a^3\}$	$\{0,1,a,a^2,a^3\}$
a^2	a^2	$\{0,1,a,a^2,a^3\}$	$\{1,a,a^2,a^3\}$	$\{a,a^2,a^3\}$	$\{1,a,a^2,a^3\}$
a^3	a^3	$\{1,a,a^2,a^3\}$	$\{0,1,a,a^2,a^3\}$	$\{1,a,a^2,a^3\}$	$\{1,a^2,a^3\}$

$G=\{1,7,9,10,12,16,26,33,34\}$ is the multiplicative subgroup of index 4 in the field $\mathbb{Z}_{37}$ and $2G$, 2^2G, 2^3G, are its cosets. Denoting by 1, a, a^2, a^3 the group G and its three cosets, respectively, we have the following Cayley Table 21 for the additive canonical hypergroups of the quotient hyperfield $\mathbb{Z}_{37}/G$:

Table 21. The Cayley table of the additive canonical hypergroup of the quotient hyperfield $\mathbb{Z}_{37}/\{1,7,9,10,12,16,26,33,34\}$.

QHF_5^9	0	1	a	a^2	a^3
0	0	1	a	a^2	a^3
1	1	$\{1,a,a^2,a^3\}$	$\{1,a,a^2,a^3\}$	$\{0,1,a,a^2,a^3\}$	$\{1,a,a^2,a^3\}$
a	a	$\{1,a,a^2,a^3\}$	$\{1,a,a^2,a^3\}$	$\{1,a,a^2,a^3\}$	$\{0,1,a,a^2,a^3\}$
a^2	a^2	$\{0,1,a,a^2,a^3\}$	$\{1,a,a^2,a^3\}$	$\{1,a,a^2,a^3\}$	$\{1,a,a^2,a^3\}$
a^3	a^3	$\{1,a,a^2,a^3\}$	$\{0,1,a,a^2,a^3\}$	$\{1,a,a^2,a^3\}$	$\{1,a,a^2,a^3\}$

$G=\{1,4,10,16,18,23,25,31,37,40\}$ is the multiplicative subgroup of index 4 in the field $\mathbb{Z}_{41}$ and $3G, 3^2G, 3^3G$, are its cosets. Denoting by $1, a, a^2, a^3$ the group G and its three cosets, respectively, we have the following Cayley Table 22 for the additive canonical hypergroup of the quotient hyperfield $\mathbb{Z}_{41}/G$:

Table 22. The Cayley table of the additive canonical hypergroup of the quotient hyperfield $\mathbb{Z}_{41}/\{1,4,10,16,18,23,25,31,37,40\}$.

QHF_5^{10}	0	1	a	a^2	a^3
0	0	1	a	a^2	a^3
1	1	$\{0,a,a^2,a^3\}$	$\{1,a,a^2,a^3\}$	$\{1,a,a^2,a^3\}$	$\{1,a,a^2,a^3\}$
a	a	$\{1,a,a^2,a^3\}$	$\{0,1,a^2,a^3\}$	$\{1,a,a^2,a^3\}$	$\{1,a,a^2,a^3\}$
a^2	a^2	$\{1,a,a^2,a^3\}$	$\{1,a,a^2,a^3\}$	$\{0,1,a,a^3\}$	$\{1,a,a^2,a^3\}$
a^3	a^3	$\{1,a,a^2,a^3\}$	$\{1,a,a^2,a^3\}$	$\{1,a,a^2,a^3\}$	$\{0,1,a,a^2\}$

5.4.2.ii. Non-Quotient Hyperfields with Multiplicative Group Being the Cyclic Group C_4

The first non-quotient hyperfield can be constructed via Theorem 10. The Cayley Table 23 presents its additive canonical hypergroup:

Table 23. The Cayley table of the additive canonical hypergroup of the non-quotient hyperfield constructed via Theorem 10.

$NQHF_5^1$	0	1	a	a^2	a^3
0	0	1	a	a^2	a^3
1	1	$\{0, a, a^2, a^3\}$	$\{1, a\}$	$\{1, a^2\}$	$\{1, a^3\}$
a	a	$\{1, a\}$	$\{0, 1, a^2, a^3\}$	$\{a, a^2\}$	$\{a, a^3\}$
a^2	a^2	$\{1, a^2\}$	$\{a, a^2\}$	$\{0, 1, a, a^3\}$	$\{a^2, a^3\}$
a^3	a^3	$\{1, a^3\}$	$\{a, a^3\}$	$\{a^2, a^3\}$	$\{0, 1, a, a^2\}$

Cayley Table 24 presents the additive canonical hypergroup of the second non-quotient hyperfield:

Table 24. The Cayley table of the additive canonical hypergroup of the second non-quotient hyperfield.

$NQHF_5^2$	0	1	a	a^2	a^3
0	0	1	a	a^2	a^3
1	1	$\{a, a^2, a^3\}$	$\{1, a, a^2, a^3\}$	$\{0, a, a^3\}$	$\{1, a, a^2, a^3\}$
a	a	$\{1, a, a^2, a^3\}$	$\{1, a^2, a^3\}$	$\{1, a, a^2, a^3\}$	$\{0, 1, a^2\}$
a^2	a^2	$\{0, a, a^3\}$	$\{1, a, a^2, a^3\}$	$\{1, a, a^3\}$	$\{1, a, a^2, a^3\}$
a^3	a^3	$\{1, a, a^2, a^3\}$	$\{0, 1, a^2\}$	$\{1, a, a^2, a^3\}$	$\{1, a, a^2\}$

We will prove that $NQHF_5^2$ is a non-quotient hyperfield. Note that the opposite of 1 is a^2, the opposite of a is a^3 and that:

$$\left(1 + a^2\right) \cap \left(a + a^3\right) = \{0, a, a^3\} \cap \{0, 1, a^2\} = \{0\}$$

Therefore, according to Lemma 1, if $NQHF_5^2$ were isomorphic to a quotient hyperfield, then the sums of any two non-opposite elements should have had the same cardinality. However, this is not the case because, for example,

$$\text{card}(1+a)=4 \quad \text{while} \quad \text{card}(1+1)=3.$$

5.4.2.iii. Non-Classified Hyperfields Having as Multiplicative Group the Cyclic Group C_4

There remain three hyperfields whose multiplicative group is the cyclic group C_4. For these hyperfields the hypercompositions are defined as shown in Table 25:

Table 25. The Cayley tables of the additive canonical hypergroup of the three non-classified hyperfields with multiplicative group C_4.

HF_5^1	0	1	a	a^2	a^3
0	0	1	a	a^2	a^3
1	1	$\{0,1,a,a^2,a^3\}$	$\{1,a\}$	$\{1,a^2\}$	$\{1,a^3\}$
a	a	$\{1,a\}$	$\{0,1,a,a^2,a^3\}$	$\{a,a^2\}$	$\{a,a^3\}$
a^2	a^2	$\{1,a^2\}$	$\{a,a^2\}$	$\{0,1,a,a^2,a^3\}$	$\{a,a^2,a^3\}$
a^3	a^3	$\{1,a^3\}$	$\{a,a^3\}$	$\{a,a^2,a^3\}$	$\{0,1,a,a^2,a^3\}$

HF_5^2	0	1	a	a^2	a^3
0	0	1	a	a^2	a^3
1	1	1	$\{1,a\}$	$\{0,1,a,a^2,a^3\}$	$\{1,a^3\}$
a	a	$\{1,a\}$	a	$\{a,a^2\}$	$\{0,1,a,a^2,a^3\}$
a^2	a^2	$\{0,1,a,a^2,a^3\}$	$\{a,a^2\}$	a^2	$\{a^2,a^3\}$
a^3	a^3	$\{1,a^3\}$	$\{0,1,a,a^2,a^3\}$	$\{a^2,a^3\}$	a^3

HF_5^3	0	1	a	a^2	a^3
0	0	1	a	a^2	a^3
1	1	$\{1,a^2\}$	$\{1,a\}$	$\{0,1,a,a^2,a^3\}$	$\{1,a^3\}$
a	a	$\{1,a\}$	$\{a,a^3\}$	$\{a,a^2\}$	$\{0,1,a,a^2,a^3\}$
a^2	a^2	$\{0,1,a,a^2,a^3\}$	$\{a,a^2\}$	$\{1,a^2\}$	$\{a^2,a^3\}$
a^3	a^3	$\{1,a^3\}$	$\{0,1,a,a^2,a^3\}$	$\{a^2,a^3\}$	$\{a,a^3\}$

From the analysis and conclusions of the previous section, it follows that the above three hyperfields cannot be derived as a quotient of finite fields by subgroups of their multiplicative group. Thus, the question of whether they are isomorphic or not to quotient hyperfields of non-finite fields by multiplicative subgroups of index 4, still remains open.

6. Discussion

Marc Krasner introduced the hyperfield in 1956, and until 1983, no hyperfields other than the residuals ones were known in the wider mathematical society, regardless of the fact that Krasner had made his associates aware of the construction of the quotient hyperfields and hyperrings, which generalize the residual hyperfields. The criticism that he received was that if all hyperrings and hyperfields could be isomorphically embedded into the quotient hyperrings, then several conclusions of their theory would have been reached in a very straightforward manner, with the use of the theories of rings, fields, and modules, and it wouldn't have been necessary to develop new techniques, methods and methodologies for their proofs. Thus, in 1983, M. Krasner published the construction of the quotient hyperfields and hyperrings and raised the questions [24]:

Are all hyperrings which are not rings isomorphic to the subhyperrings of quotient hyperrings R/G of some ring R by some of its normal multiplicative subgroups G when they are not rings? Are all hyperfields isomorphic to a quotient K/G of a field K by some of its multiplicative subgroups G?

Negative answers to these questions first came from the works in [29,35] and then from [36,47], while Theorem 12 also constructs a new class of non-quotient hyperrings and hyperfields. The constructions thought of certain hyperfields which were introduced for answering Krasner's questions gave rise to the following problem in field theory:

When does a subgroup G of the multiplicative group of a field F possess the ability to generate F via the subtraction of G from itself?

So far, we do not have a clear and complete general solution to this problem. In the finite fields, we have sharp conclusions for the subgroups of indexes 2,3,4,5,6, as described in Theorems 14 and 15. Moreover, the construction of new hyperfields (Theorem 8) and the research on whether they belong to the quotient hyperfields introduced a new problem in the theory of fields:

Under what conditions can a field F's multiplicative subgroup G generate F··G via the subtraction of G from itself?

The question of the classification of hyperfields arose naturally as a follow-up to Krasner's question, and the Table 26 below summarizes the results of the classification of finite hyperfields with 2, 3, 4, 5 elements.

Table 26. Classification of the hyperfields of order 2,3,4,5.

Order of Hyperfields	Number of Hyperfields with Cyclic Multiplicative Subgroup	Number of Hyperfields with Non-Cyclic Multiplicative Subgroup	Fields	Quotient Hyperfields	Non-Quotient Hyperfields	Unclassified Hyperfields
2	2	–	1	1	–	–
3	5	–	1	4	–	–
4	7	–	1	4	2	–
5	–	11	–	–	4	7
	16	–	1	10	2	3

Evidently, the classification of the 10 unclassified finite hyperfields remains an open problem. For the infinite non-quotient hyperfields, note that besides the construction of finite non-quotient hyperfields, Theorems 9 and 10 give the construction of infinite non-quotient hyperfields as well. Additionally, Theorem 12 presents the construction of a class of such hyperfields. Evident examples of infinite quotient hyperfields are $\mathbb{R}/\mathbb{Q}^*$, $\mathbb{R}/\mathbb{Q}^+$, $\mathbb{C}/\mathbb{Q}^*$, $\mathbb{C}/\mathbb{Q}^+$, $\mathbb{C}/\mathbb{R}^*$, $\mathbb{C}/\mathbb{R}^+$ etc.

Author Contributions: C.G.M. and G.G.M. contributed equally to this work. All authors have read and agreed to the published version of the manuscript.

Funding: This research received no external funding. The APC was funded by the MDPI journal *Mathematics*.

Institutional Review Board Statement: Not applicable.

Data Availability Statement: Not applicable.

Acknowledgments: This study was contacted within the framework of the first author's sabbatical leave supported by the National and Kapodistrian University of Athens.

Conflicts of Interest: The authors declare no conflict of interest.

References

1. Marty, F. *Sur une Généralisation de la Notion de Groupe*; Huitième Congrès des mathématiciens Scand: Stockholm, Sweden, 1934; pp. 45–49.
2. Marty, F. Rôle de la Notion de Hypergroupe dans l' étude de Groupes non Abéliens. *C. R. Acad. Sci.* **1935**, *201*, 636–638.
3. Marty, F. Sur les groupes et hypergroupes attachés à une fraction rationelle. *Ann. De L' École Norm.* **1936**, *53*, 83–123. [CrossRef]
4. Krasner, M. Sur la primitivité des corps B-adiques. *Mathematica* **1937**, *13*, 72–191.
5. Krasner, M. La loi de Jordan—Holder dans les hypergroupes et les suites generatrices des corps de nombres P–adiqes, (I). *Duke Math. J.* **1940**, *6*, 120–140, (II) *Duke Math. J.* **1940**, *7*, 121–135. [CrossRef]
6. Krasner, M. La caractérisation des hypergroupes de classes et le problème de Schreier dans ces hypergroupes. *C. R. Acad. Sci.* **1941**, *212*, 948–950.
7. Krasner, M. Hypergroupes moduliformes et extramoduliformes. *C. R. Acad. Sci.* **1944**, *219*, 473–476.
8. Krasner, M.; Kuntzmann, J. Remarques sur les hypergroupes. *C. R. Acad. Sci.* **1947**, *224*, 525–527.
9. Kuntzmann, J. Opérations multiformes. Hypergroupes. *C. R. Acad. Sci.* **1937**, *204*, 1787–1788.
10. Kuntzmann, J. Homomorphie entre systémes multiformes. *C. R. Acad. Sci.* **1937**, *205*, 208–210.
11. Wall, H.S. Hypergroups. *Am. J. Math.* **1937**, *59*, 77–98. [CrossRef]
12. Ore, O. Structures and group theory, I. *Duke Math. J.* **1937**, *3*, 149–174. [CrossRef]
13. Dresher, M.; Ore, O. Theory of multigroups. *Am. J. Math.* **1938**, *60*, 705–733. [CrossRef]
14. Eaton, E.J.; Ore, O. Remarks on multigroups. *Am. J. Math.* **1940**, *62*, 67–71. [CrossRef]
15. Eaton, E.J. Associative Multiplicative Systems. *Am. J. Math.* **1940**, *62*, 222–232. [CrossRef]
16. Griffiths, L.W. On hypergroups, multigroups, and product systems. *Am. J. Math.* **1938**, *60*, 345–354. [CrossRef]
17. Prenowitz, W. Projective Geometries as multigroups. *Am. J. Math.* **1943**, *65*, 235–256. [CrossRef]
18. Prenowitz, W. Descriptive Geometries as multigroups. *Trans. Am. Math. Soc.* **1946**, *59*, 333–380. [CrossRef]
19. Prenowitz, W. Spherical Geometries and mutigroups. *Can. J. Math.* **1950**, *2*, 100–119. [CrossRef]
20. Dietzman, A.P. On the multigroups of complete conjugate sets of a group. *CR (Doklady) Acad. Sci. URSS (N.S.)* **1946**, *49*, 315–317.
21. Massouros, C.G. On connections between vector spaces and hypercompositional structures. *Ital. J. Pure Appl. Math.* **2015**, *34*, 133–150.
22. Massouros, C.; Massouros, G. An Overview of the Foundations of the Hypergroup Theory. *Mathematics* **2021**, *9*, 1014. [CrossRef]
23. Krasner, M. *Approximation des corps valués complets de caractéristique p≠0 par ceux de caractéristique 0*; Colloque d' Algèbre Supérieure (Bruxelles, Decembre 1956); Centre Belge de Recherches Mathématiques, Établissements Ceuterick, Louvain, Librairie Gauthier-Villars: Paris, France, 1957; pp. 129–206.
24. Krasner, M. A class of hyperrings and hyperfields. *Int. J. Math. Math. Sci.* **1983**, *6*, 307–312. [CrossRef]
25. Mittas, J. Hypergroupes canoniques hypervalues. *C. R. Acad. Sci.* **1970**, *271*, 4–7.
26. Mittas, J. Hypergroupes canoniques. *Math. Balk.* **1972**, *2*, 165–179.
27. Mittas, J. Hypergroupes canoniques values et hypervalues. Hypergroupes fortement et superieurement canoniques. *Bull. Greek Math. Soc.* **1982**, *23*, 55–88.
28. Mittas, J. Hypergroupes polysymetriques canoniques. In Proceedings of the Atti del Convegno su Ipergruppi, Altre Strutture Multivoche e loro Applicazioni, Udine, Italy, 15–18 October 1985; pp. 1–25.
29. Massouros, C.G. Methods of constructing hyperfields. *Int. J. Math. Math. Sci.* **1985**, *8*, 725–728. [CrossRef]
30. Mittas, J. Sur certaines classes de structures hypercompositionnelles. *Proc. Acad. Athens* **1973**, *48*, 298–318.
31. Mittas, J. Sur les structures hypercompositionnelles. In Proceedings of the 4th International Congruence on Algebraic Hyperstructures and Applications, Xanthi, Greece, 27–30 June 1990; World Scientific: Singapore, 1991; pp. 9–31.
32. Ameri, R.; Eyvazi, M.; Hoskova-Mayerova, S. Superring of polynomials over a hyperring. *Mathematics* **2019**, *7*, 902. [CrossRef]
33. Massouros, G.G.; Massouros, C.G. Hypercompositional algebra, Computer Science and Geometry. *Mathematics* **2020**, *8*, 1338. [CrossRef]
34. Krasner, M. *Cours d' Algebre superieure, Theories des valuation et de Galois*; Cours Faculté Sciences L' Université: Paris, France, 1967–1968; pp. 1–305.
35. Massouros, C.G. On the theory of hyperrings and hyperfields. *Algebra i Logika* **1985**, *24*, 728–742. [CrossRef]
36. Nakassis, A. Recent results in hyperring and hyperfield theory. *Intern. J. Math. Math. Sci.* **1988**, *11*, 209–220. [CrossRef]
37. Mittas, J. Hyperanneaux et certaines de leurs propriétés. *C. R. Acad. Sci.* **1969**, *269*, 623–626.
38. Mittas, J. Hypergroupes et hyperanneaux polysymétriques. *C. R. Acad. Sci.* **1970**, *271*, 920–923.
39. Mittas, J. Contributions a la théorie des hypergroupes, hyperanneaux, et les hypercorps hypervalues. *C. R. Acad. Sci.* **1971**, *272*, 3–6.

40. Mittas, J. Certains hypercorps et hyperanneaux définis à partir de corps et anneaux ordonnés. *Bull. Math. de la Soc Sci. Math. de la R. S. Roum.* **1971**, *15*, 371–378.
41. Mittas, J. Sur les hyperanneaux et les hypercorps. *Math. Balk.* **1973**, *3*, 368–382.
42. Mittas, J. Contribution à la théorie des structures ordonnées et des structures valuées. *Proc. Acad. Athens* **1973**, *48*, 318–331.
43. Mittas, J. Hypercorps totalement ordonnes. *Sci. Ann. Polytech. Sch. Univ. Thessalon.* **1974**, *6*, 49–64.
44. Mittas, J. Espaces vectoriels sur un hypercorps. Introduction des hyperspaces affines et Euclidiens. *Math. Balk.* **1975**, *5*, 199–211.
45. Massouros, C.G. Free and cyclic hypermodules. *Ann. Mat. Pura Appl.* **1988**, *150*, 153–166. [CrossRef]
46. Massouros, C.G. Constructions of hyperfields. *Math. Balk.* **1991**, *5*, 250–257.
47. Massouros, C.G. A class of hyperfields and a problem in the theory of fields. *Math. Montisnigri* **1993**, *1*, 73–84.
48. Massouros, C.G. A Field Theory Problem Relating to Questions in Hyperfield Theory. In Proceedings of the ICNAAM 2011, Halkidiki, Greece, 19–25 September 2011; AIP Conference Proceedings 1389. pp. 1852–1855. [CrossRef]
49. Massouros, C.G. Algebraic Structures with Hypercomposition. Dissertation, Patras University, Patra, Greece, 1984.
50. Massouros, C.G.; Massouros, G.G. On join hyperrings. In Proceedings of the 10th International Congruence on Algebraic Hyperstructures and Applications, Brno, Czech Republic, 3–9 September 2008; pp. 203–215.
51. Massouros, G.G.; Massouros, C.G. Homomorphic relations on Hyperringoids and Join Hyperrings. *Ratio Mat.* **1999**, *13*, 61–70.
52. Massouros, G.G. The hyperringoid. *Mult. Valued Log.* **1998**, *3*, 217–234.
53. Massouros, G.G. Automata and hypermoduloids. In Proceedings of the 5th International Congruence on Algebraic Hyperstructures and Applications, Iasi, Romania, 4–10 July 1993; Hadronic Press: Palm Harbor, FL, USA, 1994; pp. 251–265.
54. Massouros, G.G. Solving equations and systems in the environment of a hyperringoid. In Proceedings of the 6th International Congruence on Algebraic Hyperstructures and Applications, Prague, Czech Republic, 1–9 September 1996; Democritus University of Thrace Press: Komotini, Greece, 1997; pp. 103–113.
55. Procesi Ciampi, R.; Rota, R. Hyperaffine planes over hyperfields. *J. Geom.* **1995**, *54*, 123–133. [CrossRef]
56. Rota, R. Hyperaffine planes over hyperrings. *Discrete Math.* **1996**, *155*, 215–223. [CrossRef]
57. Jančic-Rašović, S. About the hyperring of polynomials. *Ital. J. Pure Appl. Math.* **2007**, *21*, 223–234.
58. Cristea, I.; Jančić-Rašović, S. Composition hyperrings. *An. Stiint. Univ. Ovidius Constanta Ser. Mat.* **2013**, *21*, 81–94. [CrossRef]
59. Bordbar, H.; Jančić-Rašović, S.; Cristea, I. Regular local hyperrings and hyperdomains. *AIMS Math.* **2022**, *7*, 20767–20780. [CrossRef]
60. Bordbar, H.; Cristea, I.; Novak, M. Height of hyperideals in Noetherian Krasner hyperrings. *UPB Sci. Bull. Ser. A Appl. Math. Phys.* **2017**, *79*, 31–42.
61. Bordbar, H.; Cristea, I. Regular parameter elements and regular local hyperrings. *Mathematics* **2021**, *9*, 243. [CrossRef]
62. Cristea, I.; Kankaraš, M. The Reducibility Concept in General Hyperrings. *Mathematics* **2021**, *9*, 2037. [CrossRef]
63. Vahedi, V.; Jafarpour, M.; Aghabozorgi, H.; Cristea, I. Extension of elliptic curves on Krasner hyperfields. *Comm. Algebra* **2019**, *47*, 4806–4823. [CrossRef]
64. Vahedi, V.; Jafarpour, M.; Cristea, I. Hyperhomographies on Krasner hyperfields. *Symmetry* **2019**, *11*, 1442. [CrossRef]
65. Vahedi, V.; Jafarpour, M.; Hoskova-Mayerova, S.; Aghabozorgi, H.; Leoreanu-Fotea, V.; Bekesiene, S. Derived Hyperstructures from Hyperconics. *Mathematics* **2020**, *8*, 429. [CrossRef]
66. Iranmanesh, M.; Jafarpour, M.; Aghabozorgi, H.; Zhan, J.M. Classification of Krasner Hyperfields of Order 4. *Acta Math. Sin. (Engl. Ser.)* **2020**, *36*, 889–902. [CrossRef]
67. Connes, A.; Consani, C. From monoids to hyperstructures: In search of an absolute arithmetic. *arXiv* **2010**, arXiv:1006.4810.
68. Connes, A.; Consani, C. The hyperring of adèle classes. *J. Number Theory* **2011**, *131*, 159–194. [CrossRef]
69. Viro, O. Hyperfields for tropical geometry I. Hyperfields and dequantization. *arXiv* **2010**, arXiv:1006.3034.
70. Viro, O. On basic concepts of tropical geometry. *Proc. Steklov Inst. Math.* **2011**, *273*, 252–282. [CrossRef]
71. Ameri, R.; Eyvazi, M.; Hoskova-Mayerova, S. Advanced results in enumeration of hyperfields. *AIMS Math.* **2020**, *5*, 6552–6579. [CrossRef]
72. Baker, M.; Bowler, N. Matroids over hyperfields. In Proceedings of the ICNAAM 2017, Thessaloniki, Greece, 25–30 September 2017. AIP Conference Proceedings 1978, 340010. [CrossRef]
73. Baker, M.; Jin, T. On the Structure of Hyperfields Obtained as Quotients of Fields. *Proc. Am. Math. Soc.* **2021**, *149*, 63–70. [CrossRef]
74. Baker, M.; Lorscheid, O. Descartes' rule of signs, Newton polygons, and polynomials over hyperfields. *J. Algebra* **2021**, *569*, 416–441. [CrossRef]
75. Jun, J. Geometry of hyperfields. *arXiv* **2017**, arXiv:1707.09348. [CrossRef]
76. Lorscheid, O. Tropical geometry over the tropical hyperfield. *arXiv* **2019**, arXiv:1907.01037. [CrossRef]
77. Liu, Z. Finite Hyperfields with order n≤5. *arXiv* **2020**, arXiv:2004.07241. [CrossRef]
78. Shojaei, H.; Fasino, D. Isomorphism Theorems in the Primary Categories of Krasner Hypermodules. *Symmetry* **2019**, *11*, 687. [CrossRef]
79. Das, K.; Singha, M. Topological Krasner hyperrings with special emphasis on isomorphism theorems. *Appl. Gen. Topol.* **2022**, *23*, 201–212. [CrossRef]
80. Roberto, K.; Mariano, H. On superrings of polynomials and algebraically closed multifields. *J. Appl. Log. IFCoLog J. Log. Appl.* **2022**, *9*, 419–444.
81. Roberto, K.; Mariano, H.; Ribeiro, H. On algebraic extensions and algebraic closures of superfields. *arXiv* **2022**, arXiv:2208.08537.

82. Roberto, K.; Ribeiro, H.; Mariano, H. Quadratic Extensions of Special Hyperfields and the general Arason-Pfister Hauptsatz. *arXiv* **2022**, arXiv:2210.03784.
83. Corsini, P. *Prolegomena of Hypergroup Theory*; Aviani Editore: Udine, Italy, 1993.
84. Davvaz, B.; Leoreanu-Fotea, V. *Hyperring Theory and Applications*; International Academic Press: Palm Harber, FL, USA, 2007.
85. Mittas, J.; Yatras, C. M-polysymmetrical hyperrings. *Ratio Math.* **1997**, *12*, 45–65.
86. Yatras, C. Characteristic of M-polysymmetrical hyperrings and some properties of M- polysymmetrical hyperrings with unity. *Bull. Greek Math. Soc.* **1996**, *38*, 115–125.
87. Yatras, C. M-polysymmetrical subhyperrings and M-polysymmetrical hyperideals. In Proceedings of the 6th International Congruence on Algebraic Hyperstructures and Applications, Prague, Czech Republic, 1–9 September 1996; Democritous Univ. of Thrace Press: Komotini, Greece, 1996; pp. 103–113.
88. Atamewoue Tsafack, S.; Wen, S.; Onasanya, B.O.; Feng, Y. Skew polynomial superrings. *Soft Comput.* **2022**, *26*, 11277–11286. [CrossRef]
89. Linz, A.; Touchard, P. On the hyperfields associated to valued fields. *arXiv* **2022**, arXiv:2211.05082.
90. Creech, S. Extensions of hyperfields. *arXiv* **2019**, arXiv:1912.05919.
91. Gunn, T. Tropical Extensions and Baker-Lorscheid Multiplicities for Idylls. *arXiv* **2022**, arXiv:2211.06480.
92. Novák, M.; Křehlík, Š. *EL*-hyperstructures revisited. *Soft Comput.* **2018**, *22*, 7269–7280. [CrossRef]
93. Massouros, C.G.; Massouros, G.G. On open and closed hypercompositions. In Proceedings of the ICNAAM 2017, Thessaloniki, Greece, 25–30 September 2017. AIP Conference Proceedings 1978, 340002. [CrossRef]
94. Massouros, C.G. Hypergroups and Their Applications. Ph.D. Thesis, National Technical University of Athens, Athens, Greece, 1988.
95. Massouros, C.G. On the result of the difference of a subgroup of the multiplicative group of a field from itself. In Proceedings of the 6th International Congruence on Algebraic Hyperstructures and Applications, Prague, Czech Republic, 1–9 September 1996; Democritous Univercity of Thrace Press: Alexandroupolis, Greece, 1997; pp. 89–101.
96. Massouros, C.G. Getting a field from differences of its multiplicative subgroups. *Bull. Polytech. Inst. Iasi Sect. Math. Theor. Mech. Phys.* **2000**, *XLVI (L)*, 46.
97. Leep, D.; Shapiro, D. Multiplicative subgroups of index three in a field. *Proc. Am. Math. Soc.* **1989**, *105*, 802–807. [CrossRef]
98. Berrizbeitia, P. Additive properties of multiplicative subgroups of finite index in fields. *Proc. Am. Math. Soc.* **1991**, *112*, 365–369. [CrossRef]
99. Bergelson, V.; Shapiro, D. Multiplicative subgroups of finite index in a ring. *Proc. Am. Math. Soc.* **1992**, *116*, 885–896. [CrossRef]
100. Turnwald, G. Multiplicative subgroups of finite index in a division ring. *Proc. Am. Math. Soc.* **1994**, *120*, 377–381. [CrossRef]
101. Cameron, P.J.; Hall, J.I.; van Lint, J.H.; Springer, T.A.; van Tilborg, H.C.A. Translates of subgroups of the multiplicative group of a finite field. *Indag. Math. (Proc.)* **1975**, *78*, 285–289. [CrossRef]

Disclaimer/Publisher's Note: The statements, opinions and data contained in all publications are solely those of the individual author(s) and contributor(s) and not of MDPI and/or the editor(s). MDPI and/or the editor(s) disclaim responsibility for any injury to people or property resulting from any ideas, methods, instructions or products referred to in the content.

Article

Algebraic Hyperstructure of Multi-Fuzzy Soft Sets Related to Polygroups

Osman Kazancı [1], Sarka Hoskova-Mayerova [2,*] and Bijan Davvaz [3]

[1] Department of Mathematics, Karadeniz Technical University, 61080 Trabzon, Türkiye; kazancio@yahoo.com
[2] Department of Mathematics and Physics, University of Defence, Kounicova 65, 662 10 Brno, Czech Republic
[3] Department of Mathematics, Yazd University, Yazd 89136, Iran; davvaz@yazd.ac.ir
* Correspondence: sarka.mayerova@unob.cz; Tel.: +420-973-44-2225

Abstract: The combination of two elements in a group structure is an element, while, in a hypergroup, the combination of two elements is a non-empty set. The use of hypergroups appears mainly in certain subclasses. For instance, polygroups, which are a special subcategory of hypergroups, are used in many branches of mathematics and basic sciences. On the other hand, in a multi-fuzzy set, an element of a universal set may occur more than once with possibly the same or different membership values. A soft set over a universal set is a mapping from parameters to the family of subsets of the universal set. If we substitute the set of all fuzzy subsets of the universal set instead of crisp subsets, then we obtain fuzzy soft sets. Similarly, multi-fuzzy soft sets can be obtained. In this paper, we combine the multi-fuzzy soft set and polygroup structure, from which we obtain a new soft structure called the multi-fuzzy soft polygroup. We analyze the relation between multi-fuzzy soft sets and polygroups. Some algebraic properties of fuzzy soft polygroups and soft polygroups are extended to multi-fuzzy soft polygroups. Some new operations on a multi-fuzzy soft set are defined. In addition to this, we investigate normal multi-fuzzy soft polygroups and present some of their algebraic properties.

Keywords: multi-fuzzy soft set; multi-fuzzy soft polygroup; normal multi-fuzzy soft polygroup

MSC: 20N20; 20N25; 08A72

1. Introduction

The concept of a hyperstructure was first introduced by Marty [1], at the 8th Congress of Scandinavian Mathematicians in 1934, when he defined hypergroups and started to analyze their properties. Indeed, the notion of hypergroups is a generalization of groups. Let H be a non-empty set and $\circ$ be a function (hyperoperation) from $H \times H$ to the family of non-empty subsets of H. Then, $(H, \circ)$ is a hypergroup, if $\circ$ is associative and $a \circ H = H \circ a = H$, for all $a \in H$. The hypergroup is a very general structure. Some researchers considered hypergroups with additional axioms. One of the axioms is the transposition axiom. This axiom is considered by Prenowitz [2–4], and then Jantosciak introduced the notion of transposition hypergroups [5]. A transposition hypergroup that has a scalar identity is called a quasicanonical hypergroup [6,7] or polygroup [8–11]. One can consider the quasicanonical hypergroups as a generalization of canonical hypergroups, introduced in [12]. Examples of polygroups, such as double set algebras, Prenowitz algebras, conjugacy class polygroups and character polygroups, can be found in [11]. This book contains the principal definitions, illustrated with examples and basic results of the theory. The category of polygroups is a category between the category of groups and transposition hypergroups; see Figure 1. More precisely, each group is a polygroup, and each polygroup is a transposition hypergroup. Recently, in [13], an excellent review of the several types of hypergroups was presented. Interesting results can be also found in [14]. The theory of algebraic hyperstructures has

become a well-established branch in algebraic theory and it has extensive applications in many branches of mathematics and applied sciences; see [15–19].

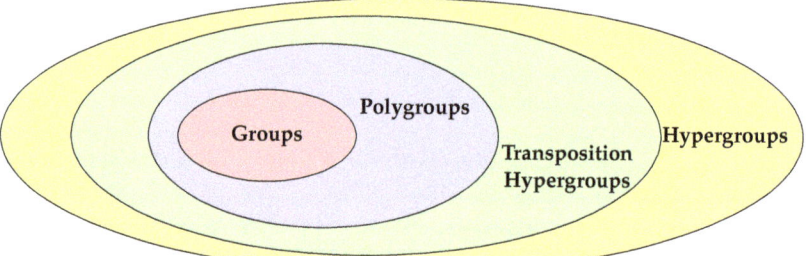

Figure 1. Each group is a polygroup, each polygroup is a transposition hypergroup, and each transposition hypergroup is a hypergroup.

The theory of fuzzy sets proposed by Zadeh [20] has achieved great success in many fields. Many researchers have applied the theory of fuzzy sets to hyperstructures. Firstly, Zahedi [21] discussed the subject of polygroups and fuzzy subpolygroups, and then Davvaz [22] presented the fuzzy subhypergroup concept, which is a generalization of Rosenfeld's fuzzy subgroup [23]. There are many articles dealing with the link between fuzzy sets and hyperstructures; see [24–26].

Soft set theory, introduced by Molodtsov [27], has been considered as an effective mathematical tool for modeling uncertainties. After Molodsov's work, different applications of soft sets were investigated in [28,29]. The idea of a fuzzy soft set, which is more general than fuzzy sets and soft sets, was first introduced by Maji et al. [30], and the algebraic properties of this concept were examined. Both of these theories have been applied to algebraic structures and algebraic hyperstructures—for instance, see [31,32].

Sebastian et al. in [33] proposed the concept of the multi-fuzzy set, which is a more general fuzzy set using ordinary fuzzy sets as building blocks; its membership function is an ordered sequence of ordinary fuzzy membership functions. Later, Yang et al. [34] introduced the concept of the multi-fuzzy soft set, which is a combination of the multi-fuzzy set and soft set, and studied its basic operations. They also introduced the application of this concept in decision making. In recent years, multi-fuzzy sets have become a subject of great interest to researchers and have been widely applied to algebraic structures. Some researchers—for instance, Onasanya and Hoskova-Mayerova [35]—studied the concept of multi-fuzzy groups, while Hoskova-Mayerova et al. [36] studied fuzzy multi-hypergroups and also fuzzy multi-polygroups in [37]. Akın [38] studied the concept of multi-fuzzy soft groups as a generalization of fuzzy soft groups, and Kazancı et al. [39] introduced a novel soft hyperstructure called the multi-fuzzy soft hyperstructure and investigated the notion of multi-fuzzy soft hypermodules and some of their structural properties on a hypermodule.

In a multi-fuzzy set, an element of a universal set U may occur more than once with possibly the same or different membership values. For example, if $U = \{x_1, x_2, x_3, x_4, x_5, x_6\}$, then the set $A = \{<x_1, (0.3, 0.8)>, <x_2, (0.5, 0.7)>, <x_3, (0.1, 0.3)>, <x_4, (0.5, 0.4)>, <x_5, (0.8, 0.6)>, <x_6, (0.4, 0.7)>\}$ is a multi-fuzzy set. A soft set over a universe U is a mapping F from parameters to $\mathcal{P}(U)$. For example, let $U = \{x_1, x_2, x_3, x_4, x_5, x_6\}$ be a set of apartments under consideration, and $A = \{e_1, e_2, e_3, e_4\}$ be a set of parameters such that e_1 = beautiful, e_2 = expensive, e_3 = a good view, and e_4 = near to the city center. If $F(e_1) = \{x_1, x_3\}$, $F(e_2) = \{x_1, x_2, x_5\}$, $F(e_3) = \{x_4, x_6\}$ and $F(e_4) = \{x_2, x_3, x_6\}$, then (F, A) is a soft set. If we substitute the set of all fuzzy subsets of U instead of crisp subsets of U, then we obtain fuzzy soft sets. Similarly, we can define multi-fuzzy soft sets.

In this paper, we combine three separated concepts: polygroups (or quasicanonical hypergroups), soft sets and multi-fuzzy sets (as a generalization of fuzzy sets). Previously, the authors have worked only on the one of these subjects or at most two of them. Indeed, we combine the multi-fuzzy soft set and polygroup structure, from which we obtain a new soft structure called the multi-fuzzy soft polygroup. The relation between the generalization of polygroups is indicated in Figure 2. To facilitate our discussion, we first review some basic concepts of the soft set, fuzzy soft set, multi-fuzzy set and polygroup in Section 2. In Section 3, we apply these to the notion of multi-fuzzy soft sets and polygroups and introduce multi-fuzzy soft polygroups. Then, we study some of their structural characterizations in Sections 4 and 5. Finally, we give the concept of a normal multi-fuzzy soft polygroup and discuss some of their structural characteristics. Finally, some conclusions are pointed out in Section 6.

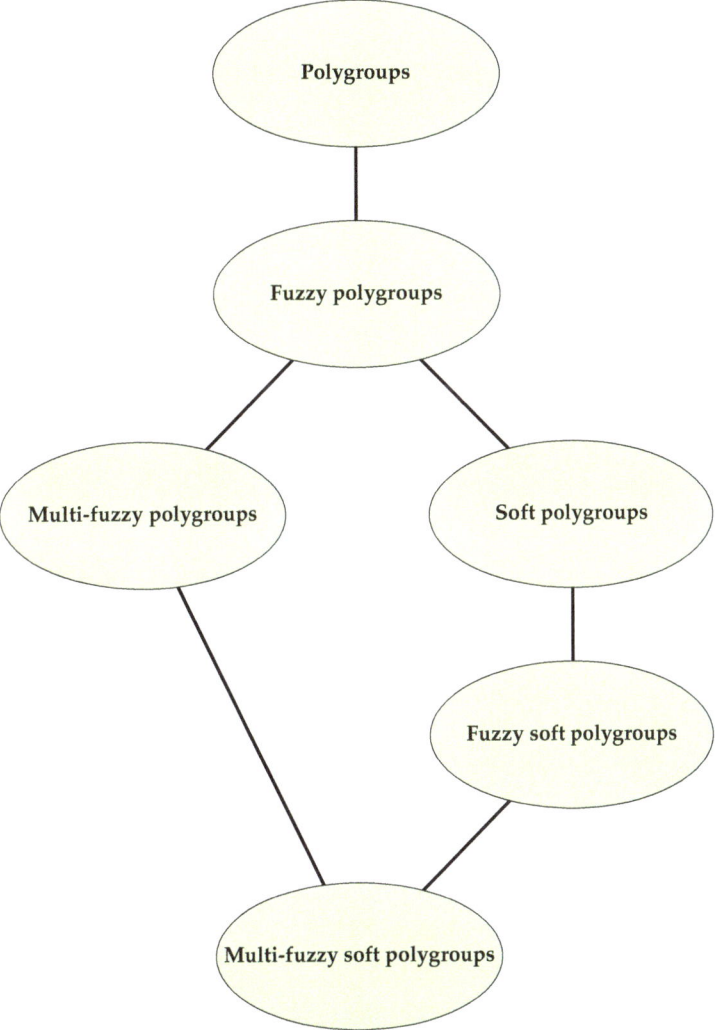

Figure 2. The relation between generalizations of polygroups.

2. Preliminaries

In this section, we provide some definitions and results of soft set theory that will help in understanding the content of the article [27,28,31,32,40]. Let $\mathcal{P}(U)$ denote the power set of U, where U is an initial universe set, E is a set of parameters and $A \subseteq E$.

Definition 1 ([27]). *Let $A \subseteq E$ and $F : A \to \mathcal{P}(U)$ be a set-valued function. Then, the pair (F, A) is called a soft set over U. For all $x \in A$ $F(x) = \{y \in U \mid (x, y) \in R\}$ and R stand for an arbitrary binary relation between an element of A and an element of U—that is, $R \subseteq A \times U$. In fact, a soft set over U is a parameterized family of subsets of the universe U.*

Definition 2 ([30,31]). *Let $A \subseteq E$ and $f : A \to FS(U)$ be a mapping. Then, the pair (f, A) is called a fuzzy soft set over U, where $FS(U)$ is the collection of all fuzzy subsets of U. That is, for each $a \in A$, $f(a)$ is a fuzzy set on U.*

Definition 3 ([33]). *A multi-fuzzy set (MF-set) $\widetilde{A}$ in U is a set of ordered sequences*

$$\widetilde{A} = \{< u, (\mu_i(u)) >: u \in U, \mu_i \in FS(U), i = 1, 2, ..., k\} \text{ and } k \text{ is a positive integer.}$$

The function $\mu_{\widetilde{A}} = (\mu_i(u))$ is said to be the multi membership function of $\widetilde{A}$ denoted by $MF_{\widetilde{A}}$, and k is called dimension of $\widetilde{A}$. The set of all MF-sets of dimension k in U is denoted by $M^k FS(U)$.

It is obvious that the one-dimensional MF-set is Zadeh's fuzzy set, and Atanassov's intuotionistic fuzzy set is a two-dimensional MF-set with $\mu_1(u) + \mu_2(u) \leq 1$.

Definition 4 ([33]). *Let $\widetilde{A} \in M^k FS(U)$. If $\widetilde{A} = \{u/(0, 0, ..., 0) : u \in U\}$, then $\widetilde{A}$ is said to be the null MF-set, defined by $\widetilde{\Phi}_k$. If $\widetilde{A} = \{u/(1, 1, ..., 1) : u \in U\}$, then $\widetilde{A}$ is said to be the absolute MF-set, denoted by $\widetilde{1}_k$.*

Definition 5 ([33]). *Let*

$$\widetilde{A} = \{< u, (\mu_i(u)) >: i = 1, 2, ..., k\} \text{ and } \widetilde{B} = \{< u, (\nu_i(u)) >: i = 1, 2, ..., k\} \in M^k FS(U).$$

Then

(i) $\widetilde{A} \sqsubseteq \widetilde{B}$ if and only if $MF_{\widetilde{A}} \leq MF_{\widetilde{B}}$, i.e $\mu_i(u) \leq \nu_i(u), \forall u \in U$ and $1 \leq i \leq k$.
(ii) $\widetilde{A} = \widetilde{B}$ if and only if $MF_{\widetilde{A}} = MF_{\widetilde{B}}$, i.e $\mu_i(u) = \nu_i(u), \forall u \in U$ and $1 \leq i \leq k$.
(iii) $\widetilde{A} \sqcup \widetilde{B} = \{< u, (\mu_i(u) \vee \nu_i(u)) >: i = 1, 2, ..., k\}$. That is $MF_{\widetilde{A} \sqcup \widetilde{B}} = MF_{\widetilde{A}} \vee MF_{\widetilde{B}}$.
(iv) $\widetilde{A} \sqcap \widetilde{B} = \{< u, (\mu_i(u) \wedge \nu_i(u)) >: i = 1, 2, ..., k\}$. That is $MF_{\widetilde{A} \sqcap \widetilde{B}} = MF_{\widetilde{A}} \wedge MF_{\widetilde{B}}$.

Definition 6 ([34]). *Let $\widetilde{f} : A \to M^k FS(U)$. Then, we call a pair $(\widetilde{f}, A)$ a multi-fuzzy soft set (MFS-set) of dimension k over U. That is, for every $a \in A$, $\widetilde{f}(a) = MF_{\widetilde{f}(a)} \in M^k FS(U)$. Here, $\widetilde{f}(a)$ may be considered a set of a-approximate elements of the multi-fuzzy soft set $(\widetilde{f}, A)$ for $a \in A$.*

Let $A \subseteq E$. Denote the set of all MFS-sets of dimension k over U by $M^k F_S^S(U, E)$

Definition 7 ([34]). *Let $A, B \subseteq E$ and $(\widetilde{f}, A), (\widetilde{g}, B) \in M^k F_S^S(U, E)$. Then, $(\widetilde{f}, A) \sqsubseteq (\widetilde{g}, B)$ if and only if $A \subseteq B$ and $MF_{\widetilde{f}(a)} \sqsubseteq MF_{\widetilde{g}(a)}$ for all $a \in A$.*

Definition 8 ([34]). *Let $(\widetilde{f}, A) \in M^k F_S^S(U, E)$. Then, $(\widetilde{f}, A)$ is said to be a null MFS-set, denoted by $\widetilde{\Phi}_A^k$, if $MF_{\widetilde{f}(a)} = \widetilde{\Phi}_k$ for all $a \in A$.*

$(\widetilde{f}, A)$ is said to be an absolute MFS-set defined by $\widetilde{U}_A^k$ if $MF_{\widetilde{f}(a)} = \widetilde{1}_k$ for each $a \in A$.

Definition 9 ([34]). *Let $(\widetilde{f}, A), (\widetilde{g}, B) \in M^k F_S^S(U, E)$.*

(i) *The $\widetilde{\wedge}$-intersection $(\widetilde{f}, A) \widetilde{\wedge} (\widetilde{g}, B)$ is defined as $(\widetilde{h}, A \times B)$, where $\widetilde{h}(a, b) = \widetilde{f}(a) \sqcap \widetilde{g}(b)$, for all $(a, b) \in A \times B$.*
(ii) *The $\widetilde{\vee}$-union $(\widetilde{f}, A) \widetilde{\vee} (\widetilde{g}, B)$ is defined as $(\widetilde{h}, A \times B)$, where $\widetilde{h}(a, b) = \widetilde{f}(a) \sqcup \widetilde{g}(b)$, for all $(a, b) \in A \times B$.*

(iii) The union $(\widetilde{f}, A) \widetilde{\sqcup} (\widetilde{g}, B)$ is defined as $(\widetilde{h}, C)$, where $C = A \cup B$ and for all $c \in C$ $\widetilde{h}(c) = \widetilde{f}(c)$ if $c \in A - B$, $\widetilde{h}(c) = \widetilde{g}(c)$ if $c \in B - A$ and $\widetilde{h}(c) = \widetilde{f}(c) \sqcup \widetilde{g}(c)$ if $c \in A \cap B$.

Definition 10. *Let $(\widetilde{f}, A), (\widetilde{g}, B) \in M^k F_S^S(U, E)$.*

(i) *The restricted intersection of $(\widetilde{f}, A)$ and $(\widetilde{g}, B)$ is the MFS-set $(\widetilde{h}, C)$ with $A \cap B \neq \emptyset$ where $C = A \cap B$, and for all $c \in C$, $\widetilde{h}(c) = \widetilde{f}(c) \sqcap \widetilde{g}(c)$. The situation is denoted by $(\widetilde{f}, A) \sqcap_{\mathbb{R}} (\widetilde{g}, B) = (\widetilde{h}, C)$.*

(ii) *The extended intersection of $(\widetilde{f}, A)$ and $(\widetilde{g}, B)$ is the MFS-set $(\widetilde{h}, C)$, where $C = A \cup B$ and for all $c \in C$, $\widetilde{h}(c) = \widetilde{f}(c)$ if $c \in A - B$, $\widetilde{h}(c) = \widetilde{g}(c)$ if $c \in B - A$ and $\widetilde{h}(c) = \widetilde{f}(c) \sqcap \widetilde{g}(c)$ if $c \in A \cap B$. In this case, we write $(\widetilde{f}, A) \sqcap_{\mathfrak{E}} (\widetilde{g}, B) = (\widetilde{h}, C)$.*

Definition 11. *Let H be a non-empty set and let $\mathcal{P}^*(H)$ be the set of all non-empty subsets of H. A hyperoperation on H is a map $\circ : H \times H \to \mathcal{P}^*(H)$ and the pair $(H, \circ)$ is called a hypergroupoid.*

Definition 12 ([11,21]). *A multi-valued system $P = <P, \circ, e, ^{-1}>$ is called a polygroup where $e \in P$, $^{-1} : P \to P$, $\circ : P \times P \to \mathcal{P}^*(P)$ if the following axioms hold for all x, y, z in P.*

(i) $x \circ (y \circ z) = (x \circ y) \circ z$,
(ii) $x \circ e = e \circ x = x$,
(iii) $x \in y \circ z$ implies $y \in x \circ z^{-1}$ and $z \in y^{-1} \circ x$.

The following elementary properties follow from the axioms:

$$e \in x \circ x^{-1} \cap x^{-1} \circ x, \quad e^{-1} = e, \quad (x^{-1})^{-1} = x, \text{ and } (x \circ y)^{-1} = y^{-1} \circ x^{-1},$$

where $A^{-1} = \{a^{-1} \mid a \in A\}$.

Let P be a polygroup and K a non-empty subset of P; then, K is called a *subpolygroup* of P if $e \in K$ and $<K, \circ, e, ^{-1}>$ is a polygroup.

A subhypergroup N of a hypergroup is normal if $aN = Na$ [5]. According to [7], a quasicanonical subhypergroup N of a quasicanonical hypergroup H is called *normal* if and only if it is a member of an appreciated quotient system of H by some congruence relation.

Example 1. *Suppose that H is a subgroup of a group G. Define a system $G//H = <\{HgH \mid g \in G\}, *, H, ^{-1}>$, where $(HgH)^{-1} = Hg^{-1}H$ and*

$$(Hg_1 H) * (Hg_2 H) = \{Hg_1 hg_2 H \mid h \in H\}.$$

The algebra of double cosets $G//H$ is a polygroup introduced in (Dresher and Ore [11]).

Example 2. *Consider $P = \{0, 1, 2, a, b\}$ and define $\circ$ on P by the following table:.*

$\circ$	0	1	2	a	b
0	0	1	2	a	b
1	1	$\{0,2\}$	$\{1,2\}$	a	b
2	2	$\{1,2\}$	$\{0,1\}$	a	b
a	a	a	a	$\{0,1,2,b\}$	$\{a,b\}$
b	b	b	b	$\{a,b\}$	$\{0,1,2,a\}$

Then, P is a canonical hypergroup. Suppose that S_3 is the symmetric group on a set with three elements. We consider

$$P \times S_3 = \{(p, x) \mid p \in P \text{ and } x \in S_3\},$$

with the usual hyperoperation

$$(p_1, x_1) \odot (p_2, x_2) \mid p \in p_1 \circ p_2 \text{ and } x = x_1 \cdot x_2\},$$

for all $(p_1, x_1), (p_2, x_2) \in P \times S_3$. Then, $P \times S_3$ is a non-commutative polygroup or quasicanonical hypergroup.

3. Multi-Fuzzy Soft Polygroups

The concept of the MF-set was introduced by Sebastian et al. in [33]. By combining the MF-set and soft set, Yang et al. introduced the concept of the MFS-set [34]. Both of these theories have been applied to algebraic structures. At this point, we give a new type of polygroup named the multi-fuzzy soft polygroup (MFS-polygroup). Since the concepts of uncertainty and fuzziness can be better expressed with MFS-sets, their applications in hyperalgebraic structures are extremely important. Thus, in this section, we provide a new connection between the polygroup structure and MFS-set.

Definition 13. *Let P be a polygroup and $(\tilde{f}, A) \in M^k F_S^S(P, E)$. Then, $(\tilde{f}, A)$ is said to be an MFS-polygroup of dimension k over P if and only if, for all $a \in A$ and $x, y \in P$,*

(i) $\min\{MF_{\tilde{f}(a)}(x), MF_{\tilde{f}(a)}(y)\} \leq \inf_{z \in x \circ y}\{MF_{\tilde{f}(a)}(z)\}$,

(ii) $MF_{\tilde{f}(a)}(x) \leq MF_{\tilde{f}(a)}(x^{-1})$.

That is, for each $a \in A$, $MF_{\tilde{f}(a)}$ is a multi-fuzzy subpolygroup.

The first condition requires that the polygroup is closed under multi-fuzzy soft hyperoperation $\circ$ and the second condition is a generalization of the inverse element under $\circ$.

To better understand this new algebraic structure, consider the following examples.

Example 3. *Let $P = \{e, a, b, c\}$ be a polygroup with the Cayley table:*

$\circ$	e	a	b	c
e	e	a	b	c
a	a	a	$\{e,a,b,c\}$	c
b	b	$\{e,a,b\}$	b	$\{b,c\}$
c	c	$\{a,c\}$	c	$\{e,a,b,c\}$

Let $A = \{e_1, e_2, e_3\}$ be the set of parameters.
Consider the MF-set $\tilde{f}: A \to M^3 FS(P)$ defined as follows. $\tilde{f}: A \to M^3 FS(P)$ as follows.

$MF_{\tilde{f}(e_1)} = \{e/(0.9, 0.8, 0.7), a/(0.6, 0.5, 0.6), b/(0.4, 0.1, 0.2), c/(0.4, 0.1, 0.2)\}$,

$MF_{\tilde{f}(e_2)} = \{e/(0.8, 0.5, 0.6), a/(0.7, 0.4, 0.5), b/(0.6, 0.3, 0.1), c/(0.6, 0.3, 0.1)\}$,

$MF_{\tilde{f}(e_3)} = \{e/(0.8, 0.8, 0.7), a/(0.5, 0.6, 0.3), b/(0.3, 0.6, 0.2), c/(0.2, 0.5, 0.1)\}$.

Then, $(\tilde{f}, A)$ is not an MFS-polygroup of dimension 3 over P since

$$\inf_{c \in c \circ b}\{MF_{\tilde{f}(e_3)}(c)\} \not\geq \min\{MF_{\tilde{f}(e_3)}(c), MF_{\tilde{f}(e_3)}(b)\}.$$

Example 4. *Consider the polygroup given in Example 3 and define the MF-set $\tilde{f}: A \to M^3 FS(P)$ as follows.*

$MF_{\tilde{f}(e_1)} = \{e/(0.8, 0.6, 0.7), a/(0.4, 0.5, 0.6), b/(0.3, 0.4, 0.2), c/(0.3, 0.4, 0.2)\}$,

$MF_{\tilde{f}(e_2)} = \{e/(0.8, 0.5, 0.6), a/(0.6, 0.4, 0.5), b/(0.4, 0.3, 0.4), c/(0.4, 0.3, 0.4)\}$,

$MF_{\tilde{f}(e_3)} = \{e/(0.8, 0.8, 0.7), a/(0.5, 0.6, 0.3), b/(0.3, 0.6, 0.2), c/(0.3, 0.6, 0.2)\}$.

Then, for all $a \in A$, $MF_{\tilde{f}(a)}$ is an MF-subpolygroup of P. By Definition 13, $(\tilde{f}, A)$ is an MFS-polygroup of dimension 3 over P.

Example 5. *Consider the polygroup given in Example 3 and define the MF-set $\widetilde{f} : A \to M^3FS(P)$ as follows.*

$$MF_{\widetilde{f}(e_1)} = \{e/(0.9, 0.8, 0.6), a/(0.8, 0.7, 0.6), b/(0.7, 0.6, 0.5), c/(0.7, 0.6, 0.5)\},$$
$$MF_{\widetilde{f}(e_2)} = \{e/(0.8, 0.8, 0.6), a/(0.7, 0.6, 0.5), b/(0.4, 0.5, 0.2), c/(0.4, 0.5, 0.2)\},$$
$$MF_{\widetilde{f}(e_3)} = \{e/(0.6, 0.7, 0.5), a/(0.5, 0.6, 0.4), b/(0.4, 0.3, 0.1), c/(0.3, 0.4, 0.1)\}.$$

Then, it is clear to see that $MF_{\widetilde{f}(e_1)}$ and $MF_{\widetilde{f}(e_2)}$ are MF-subpolygroups of P. However, $MF_{\widetilde{f}(e_3)}$ is not an MF-subpolygroup of P since

$$\inf_{b \in c \circ c} \{MF_{\widetilde{f}(e_3)}(b)\} \not\geq \min\{MF_{\widetilde{f}(e_3)}(c), MF_{\widetilde{f}(e_3)}(c)\} = MF_{\widetilde{f}(e_3)}(c).$$

By Definition 13 $(\widetilde{f}, A)$ is not an MFS-polygroup of dimension 3 over P.

The following example shows that every soft set (F, A) over P can be seen as an MFS-set of dimension k over P.

Example 6. *Let $A \subset E$ and (F, A) be a soft set over P. For all $a \in A$, the MF-set $\widetilde{\chi}_{F(a)} : A \to M^k FS(P)$ defined by*

$$MF_{\widetilde{\chi}_{F(a)}}(b) = \begin{cases} \widetilde{1}_k & \text{if } b \in F(a) \\ \widetilde{\Phi}_k, & \text{otherwise} \end{cases}$$

for all $b \in A$. Then, $(\widetilde{\chi}_{F(a)}, A) \in M^k F^S_S(P, E)$.

Proposition 1. *Let $(\widetilde{f}, A) \in M^k F^S_S(P, E)$. If $(\widetilde{f}, A)$ is an MFS-polygroups, then, for all $a \in A$ and $x, y \in P$,*
(i) $MF_{\widetilde{f}(a)}(x^{-1}) = MF_{\widetilde{f}(a)}(x),$
(ii) $\inf_{e \in x \circ x^{-1}} \{MF_{\widetilde{f}(a)}(e)\} \geq MF_{\widetilde{f}(a)}(x).$

Proof. (i) By Definition 13, $MF_{\widetilde{f}(a)}(x) \leq MF_{\widetilde{f}(a)}(x^{-1})$ for all $a \in A$ and $x \in P$. Moreover, $MF_{\widetilde{f}(a)}(x) = MF_{\widetilde{f}(a)}(x^{-1})^{-1} \leq MF_{\widetilde{f}(a)}(x^{-1})$. This completes the proof of (i).

(ii) Suppose that $x \in P$. Since $e \in x \circ x^{-1}$ and $(\widetilde{f}, A)$ is an MFS-polygroup, then, for all $a \in A$, we obtain

$$\inf_{e \in x \circ x^{-1}} \{MF_{\widetilde{f}(a)}(e)\} \geq \min\{MF_{\widetilde{f}(a)}(x), MF_{\widetilde{f}(a)}(x^{-1})\}$$
$$= MF_{\widetilde{f}(a)}(x).$$

□

The relationship between soft polygroups and MFS-polygroups is given in the following theorem.

Theorem 1. *Let $F : A \to \mathcal{P}^*(P)$ be a soft set over P. Then, (F, A) is a soft polygroup over P if and only if $(\widetilde{\chi}_{F(a)}, A) \in M^k F^S_S(P, E)$ is an MFS-polygroup.*

Proof. The proof follows by Example 6. □

In Theorem 2, we show that the restricted intersection and the extended intersection of two MFS-polygroups are also an MFS-polygroup.

Theorem 2. Let $(\widetilde{f}, A), (\widetilde{g}, B) \in M^k F_S^S(P, E)$ be two MFS-polygroups.

(i) $(\widetilde{f}, A) \sqcap_{\Re} (\widetilde{g}, B) \in M^k F_S^S(P, E)$ is an MFS-polygroup.

(ii) $(\widetilde{f}, A) \sqcap_{\Im} (\widetilde{g}, B) \in M^k F_S^S(P, E)$ is an MFS-polygroup.

Proof. (i) By Definition 10 (i), let $(\widetilde{f}, A) \sqcap_{\Re} (\widetilde{g}, B) = (\widetilde{h}, C)$, where $C = A \cap B$ and for all $c \in C$, $\widetilde{h}(c) = \widetilde{f}(c) \sqcap \widetilde{g}(c)$. Since $(\widetilde{f}, A)$ and $(\widetilde{g}, B)$ are MFS-polygroups, we have for arbitrary $c \in C$ and for all $x, y \in P$

$$\inf_{z \in x \circ y} \{MF_{\widetilde{f}(c)}(z)\} \geq \min\{MF_{\widetilde{f}(c)}(x), MF_{\widetilde{f}(c)}(y)\},$$

$$MF_{\widetilde{f}(c)}(x) \leq MF_{\widetilde{f}(c)}(x^{-1}) \text{ and}$$

$$\inf_{z \in x \circ y} \{MF_{\widetilde{g}(c)}(z)\} \geq \min\{MF_{\widetilde{g}(c)}(x), MF_{\widetilde{g}(c)}(y)\},$$

$$MF_{\widetilde{g}(c)}(x) \leq MF_{\widetilde{g}(c)}(x^{-1}).$$

For arbitrary $c \in C$ and for all $x, y \in P$,

$$\begin{aligned}
\inf_{z \in x \circ y} \{MF_{\widetilde{h}(c)}(z)\} &= \inf_{z \in x \circ y} \{MF_{\widetilde{f}(c) \sqcap \widetilde{g}(c)}(z)\} \\
&= \inf_{z \in x \circ y} \{MF_{\widetilde{f}(c)}(z) \wedge MF_{\widetilde{g}(c)}(z)\} \\
&= \inf_{z \in x \circ y} \{MF_{\widetilde{f}(c)}(z)\} \wedge \inf_{z \in x \circ y} \{MF_{\widetilde{g}(c)}(z)\} \\
&\geq \min\{MF_{\widetilde{f}(c)}(x), MF_{\widetilde{f}(c)}(y)\} \wedge \min\{MF_{\widetilde{g}(c)}(x), MF_{\widetilde{g}(c)}(y)\} \\
&= \min\{MF_{\widetilde{f}(c)}(x), MF_{\widetilde{g}(c)}(x)\} \wedge \min\{MF_{\widetilde{f}(c)}(y), MF_{\widetilde{g}(c)}(y)\} \\
&= \min\{MF_{\widetilde{f}(c) \sqcap \widetilde{g}(c)}(x), MF_{\widetilde{f}(c) \sqcap \widetilde{g}(c)}(y)\} \\
&= \min\{MF_{\widetilde{h}(c)}(x), MF_{\widetilde{h}(c)}(y)\}.
\end{aligned}$$

Moreover,

$$\begin{aligned}
MF_{\widetilde{h}(c)}(x) &= MF_{\widetilde{f}(c) \sqcap \widetilde{g}(c)}(x) \\
&= \min\{MF_{\widetilde{f}(c)}(x), MF_{\widetilde{g}(c)}(x)\} \\
&\leq \min\{MF_{\widetilde{f}(c)}(x^{-1}), MF_{\widetilde{g}(c)}(x^{-1})\} \\
&= MF_{\widetilde{f}(c) \sqcap \widetilde{g}(c)}(x^{-1}) \\
&= MF_{\widetilde{h}(c)}(x^{-1}).
\end{aligned}$$

Therefore, $(\widetilde{f}, A) \sqcap_{\Re} (\widetilde{g}, B)$ is an MFS-polygroup of dimension k over P.

(ii) According to Definition 10 (ii), we can write $(\widetilde{f}, A) \sqcap_{\Im} (\widetilde{g}, B) = (\widetilde{h}, C)$, $C = A \cup B$. If $c \in A - B$, then $\widetilde{h}(c) = \widetilde{f}(c)$ is an MF-subpolygroup of P, since $(\widetilde{f}, A)$ is an MFS-polygroup over P; if $c \in B - A$, then $\widetilde{h}(c) = \widetilde{g}(c)$ is an MF-subpolygroup of P, since $(\widetilde{g}, B)$ is an MFS-polygroup over P; if $c \in A \cap B$, then $\widetilde{h}(c) = \widetilde{f}(c) \sqcap \widetilde{g}(c)$ is an MF-subpolygroup of P by (i). Therefore, $(\widetilde{f}, A) \sqcap_{\Im} (\widetilde{g}, B)$ is an MFS-polygroup of dimension k over P. □

The following corollary follows from Theorem 2.

Corollary 1. Let $\{(\widetilde{f}_i, A_i) \mid i \in I\} \in M^k F_S^S(P, E)$ be a family of MFS-polygroups. If $\cap_{i \in I} A_i \neq \emptyset$. Then,

(i) $(\sqcap_{\Re})_{i \in I} (\widetilde{f}_i, A_i) \in M^k F_S^S(P, E)$ is an MFS-polygroup.

(ii) $(\sqcap_{\Im})_{i \in I} (\widetilde{f}_i, A_i) \in M^k F_S^S(P, E)$ is an MFS-polygroup.

The union of two MFS-polygroups is not an MFS-polygroup. In Theorem 3, we provide a condition for the union to be an MFS-polygroup as well.

Theorem 3. *Let $(\tilde{f}, A), (\tilde{g}, B) \in M^k F_S^S(P, E)$ be two MFS-polygroups. If $A \cap B = \emptyset$, then $(\tilde{f}, A) \tilde{\sqcup} (\tilde{g}, B) \in M^k F_S^S(P, E)$ is an MFS-polygroup.*

Proof. By Definition 9(iii), we can write $(\tilde{f}, A) \tilde{\sqcup} (\tilde{g}, B) = (\tilde{h}, C)$, where $C = A \cup B$. Since $A \cap B = \emptyset$, it follows that either $c \in A - B$ or $c \in B - A$ for all $c \in C$. If $c \in A - B$, then $\tilde{h}(c) = \tilde{f}(c)$ is an MF-subpolygroup of P and if $c \in B - A$, then $\tilde{h}(c) = \tilde{g}(c)$ is an MF-subpolygroup of P. Therefore, $(\tilde{f}, A) \tilde{\sqcup} (\tilde{g}, B)$ is an MFS-polygroup of dimension k over P. □

Theorem 4. *Let $(\tilde{f}, A), (\tilde{g}, B) \in M^k F_S^S(P, E)$ be two MFS-polygroups. Then, $(\tilde{f}, A) \tilde{\wedge} (\tilde{g}, B) \in M^k F_S^S(P, E)$ is an MFS-polygroup.*

Proof. By Definition 9(i), let $(\tilde{f}A) \tilde{\wedge} (\tilde{g}, B) = (\tilde{h}, A \times B)$. We know that for all $a \in A$, $\tilde{f}(a)$ is an MF-subpolygroup of P and for all $b \in B$, $\tilde{g}(b)$ is an MF-subpolygroup of P and so is $\tilde{h}(a, b) = MF_{\tilde{h}(a,b)} = MF_{\tilde{f}(a) \cap \tilde{g}(b)}$ for all $(a, b) \in A \times B$, because the intersection of two multi-fuzzy subpolygroups is also an MF-subpolygroup. Hence, $(\tilde{f}, A) \tilde{\wedge} (\tilde{g}, B)$ is an MFS-polygroup of dimension k over P. □

By Theorems 3 and 4, we obtain the following corollary.

Corollary 2. *Let $\{(\tilde{f}_i, A_i) \mid i \in I\} \in M^k F_S^S(P, E)$ be a family of MFS-polygroups.*

(i) *If $A_i \cap A_j = \emptyset$ for all $i, j \in I$ and $i \neq j$, then $\tilde{\sqcup}_{i \in I}(\tilde{f}_i, A_i) \in M^k F_S^S(P, E)$ is an MFS-polygroup.*

(ii) *$\tilde{\wedge}_{i \in I}(\tilde{f}_i, A_i) \in M^k F_S^S(P, E)$ is an MFS-polygroup.*

The following theorem gives a condition for the $\tilde{\vee}$-union of two MFS-polygroups to be an MFS-polygroup.

Theorem 5. *Let $(\tilde{f}, A), (\tilde{g}, B) \in M^k F_S^S(P, E)$ be two MFS-polygroups. If $(\tilde{f}, A) \sqsubset (\tilde{g}, B)$ or $(\tilde{g}, B) \sqsubset (\tilde{f}, A)$, then $(\tilde{f}, A) \tilde{\vee} (\tilde{g}, B) \in M^k F_S^S(P, E)$ is an MFS-polygroup.*

Proof. Suppose that $(\tilde{f}, A)$ and $(\tilde{g}, B)$ are MFS-polygroups of dimension k over P. By Definition 9 (ii), we can write $(\tilde{f}, A) \tilde{\vee} (\tilde{g}, B) = (\tilde{h}, C)$, where $C = A \times B$, and $\tilde{h}(a, b) = \tilde{f}(a) \sqcup \tilde{g}(b)$ for all $(a, b) \in C$. Since $(\tilde{f}, A)$ and $(\tilde{g}, B)$ are MFS-polygroups of dimension k over P, we obtain that for all $a \in A$, $\tilde{f}(a)$ is an MF-subpolygroup of P and for all $b \in B$, $\tilde{g}(b)$ is an MF-subpolygroup of P. By assumption, $\tilde{h}(a, b) = \tilde{f}(a) \sqcup \tilde{g}(b)$ is an MF-subpolygroup of P for all $(a, b) \in C$. Hence, $(\tilde{f}, A) \tilde{\vee} (\tilde{g}, B)$ is an MFS-polygroup. □

Definition 14. *The sum of two MFS-sets $(\tilde{f}, A)$ and $(\tilde{g}, B)$ of dimension k over P, denoted by $(\tilde{f}, A) \oplus (\tilde{g}, B)$, is the MFS-set $(\tilde{h}, C)$, where $C = A \cup B$ and for all $c \in C$,*

$$\tilde{h}(c) = \begin{cases} \tilde{f}(c) & \text{if } c \in A \setminus B \\ \tilde{g}(c) & \text{if } c \in B \setminus A \\ \tilde{f}(c) \oplus \tilde{g}(c) & \text{if } c \in A \cap B \end{cases}$$

For every $z \in P$,

$$(\tilde{f}(c) \oplus \tilde{g}(c))(z) = \bigvee \{MF_{\tilde{f}(c)}(x) \wedge MF_{\tilde{g}(c)}(y), x, y \in P, z \in x \circ y\}.$$

The next theorem gives a condition for the sum of two MFS-polygroups to be an MFS-polygroup.

Theorem 6. Let $(\tilde{f}, A), (\tilde{g}, B) \in M^k F_S^S(P, E)$ be two MFS-polygroups. If $(\tilde{f}, A) \oplus (\tilde{g}, B) = (\tilde{g}, B) \oplus (\tilde{f}, A)$, then $(\tilde{f}, A) \oplus (\tilde{g}, B) \in M^k F_S^S(P, E)$ is an MFS-polygroup.

Proof. The proof is straightforward. □

Definition 15. Let $(\tilde{f}, A) \in M^k F_S^S(P, E)$. The soft set

$$(\tilde{f}, A)_t = \{(MF_{\tilde{f}(a)})_t \mid a \in A\} \text{ where } (MF_{\tilde{f}(a)})_t = \{x \in P \mid MFS_{\tilde{f}(a)}(x) \geq t\},$$

for all $t = (t_1, t_2, ..., t_k)$, $t_i \in (0, 1]$ $1 \leq i \leq k$, is called a t-level soft set of the MFS-set $(\tilde{f}, A)$, where $(MF_{\tilde{f}(a)})_t$ is a t-level subset of the MF-set $MF_{\tilde{f}(a)}$.

The following theorem explores the relation between MFS-polygroups and t-level soft sets.

Theorem 7. Let $(\tilde{f}, A) \in M^k F_S^S(P, E)$. Then, $(\tilde{f}, A)$ is an MFS-polygroup if and only, if for all $a \in A$ and for arbitrary $t \in (0, 1]$ with $(MF_{\tilde{f}(a)})_t \neq \emptyset$, the t-level soft set $(\tilde{f}, A)_t$ is a soft polygroup over P in Wanga's sense [40].

Proof. Let $(\tilde{f}, A) \in M^k F_S^S(P, E)$ be an MFS-polygroup. Then, for each $a \in A$, $MF_{\tilde{f}(a)}$ is an MF-subpolygroup of P. Suppose that $t \in (0, 1]$ with $(MF_{\tilde{f}(a)})_t \neq \emptyset$ and $x, y \in (MF_{\tilde{f}(a)})_t$. Then, $MF_{\tilde{f}(a)}(x) \geq t$, $MF_{\tilde{f}(a)}(y) \geq t$. Thus,

$$t \leq \min\{MF_{\tilde{f}(a)}(x), MF_{\tilde{f}(a)}(y)\} \leq \inf_{z \in x \circ y}\{MF_{\tilde{f}(a)}(z)\}.$$

which implies $MF_{\tilde{f}(a)}(z) \geq t$ for all $z \in x \circ y$. Therefore, $x \circ y \subseteq (MF_{\tilde{f}(a)})_t$. Moreover, for $x \in (MF_{\tilde{f}(a)})_t$, we have $MF_{\tilde{f}(a)}(x^{-1}) \geq MF_{\tilde{f}(a)}(x) \geq t$. It follows that $x^{-1} \in (MF_{\tilde{f}(a)})_t$. we obtain that $(MF_{\tilde{f}(a)})_t$ is a subpolygroup of P for all $a \in A$. Consequently, $(\tilde{f}, A)_t$ is a soft polygroup over P. Conversely, let $(\tilde{f}, A)_t$ be a soft polygroup over P for all $t \in (0, 1]$. Let $t_0 = \min\{MF_{\tilde{f}(a)}(x), MF_{\tilde{f}(a)}(y)\}$ for some $x, y \in P$. Then, obviously, $x, y \in (MF_{\tilde{f}(a)})_{t_0}$; consequently, $x \circ y \subseteq (MF_{\tilde{f}(a)})_{t_0}$. Thus,

$$\min\{MF_{\tilde{f}(a)}(x), MFS_{\tilde{f}(a)}(y)\} = t_0 \leq \inf_{z \in x \circ y}\{MF_{\tilde{f}(a)}(z)\}.$$

Now, $t_0 = MF_{\tilde{f}(a)}(x)$ for some $x \in P$. Since, by the assumption, every non-empty t-level soft set $(\tilde{f}, A)_t$ is a soft polygroup over P, $x^{-1} \in (MF_{\tilde{f}(a)})_{t_0}$. Hence, $MF_{\tilde{f}(a)}(x^{-1}) \geq t_0 = MF_{\tilde{f}(a)}(x)$. As a result, we obtain that $MF_{\tilde{f}(a)}$ is an MF-subpolygroup of P for all $a \in A$. Consequently, $(\tilde{f}, A)$ is an MFS-polygroup of dimension k over P. □

4. The Behavior Image and Inverse Image of MFS-Polygroups

Definition 16. A pair (φ, ψ) is called an MF-soft function from P_1 to P_2, where $\varphi : P_1 \to P_2$ and $\psi : E_1 \to E_2$ are functions.

Definition 17. Let $(\tilde{f}, A) \in M^k F_S^S(P_1, E_1), (\tilde{g}, B) \in M^k F_S^S(P_2, E_2)$ and (φ, ψ) be an MF- soft function from P_1 to P_2.

(i) The image of $(\widetilde{f}, A)$ under the MF-soft function (φ, ψ), denoted by $(\varphi, \psi)(\widetilde{f}, A)$, is the MFS-set $(\varphi(\widetilde{f}), \psi(A))$ such that the MF-set $\varphi(\widetilde{f})(t)$ for any $t \in \psi(A)$ is characterized by the following MF-membership function:

$$MF_{\varphi(\widetilde{f})(t)}(y) = \begin{cases} \bigvee_{\varphi(x)=y} \bigvee_{\psi(a)=t} MF_{\widetilde{f}(a)}(x) & \text{if } \exists x \in \varphi^{-1}(y) \\ 0, & \text{otherwise} \end{cases}$$

for all $y \in P_2$.

(ii) The pre-image of $(\widetilde{g}, B)$ under the MF-soft function (φ, ψ), denoted by $(\varphi, \psi)^{-1}(\widetilde{g}, B)$, is the MFS-set $(\varphi^{-1}(\widetilde{g}), \psi^{-1}(B))$ such that the MF-set $\varphi^{-1}(\widetilde{g})(a)$ is characterized by the following MF-membership function:

$$MF_{\varphi^{-1}(\widetilde{g})(a)}(x) = MF_{\widetilde{g}(\psi(a))}(\varphi(x))$$

for all $a \in \psi^{-1}(B)$ and $x \in P_1$.

If φ and ψ are injective (surjective), then (φ, ψ) is said to be injective (surjective).

Definition 18. *Let P_1, P_2, be two polygroups and (φ, ψ) be an MF-soft function from P_1 to P_2. If φ is a strong homomorphism of polygroups, then the pair (φ, ψ) is called an MF-soft homomorphism. If φ is an isomorphism and ψ is a one-to-one mapping, then (φ, ψ) is said to be an MF-soft isomorphism.*

Theorem 8. *Let P_1, P_2 be two polygroups and (φ, ψ) be an MF-soft homomorphism from P_1 to P_2. If $(\widetilde{f}, A) \in M^k F_S^S(P_1, E_1)$ is an MFS-polygroup, then $(\varphi, \psi)(\widetilde{f}, A) \in M^k F_S^S(P_2, E_2)$ is an MFS-polygroup.*

Proof. Let $k \in \psi(A)$, $u, v \in P_2$. If $\varphi^{-1}(u) = \emptyset$ or $\varphi^{-1}(v) = \emptyset$, the proof is straightforward. Assume that there exists $x, y \in P_1$, such that $\varphi(x) = u$ and $\varphi(y) = v$. Since $(\widetilde{f}, A) \in M^k F_S^S(P_1, E_1)$ is an MFS-polygroup, it follows that for each $a \in A$

$$\min\{MF_{\widetilde{f}(a)}(x), MF_{\widetilde{f}(a)}(y)\} \leq MF_{\widetilde{f}(a)}(z)$$

for all $z \in x \circ y$. Let $z^* \in u \circ v = \varphi(x \circ y)$. We obtain $z^* = \varphi(z)$. Then, we have

$$\min\{\bigvee_{\varphi(x)=u} MF_{\widetilde{f}(a)}(x), \bigvee_{\varphi(y)=v} MF_{\widetilde{f}(a)}(y)\} \leq \bigvee_{\varphi(x)=u} \bigvee_{\varphi(y)=v} MF_{\widetilde{f}(a)}(z).$$

Hence,

$$\min\{MF_{\varphi(\widetilde{f})(t)}(u), MF_{\varphi(\widetilde{f})(t)}(v)\} \leq \bigvee_{\psi(a)=t} \bigvee_{\varphi(x)=u} \bigvee_{\varphi(y)=v} MF_{\widetilde{f}(a)}(z)$$

$$= \bigvee_{\psi(a)=t} \bigvee_{\varphi(z)=z^*} MF_{\varphi(\widetilde{f})(t)}(z)$$

for all $z^* \in u \circ v$. Then, we have

$$\inf_{z^* \in u \circ v}\{MF_{\varphi(\widetilde{f})(t)}(z^*)\} \geq \min\{MF_{\varphi(\widetilde{f})(t)}(u), MF_{\varphi(\widetilde{f})(t)}(v)\}$$

Moreover, for all $u \in P_2$ where $\varphi(x) = u$ and $x \in P_1$, we have

$$MF_{\varphi(\widetilde{f})(t)}(u^{-1}) = \bigvee_{\varphi(x^{-1})=u^{-1}} \bigvee_{\psi(a)=k} MF_{\widetilde{f}(a)}(x^{-1})$$
$$\geq \bigvee_{\varphi(x)=u} \bigvee_{\psi(a)=t} MF_{\widetilde{f}(a)}(x)$$
$$= MF_{\varphi(\widetilde{f})(t)}(u)$$

Consequently, $(\varphi, \psi)(\widetilde{f}, A) \in M^k F_S^S(P_2, E_2)$ is an MFS-polygroup. □

Theorem 9. *Let P_1, P_2 be two polygroups and (φ, ψ) be an MF-soft homomorphism from P_1 to P_2. If $(\widetilde{g}, B) \in M^k F_S^S(P_2, E_2)$ is an MFS-polygroup, then $(\varphi^{-1}(\widetilde{g}), \psi^{-1}(B)) \in M^k F_S^S(P_1, E_1)$ is an MFS-polygroup.*

Proof. Let $a \in \psi^{-1}(B)$, $x, y \in P_1$. For all $z \in x \circ y$, we have

$$\inf_{z \in x \circ y}\{MF_{\varphi^{-1}(\widetilde{g})(a)}(z)\} = \inf_{z \in x \circ y}\{MF_{\widetilde{g}(\psi(a))}(\varphi(z))\}$$
$$\geq \min\{MF_{\widetilde{g}(\psi(a))}(\varphi(x)), MF_{\widetilde{g}(\psi(a))}(\varphi(y))\}$$
$$= \min\{MF_{(\varphi^{-1}\widetilde{g})(a)}(x), MF_{(\varphi^{-1}\widetilde{g})(a)}(y)\}$$

Similarly, we obtain $MF_{(\varphi^{-1}\widetilde{g})(a)}(x^{-1}) \geq MF_{(\varphi^{-1}\widetilde{g})(a)}(x)$. Therefore, we conclude that $(\varphi^{-1}(\widetilde{g}), \psi^{-1}(B)) \in M^k F_S^S(P_1, E_1)$ is an MFS-polygroup. □

5. Normal MFS-Polygroups

In this section, we define normal MFS-polygroups and study some of their basic properties. We proved that the images of normal MFS-polygroups are the normal MFS-polygroups under some conditions.

Definition 19. *Let $(\widetilde{f}, A) \in M^k F_S^S(P, E)$ be an MFS-polygroup. Then, $(\widetilde{f}, A)$ is said to be normal if and only if*

$$\inf_{z \in x \circ y}\{MF_{\widetilde{f}(a)}(z)\} = \inf_{z' \in y \circ x}\{MF_{\widetilde{f}(a)}(z')\},$$

for all $a \in A$ and $x, y \in P$.

It is obvious that if $(\widetilde{f}, A)$ is a normal MFS-polygroup, then

$$\inf_{z \in x \circ y}\{MF_{\widetilde{f}(a)}(z) = \inf_{z' \in x \circ y}\{MF_{\widetilde{f}(a)}(z')\},$$

for all $a \in A$ and $x, y \in P$.

Theorem 10. *Let $(\widetilde{f}, A) \in M^k F_S^S(P, E)$ be an MFS-polygroup. Then, the following conditions are equivalent:*

(i) *$(\widetilde{f}, A)$ is a normal MFS-polygroup,*
(ii) $\inf_{z \in x \circ y \circ x^{-1}}\{MF_{\widetilde{f}(a)}(z)\} = MF_{\widetilde{f}(a)}(y)$, *for all $a \in A$ and $x, y \in P$,*
(iii) $\inf_{z \in x \circ y \circ x^{-1}}\{MF_{\widetilde{f}(a)}(z)\} \geq MF_{\widetilde{f}(a)}(y)$, *for all $a \in A$ and $x, y \in P$,*
(iv) $\inf_{z \in y^{-1} \circ x^{-1} \circ y \circ x}\{MF_{\widetilde{f}(a)}(z)\} \geq MF_{\widetilde{f}(a)}(y)$, *for all $a \in A$ and $x, y \in P$.*

Proof. $(i) \Rightarrow (ii)$: For any $a \in A$, suppose that $x, y \in P$ and $z \in x \circ y \circ x^{-1}$. Then, $z \in x \circ s$, where $s \in y \circ x^{-1}$. Since $s \in y \circ x^{-1}$, then $y \in s \circ (x^{-1})^{-1} = s \circ x$. Thus, by hypothesis, we obtain

$$\inf_{z \in x \circ s}\{MF_{\tilde{f}(a)}(z)\} = \inf_{y \in s \circ x}\{MF_{\tilde{f}(a)}(y)\} = MF_{\tilde{f}(a)}(y).$$

That is, $\inf_{z \in x \circ y \circ x^{-1}}\{MF_{\tilde{f}(a)}(z)\} = MF_{\tilde{f}(a)}(y).$

$(ii) \Rightarrow (iii)$: The proof is trivial.

$(iii) \Rightarrow (iv)$: For any $a \in A$, suppose that $x, y \in P$ and $z \in y^{-1} \circ x^{-1} \circ y \circ x$. Then, $z \in y^{-1} \circ s$, where $s \in x^{-1} \circ y \circ x$. By (iii), we obtain $\inf_{s \in x^{-1} \circ y \circ x}\{MF_{\tilde{f}(a)}(s)\} \geq MF_{\tilde{f}(a)}(y).$ Since $z \in y^{-1} \circ s$ and letting $(\tilde{f}, A) \in M^k FS(P)$ be a MFS-polygroup, then we have

$$\inf_{z \in y^{-1} \circ s}\{MF_{\tilde{f}(a)}(z)\} \geq \min\{MF_{\tilde{f}(a)}(y^{-1}), MF_{\tilde{f}(a)}(s)\} = MF_{\tilde{f}(a)}(y).$$

That is, $\inf_{z \in y^{-1} \circ x^{-1} \circ y \circ x}\{MF_{\tilde{f}(a)}(z)\} \geq MF_{\tilde{f}(a)}(y)$ for all $a \in A$ and $x, y \in P$.

$(iv) \Rightarrow (i)$: For any $a \in A$, suppose that $x, y \in P$ and $u \in x^{-1} \circ y \circ x$. Then, $u \in x^{-1} \circ y \circ x \subset y \circ y^{-1} \circ x^{-1} \circ y \circ x$. Thus, $u \in y \circ s$, where $s \in y^{-1} \circ x^{-1} \circ y \circ x$. By (iv), we obtain $\inf_{s \in y^{-1} \circ x^{-1} \circ y \circ x}\{MF_{\tilde{f}(a)}(s)\} \geq MF_{\tilde{f}(a)}(y).$ On the other hand,

$$\inf_{u \in y \circ s}\{MF_{\tilde{f}(a)}(u)\} \geq \min\{MF_{\tilde{f}(a)}(y), MF_{\tilde{f}(a)}(s)\} = MF_{\tilde{f}(a)}(y).$$

Now, let $\omega \in x \circ y$ and $v \in y \circ x$. Then, $y \in v \circ x^{-1}$ and so $\omega \in x \circ y \subset x \circ v \circ x^{-1}$. By the above result, $MF_{\tilde{f}(a)}(\omega) \geq MF_{\tilde{f}(a)}(v)$. Similarly, we obtain $MF_{\tilde{f}(a)}(v) \geq MF_{\tilde{f}(a)}(\omega)$. Therefore,

$$\inf_{\omega \in x \circ y}\{MF_{\tilde{f}(a)}(\omega)\} = \inf_{v \in y \circ x}\{MF_{\tilde{f}(a)}(v)\},$$

for all $a \in A$ and $x, y \in P$. Hence, $(\tilde{f}, A)$ is a normal MFS-polygroup. □

Lemma 1. *Let $(\tilde{f}, A) \in M^k F_S^S(P, E)$ be an MFS-polygroup. If $MF_{\tilde{f}(a)}(x) < MF_{\tilde{f}(a)}(y)$ for all $a \in A$ and $x, y \in P$, then*

$$\inf_{z \in x \circ y}\{MF_{\tilde{f}(a)}(z)\} = \inf_{z' \in y \circ x}\{MFS_{\tilde{f}(a)}(z')\} = MF_{\tilde{f}(a)}(x).$$

Proof. Let $x, y \in P$ and $z \in x \circ y$. Then,

$$MF_{\tilde{f}(a)}(z) \geq \min\{MF_{\tilde{f}(a)}(x), MF_{\tilde{f}(a)}(y)\} = MF_{\tilde{f}(a)}(x)$$

for all $a \in A$. Since $z \in x \circ y$, then $x \in z \circ y^{-1}$. Thus,

$$\inf_{x \in z \circ y^{-1}}\{MF_{\tilde{f}(a)}(x)\} \geq \min\{MF_{\tilde{f}(a)}(z), MF_{\tilde{f}(a)}(y^{-1})\}$$
$$= \min\{MF_{\tilde{f}(a)}(z), MF_{\tilde{f}(a)}(y)\}.$$

If $\min\{MF_{\tilde{f}(a)}(z), MF_{\tilde{f}(a)}(y)\} = MF_{\tilde{f}(a)}(y)$, then $MF_{\tilde{f}(a)}(x) \geq MF_{\tilde{f}(a)}(y)$, a contradiction. Thus, $\min\{MF_{\tilde{f}(a)}(z), MF_{\tilde{f}(a)}(y)\} = MF_{\tilde{f}(a)}(z)$. Hence, $MF_{\tilde{f}(a)}(x) \geq MF_{\tilde{f}(a)}(z)$. Consequently, $\inf_{z \in x \circ y}\{MF_{\tilde{f}(a)}(z)\} = MF_{\tilde{f}(a)}(x)$ for all $a \in A$ and $x, y \in P$. Similarly, we obtain $\inf_{z' \in y \circ x}\{MF_{\tilde{f}(a)}(z')\} = MF_{\tilde{f}(a)}(x)$ for all $a \in A$ and $x, y \in P$. □

Theorem 11. Let $(\tilde{f}, A) \in M^k F_S^S(P, E)$ be an MFS-polygroup. Then, $(\tilde{f}, A)$ is normal if and only if

$$MF_{\tilde{f}(a)}(x) = MF_{\tilde{f}(a)}(y) \Rightarrow \inf_{z \in x \circ y}\{MF_{\tilde{f}(a)}(z)\} = \inf_{z' \in y \circ x}\{MF_{\tilde{f}(a)}(z')\},$$

for all $a \in A$ and $x, y \in P$.

Proof. The proof of Theorem 11 follows from Lemma 1. □

Theorem 12. Let $(\tilde{f}, A) \in M^k F_S^S(P, E)$. Then, $(\tilde{f}, A)$ is a normal MFS-polygroup if and only if each of its non-empty level subsets is a normal soft polygroup over P.

Proof. Let $(\tilde{f}, A) \in M^k F_S^S(P, E)$ be a normal MFS-polygroup. By Theorem 7, $(MF_{\tilde{f}(a)})_t$ is a soft polygroup over P for all $a \in A$. Now, we will show that $(MF_{\tilde{f}(a)})_t$ is normal. Suppose that $y \in (MF_{\tilde{f}(a)})_t$ and $x \in P$. Then, we have

$$\inf_{z \in x \circ y \circ x^{-1}}\{MF_{\tilde{f}(a)}(z)\} \geq MF_{\tilde{f}(a)}(y) \geq t.$$

It follows that $MF_{\tilde{f}(a)}(z) \geq t$ for all $z \in x \circ y \circ x^{-1}$. That is, $x \circ y \circ x^{-1} \subset (MF_{\tilde{f}(a)})_t$. We obtain that $(MF_{\tilde{f}(a)})_t$ is a normal subpolygroup of P for all $a \in A$. Consequently, $(\tilde{f}, A)_t$ is a normal soft polygroup over P. Conversely, let $(\tilde{f}, A)_t$ be a normal soft polygroup over P for all $t \in [0, 1]$. By Theorem 7, $(\tilde{f}, A) \in M^k F_S^S(P, E)$ is an MFS-polygroup. That is, $MF_{\tilde{f}(a)}$ is an MF-subpolygroup of P for all $a \in A$. We will show that $MF_{\tilde{f}(a)}$ is normal. Assume that $x, y \in P$, $t_0 = MF_{\tilde{f}(a)}(y)$. Then, $MF_{\tilde{f}(a)}(y) \geq t_0$. Since $(\tilde{f}, A)_{t_0}$ is normal, we have $x \circ y \circ x^{-1} \subset (MFS_{\tilde{f}(a)})_{t_0}$. Thus, $z \in (MF_{\tilde{f}(a)})_{t_0}$ for all $z \in x \circ y \circ x^{-1}$. Therefore,

$$\inf_{z \in x \circ y \circ x^{-1}}\{MF_{\tilde{f}(a)}(z)\} \geq t_0 = MF_{\tilde{f}(a)}(y).$$

We obtain that $MF_{\tilde{f}(a)}$ is a normal MF-subpolygroup of P for all $a \in A$. Consequently, $(\tilde{f}, A)$ is a normal MFS-polygroup. □

Theorem 13. Let $(\tilde{f}, A), (\tilde{g}, B) \in M^k F_S^S(P, E)$ be two normal MFS-polygroups. Then,
(i) $(\tilde{f}, A) \sqcap_{\Re} (\tilde{g}, B)$ is a normal MFS-polygroup.
(ii) $(\tilde{f}, A) \sqcap_{\Im} (\tilde{g}, B)$ is a normal MFS-polygroup.
(iii) If $A \cap B = \emptyset$, then $(\tilde{f}, A) \sqcup (\tilde{g}, B)$ is a normal MFS-polygroup.
(iv) $(\tilde{f}, A) \widetilde{\wedge} (\tilde{g}, B)$ is a normal MFS-polygroup.

Theorem 14. Let P_1, P_2 be two polygroups and (φ, ψ) be a surjective multi-fuzzy soft homomorphism from P_1 to P_2. If $(\tilde{f}, A) \in M^k F_S^S(P_1, E_1)$ is a normal MFS-polygroup, then $(\varphi, \psi)(\tilde{f}, A) \in M^k F_S^S(P_2, E_2)$ is a normal MFS-polygroup.

Proof. For each $t \in \psi(A)$ and $u, v \in P_2$, there exists $x, y \in P_1$, such that $\varphi(x) = u$ and $\varphi(y) = v$. Since $(\tilde{f}, A) \in M^k F_S^S(P, E)$ is a normal MFS-polygroup, it follows that for each $a \in A$

$$MF_{\tilde{f}(a)}(y) \leq MF_{\tilde{f}(a)}(z)$$

for all $z \in x \circ y \circ x^{-1}$. Let $z^* \in u \circ v \circ u^{-1} = \varphi(x \circ y \circ x^{-1})$. We obtain $z^* = \varphi(z)$. Then, we have

$$\bigvee_{\varphi(y)=v} MF_{\tilde{f}(a)}(y) \leq \bigvee_{\varphi(x)=u} \bigvee_{\varphi(y)=v} \bigvee_{\varphi(x^{-1})=u^{-1}} MF_{\tilde{f}(a)}(z).$$

Hence,

$$MF_{\varphi(\tilde{f})(t)}(v) \leq \bigvee_{\psi(a)=t} \bigvee_{\varphi(x)=u} \bigvee_{\varphi(y)=v} \bigvee_{\varphi(x^{-1})=u^{-1}} MF_{\tilde{f}(a)}(z)$$
$$= \bigvee_{\psi(a)=t} \bigvee_{\varphi(z)=z^*} MF_{\varphi(\tilde{f})(t)}(z)$$

for all $z^* \in u \circ v \circ u^{-1}$. Then, we have

$$\inf_{z^* \in u \circ v \circ u^{-1}} \{MF_{\varphi(\tilde{f})(t)}(z^*)\} \geq MF_{\varphi(\tilde{f})(t)}(v)\}$$

Consequently, $(\varphi, \psi)(\tilde{f}, A)$ is a normal MFS-polygroup. □

Theorem 15. *Let P_1, P_2 be two polygroups and (φ, ψ) be an MF-soft homomorphism from P_1 to P_2. If $(\tilde{g}, B) \in M^k F_S^S(P_2, E_2)$ is a normal MFS-polygroup, then $(\varphi^{-1}(\tilde{g}), \psi^{-1}(B)) \in M^k F_S^S(P_1, E_1)$ is a normal MFS-polygroup.*

Proof. Let $a \in \psi^{-1}(B)$, $x, y \in P_1$. For all $z \in x \circ y \circ x^{-1}$, we have

$$\inf_{z \in x \circ y \circ x^{-1}} \{MF_{(\varphi^{-1}(\tilde{g}))(a)}(z)\} = \inf_{z \in x \circ y \circ x^{-1}} \{MF_{\tilde{g}(\psi(a))}(\varphi(z))\}$$
$$\geq MF_{\tilde{g}(\psi(a))}(\varphi(y))$$
$$= MF_{(\varphi^{-1}(\tilde{g}))(a)}(y).$$

Therefore, $(\varphi^{-1}(\tilde{g}), \psi^{-1}(B))$ is a normal MFS-polygroup. □

6. Conclusions

In real life, many problems often involve uncertainties that are difficult to describe and solve with traditional mathematical tools. To investigate these uncertainties, many researchers have proposed mathematical theory to address the problem of uncertainty. Currently, mathematical theories dealing with the problem of uncertainty include fuzzy set theory, soft set theory, multi-fuzzy set theory, probability theory and so on. The purpose of this paper is to apply the MFS-set theory to algebraic hyperstructures, motivated by the study of the algebraic structures of MF-sets. We generalized the concept of fuzzy polygroups and studied the algebraic properties of MFS-sets in polygroup structures. Thus, this paper provides a new connection between polygroup structures and MFS-sets. We hope that our work enhances the understanding of MFS-polygroups for future researchers. To extend this work, one should study the MFS-sets related to various hyperrings, which can be researched further. A solution to a decision-making problem can be investigated using a different algorithm in the future as well.

Author Contributions: Conceptualization, O.K. and B.D.; methodology, O.K., S.H.-M. and B.D.; formal analysis, O.K.; investigation, O.K.; resources, O.K., S.H.-M. and B.D.; writing—original draft preparation, O.K.; writing—review and editing, O.K., S.H.-M. and B.D.; supervision, B.D.; project administration, S.H.-M.; funding acquisition, S.H.-M. All authors have read and agreed to the published version of the manuscript.

Funding: The research was supported by VAROPS, granted by the Ministry of Defence of the Czech Republic.

Informed Consent Statement: Not applicable.

Data Availability Statement: Not applicable.

Acknowledgments: The authors are highly grateful to the referees for their constructive suggestions for improving the paper.

Conflicts of Interest: The authors declare no conflict of interest.

References

1. Marty, F. Sur une generalization de la notion de group. In Proceedings of the 8th Congress on Mathmatics Scandenaves, Stockholm, Sweden, 14–18 August 1934; pp. 45–49.
2. Prenowitz, W. Projective geometries as multigroups. *Am. J. Math.* **1943**, *65*, 235–256. [CrossRef]
3. Prenowitz, W. Descriptive geometries as multigroups. *Trans. Am. Math. Soc.* **1946**, *59*, 333–380. [CrossRef]
4. Prenowitz, W. Spherical geometries and mutigroups. *Can. J. Math.* **1950**, *2*, 100–119. [CrossRef]
5. Jantosciak, J. Transposition hypergroups, Noncommutative Join Spaces. *J. Algebra* **1997**, *187*, 97–119. [CrossRef]
6. Bonansinga, P. Quasicanonical hypergroups. *Atti Soc. Peloritana Sci. Fis. Mat. Natur.* **1981**, *27*, 9–17. (In Italian)
7. Massouros, C.G. Quasicanonical hypergroups. In Proceedings of the 4th Internation Congress, on Algebraic Hyperstructures and Applications, Xanthi, Greece, 27–30 June 1990; World Scientific: Singapore, 1991; pp. 129–136.
8. Ioulidis, S. Polygroups et certains de leurs properietes. *Bull. Greek Math. Soc.* **1981**, *22*, 95–104.
9. Comer, S.D. Polygroups derived from cogroups. *J. Algebra* **1984**, *89*, 397–405. [CrossRef]
10. Comer, S.D. *Extension of Polygroups by Polygroups and their Representations Using Color Schemes*; Lecture Notes in Mathematics, No 1004, Universal Algebra and Lattice Theory; Springer: Berlin/Heidelberg, Germany, 1982; pp. 91–103.
11. Davvaz, B. *Polygroup Theory and Related Systems*; World Scientific Publishing Co. Pte. Ltd.: Hackensack, NJ, USA, 2013.
12. Mittas, J. Hypergroupes canoniques. *Math. Balk.* **1972**, *2*, 165–179.
13. Massouros, C.; Massouros, G. An overview of the foundations of the hypergroup theory. *Mathematics* **2021**, *9*, 1014. [CrossRef]
14. Massouros, C.; Cristea, I. 1st Symposium on "Hypercompositional Algebra—New Developments and Applications (HAnDA)". *AIP Conf. Proc.* **2018**, 340001. [CrossRef]
15. Corsini, P. *Prolegomena of Hypergroup Theory*, 2nd ed.; Aviani Editor: Tricesimo, Italy, 1993.
16. Corsini, P.; Leoreanu-Fotea, V. Applications of hyperstructures theory. In *Advanced in Mathematics*; Kluwer: Dordrecht, The Netherlands, 2003.
17. Davvaz, B.; Leoreanu-Fotea, V. *Hyperring Theory and Applications*; Hadronic Press, Inc.: Palm Harber, FL, USA, 2007.
18. Davvaz, B.; Cristea, I. *Fuzzy Algebraic Hyperstructures*; Studies in Fuzziness and Soft Computing 321; Springer International Publishing: Cham, Switzerland, 2015.
19. Vougiouklis, T. *Hyperstructures and Their Representations*; Hadronic Press: Palm Harbor, FL, USA, 1994.
20. Zadeh, L.A. Fuzzy sets. *Inf. Control* **1965**, *8*, 338–353. [CrossRef]
21. Zahedi, M.M.; Bolurian, M.; Hasankhani, A. On polygroups and fuzzy subpolygroups. *J. Fuzzy Math.* **1995**, *3*, 1–15.
22. Davvaz, B. Fuzzy H_v-groups. *Fuzzy Sets Syst.* **1999**, *101*, 191–195. [CrossRef]
23. Rosenfeld, A. Fuzzy groups. *J. Math. Anal. Appl.* **1971**, *35*, 512–517. [CrossRef]
24. Corsini, P. A new connection between hypergroups and fuzzy sets. *Southeast Asian Bull. Math.* **2003**, *27*, 221–229.
25. Davvaz, B.; Corsini, P. Generalized fuzzy polygroups. *Iran. J. Fuzzy Syst.* **2006**, *3*, 59–75.
26. Kazancı, O.; Davvaz, B.; Yamak, S. A new characterization of fuzzy n-ary polygroups. *Neural Comput. Appl.* **2012**, *19*, 649–655. [CrossRef]
27. Molodtsov, D. Soft set theory first results. *Comp. Math. Appl.* **1999**, *37*, 19–31. [CrossRef]
28. Aktaş, H.; Çağman, N. Soft sets and soft groups. *Inf. Sci.* **2007**, *177*, 2726–2735. [CrossRef]
29. Maji, P.K.; Biswas, R.; Roy, A.R. Soft set theory. *Comp. Math. Appl.* **2003**, *45*, 555–562. [CrossRef]
30. Maji, P.K.; Biswas, R.; Roy, A.R. Fuzzy soft sets. *J. Fuzzy Math.* **2001**, *9*, 589–602.
31. Aygunoğlu, A.; Aygun, H. Introduction to fuzzy soft groups. *Comp. Math. Appl.* **2009**, *58*, 1279–1286. [CrossRef]
32. Leoreanu-Fotea, V.; Feng, F.; Zhan, J. Fuzzy soft hypergroup. *Int. J. Comp. Math.* **2012**, *89*, 1–12. [CrossRef]
33. Sebastian, S.; Ramakrishnan, T.V. Multi-fuzzy sets: An extension of fuzzy sets. *Fuzzy Inf. Eng.* **2011**, *3*, 35–43. [CrossRef]
34. Yang, Y.; Tan, X.; Meng, C. The multi-fuzzy soft set and its application in decision making. *Appl. Math. Modell.* **2013**, *37*, 4915–4923. [CrossRef]
35. Onasanya, B.O.; Hoskova-Mayerova, S. Multi-fuzzy group induced by multisets. *Ital. J. Pure Appl. Math.* **2019**, *41*, 597–604.
36. Hoskova-Mayerova, S.; Al-Tahan, M.; Davvaz, B. Fuzzy multi-Hypergroups. *Mathematics* **2020**, *8*, 244. [CrossRef]
37. Al-Tahan, M.; Hoskova-Mayerova, S.; Davvaz, B. Fuzzy multi-polygroups. *J. Intell. Fuzzy Syst.* **2019**, *38*, 2337–2345. [CrossRef]
38. Akın, C. Multi-fuzzy soft groups. *Soft Comput.* **2021**, *25*, 137–145. [CrossRef]
39. Kazancı, O.; Hoskova-Mayerova, S.; Davvaz, B. Application multi-fuzzy soft sets in hypermodules. *Mathematics* **2021**, *9*, 2182. [CrossRef]
40. Wanga, J.; Yin, M.; Gua, W. Soft polygroups. *Comp. Math. Appl.* **2011**, *62*, 3529–3537. [CrossRef]
41. Dresher, M.; Ore, O. Theory of Multigroups. *Am. J. Math.* **1938**, *60*, 705–733. [CrossRef]

Article

A Combinatorial Characterization of $H(4, q^2)$ †

Stefano Innamorati and Fulvio Zuanni *

Department of Industrial and Information Engineering and Economics, University of L'Aquila, Piazzale Ernesto Pontieri, 1, 67100 L'Aquila, Italy; stefano.innamorati@univaq.it
* Correspondence: fulvio.zuanni@univaq.it
† Dedicated to Prof. Franco Eugeni on the occasion of their 80th birthday.

Abstract: In this paper, we remove the solid incidence assumption in a characterization of $H(4, q^2)$ by J. Schillewaert and J. A. Thasby proving that Hermitian plane incidence numbers imply Hermitian solid incidence numbers, except for a few possible small cases.

Keywords: Hermitian variety; three character sets; intersection number

MSC: 51E20

1. Introduction and Motivation

Let q denote a prime power p^h with exponent $h \geq 1$. In $PG(r,q)$, the projective space of dimension r and order q, let K denote a k-set, i.e., a set of k points. For each integer i such that $0 \leq i \leq \theta_d := \sum_{j=0}^{d} q^j$, let us denote by $t_i^d = t_i^d(K)$ the number of d-subspaces of $PG(r,q)$ meeting K in exactly i points. The nonnegative integers t_i^d are called the characters of K with respect to the dimension d, as can be seen in [1–3]. Let $m_1, m_2, \ldots, m_s$ be s integers such that $0 \leq m_1 < m_2 < \cdots < m_s \leq \theta_d$. A set K is said to be of class $[m_1, m_2, \ldots, m_s]_d$ if $t_i^d > 0$ only if $i \in \{m_1, m_2, \ldots, m_s\}$. Moreover, K is said to be of type $(m_1, m_2, \ldots, m_s)_d$ if $t_i^d > 0$ if and only if $i \in \{m_1, m_2, \ldots, m_s\}$. The nonnegative integers $m_1, m_2, \ldots, m_s$ are called intersection numbers with respect to the dimension d. Intersection numbers with respect to dimensions 2 and 3 will be called plane and solid intersection numbers, respectively. A full swing research topic is to recognize algebraic varieties by intersection numbers, as can be seen in [4–6]. The Hermitian variety $H(4, q^2)$ is the set of all absolute points of a non-degenerate unitary polarity in $PG(4, q^2)$; it is a non-singular algebraic hypersurface of degree $q + 1$ in $PG(4, q^2)$ with three plane intersection numbers and two solid intersection numbers (for more details, we refer the reader to Chapter 23 of [1]). The size and the solid intersection numbers are generally not sufficient to characterize Hermitian varieties due to the existence of quasi-Hermitian varieties, as can be seen in [7,8]. In [9], Theorem 4.2, J. Schillewaert and J. A. Thas proved the following

Result 1. *In $PG(4, q^2)$, any set of class $[q^2 + 1, q^3 + 1, q^3 + q^2 + 1]_2$ and of class $[q^5 + q^2 + 1, q^5 + q^3 + q^2 + 1]_3$ is the Hermitian variety $H(4, q^2)$.*

In this paper, we remove the solid incidence assumption of Result 1 by proving the following

Theorem 1. *In $PG(4, q^2)$, apart from possible cases with $q \in \{2, 3, 5\}$, any set of class $[q^2 + 1, q^3 + 1, q^3 + q^2 + 1]_2$ is the Hermitian variety $H(4, q^2)$.*

In order to remove the solid incidence assumption, we have to calculate the solid intersection numbers of a set of class $[q^2 + 1, q^3 + 1, q^3 + q^2 + 1]_2$ in $PG(4, q^2)$. To do this, in Section 2, we analyze the possible sizes of a set that have the same plane intersection numbers in $PG(3, q^2)$.

2. Sets of Class $[q^2 + 1, q^3 + 1, q^3 + q^2 + 1]_2$ in $PG(3, q^2)$

We start by recalling the following

Result 2 (see [10] Lemma 2.2). *In $PG(r, q)$ with $r \geq 2$, let K be a k-set of class $[m_1, m_2, \ldots, m_s]_d$ and of class $[n_1, n_2, \ldots, n_u]_{d+1}$ with $1 \leq d < d+1 \leq r$. If there is an integer x such that for any $m_i \in \{m_1, m_2, \ldots, m_s\}$, we have $m_i \equiv x \bmod q$; then, for any $n_j \in \{n_1, n_2, \ldots, n_u\}$, we have $n_j \equiv x \bmod q$. Thus, $k \equiv x \bmod q$ as well, since K is of type $(k)_r$.*

In this section, we will prove the following:

Theorem 2. *In $PG(3, q^2)$, with $q = p^h$ a prime power, let K be a k-set of class $[q^2 + 1, q^3 + 1, q^3 + q^2 + 1]_2$. Then, there is an integer a such that $k = aq^2 + 1$ with either $a \equiv 0 \pmod{q}$ or $a \equiv 1 \pmod{q}$. Furthermore:*

1. $t^2_{q^2+1} = 0$ *if and only if* $k = q^5 + q^3 + q^2 + 1$;
   furthermore, K is of type $(q^3 + 1, q^3 + q^2 + 1)_2$;
2. *If $t^2_{q^2+1} \geq 1$, then*

   - $q = 2$ and $k \in \{25, 33, 49\}$; *furthermore:*
     - *If $k = 25$, then K is a set of type $(5, 9)_2$ of $PG(3, 4)$;*
     - *If $k \in \{33, 49\}$, then K is a set of type $(5, 9, 13)_2$ of $PG(3, 4)$;*
     - *If $k = 49$, then a line meets K in at most 4 points and therefore K contains no line;*
   - $q = 3$ and $k = 244$; *furthermore:*
     - *K is a set of type $(10, 28, 37)_2$ of $PG(3, 9)$;*
     - *A line meets K in at most 8 points and therefore K contains no line;*
   - $q = 5$ and $k = 3126$; *furthermore:*
     - *K is a set of type $(26, 126, 151)_2$ of $PG(3, 25)$;*
     - *A line meets K in at most 12 points and therefore K contains no line;*
   - $k = q^5 + q^2 + 1$ *for any $q \geq 2$;*
     furthermore, K is a set of type $(q^2 + 1, q^3 + 1, q^3 + q^2 + 1)_2$;
   - $k = q^5 + q^3 + 1$ *for any $q \geq 2$; furthermore:*
     - *K is a set of type $(q^2 + 1, q^3 + 1, q^3 + q^2 + 1)_2$;*
     - *If α is a $(q^2 + 1)$-plane, then $\alpha \cap K$ is not a line;*
   - $k = q^5 + q^4 - q^3 + q^2 + 1$ *for any $q \geq 3$; furthermore:*
     - *K is a set of type $(q^2 + 1, q^3 + 1, q^3 + q^2 + 1)_2$;*
     - *A line meets K in at most $2q + 1$ points and therefore K contains no line.*

Theorem 2 will be a consequence of Lemmas 2–5.

Now let K be a k-set of $PG(3, q^2)$ of class $[l, m, n]_2$. Thus, by definition, $l < m < n$. By double counting the number of planes, the number of pairs (P, α) where $P \in K$ and α is a plane through P, and the number of pairs $((P, Q), \alpha)$ where P and Q are two distinct points of K and α is a plane through P and Q, we obtain the following equations on the integers $t_i = t_i^2(K)$

$$t_l + t_m + t_n = (q^2 + 1)(q^4 + 1) \tag{1}$$

$$lt_l + mt_m + nt_n = k(q^4 + q^2 + 1) \tag{2}$$

$$l(l-1)t_l + m(m-1)t_m + n(n-1)t_n = k(k-1)(q^2 + 1) \tag{3}$$

Lemma 1. *Let K be a k-set of class $[q^2 + 1, q^3 + 1, q^3 + q^2 + 1]_2$ in $PG(3, q^2)$ and let r_h be a line meeting K in exactly h points. Then:*

1. $k \equiv 1 \pmod{q^2}$;
2. $h \leq q^3 + q^2 + q + 2 - \frac{k-1}{q^2} = [q^3 + q + 1 - \frac{k-1}{q^2}] + q^2 + 1$;

3. $h \leq [q^3 + q + 1 - \frac{k-1}{q^2}] + t_{q^3+q^2+1}$;
4. If $t_{q^2+1} \geq 2$, then $k \leq q^5 + q^4 - q^3 + 2q^2 + 1$.

Proof. By Result 2, we immediately have that $k \equiv 1 \pmod{q^2}$.

Now let r_h be a line meeting K in exactly h points and let us denote by u_i^h the number of i-planes passing through r_h with $i \in \{q^2 + 1, q^3 + 1, q^3 + q^2 + 1\}$. Counting the number of points of $K \setminus r_h$ via the planes through r_h, we obtain

$$k - h = (q^2 + 1 - h)u_{q^2+1}^h + (q^3 + 1 - h)u_{q^3+1}^h + (q^3 + q^2 + 1 - h)u_{q^3+q^2+1}^h \tag{4}$$

Since $u_{q^2+1}^h + u_{q^3+1}^h + u_{q^3+q^2+1}^h = q^2 + 1$, we have that

$$h + u_{q^3+1}^h + qu_{q^2+1}^h = q^3 + q^2 + q + 2 - \frac{k-1}{q^2} \tag{5}$$

$$h + (q-1)u_{q^2+1}^h = q^3 + q + 1 - \frac{k-1}{q^2} + u_{q^3+q^2+1}^h \tag{6}$$

By (5), we immediately have that $h \leq q^3 + q^2 + q + 2 - \frac{k-1}{q^2}$. Since $u_{q^3+q^2+1}^h \leq t_{q^3+q^2+1}$, by (6), we have that $h \leq q^3 + q + 1 - \frac{k-1}{q^2} + t_{q^3+q^2+1}$.

Now let us suppose that $t_{q^2+1} \geq 2$. Let α and β be two $(q^2 + 1)$-planes and let r_h be the line $\alpha \cap \beta$. Equation (5) can be rewritten in the following way

$$q^3 + q^2 - q + 2 - \frac{k-1}{q^2} = h + q(u_{q^2+1}^h - 2) + u_{q^3+1}^h \tag{7}$$

Since $u_{q^2+1}^h - 2 \geq 0$, by (7), we have that $q^3 + q^2 - q + 2 - \frac{k-1}{q^2} \geq 0$ from which it immediately follows that $k \leq q^5 + q^4 - q^3 + 2q^2 + 1$. □

Lemma 2. *If K is a k-set of $PG(3, q^2)$ of class $[q^3 + 1, q^3 + q^2 + 1]_2$, then $k = q^5 + q^3 + q^2 + 1$.*

Proof. A set of class $[m, n]_2$ is a set of class $[l, m, n]_2$ having $t_l = 0$. Putting $t_l = 0$, $m = q^3 + 1$ and $n = q^3 + q^2 + 1$ in Equations (1)–(3), we obtain

$$[k - (q^5 + q^3 + q^2 + 1)][k - (q^5 + q^4 - q^3 + 2q + 1 - \frac{2q}{q^2+1})] = 0 \tag{8}$$

Therefore, $k = q^5 + q^3 + q^2 + 1$ necessarily, since $q \geq 2$ and k is an integer. □

From now on, K will ever be a k-set of class $[q^2 + 1, q^3 + 1, q^3 + q^2 + 1]_2$ having $t_{q^2+1} \geq 1$. By Lemma 1, there is an integer a such that $k = aq^2 + 1$.

Lemma 3. *We have that either $a \equiv 0 \pmod{q}$ or $a \equiv 1 \pmod{q}$.*

Proof. Putting $l = q^2 + 1$, $m = q^3 + 1$, $n = q^3 + q^2 + 1$ and $k = aq^2 + 1$ into Equations (1)–(3), we obtain

$$t_{q^2+1} = H + 3q^4 + (6 - 2a)q^3 - 3aq^2 - 7a - \alpha + \beta \tag{9}$$

$$t_{q^3+1} = -H + (a-4)q^4 + (3a-7)q^3 + 4aq^2 - (a-1)^2 q + 8a - \beta \tag{10}$$

$$t_{q^3+q^2+1} = q^6 + (2-a)q^4 + (1-a)q^3 + (1-a)q^2 + (a-1)^2 q - a + 1 + \alpha \tag{11}$$

with $H = q^6 + 2q^5 + 8q^2 + (9 - 5a)q + a^2 + 11$, $\alpha = \frac{a(a-1)}{q}$ and $\beta = \frac{2(a-2)(a-3)}{q-1}$. Since $\alpha = \frac{a(a-1)}{q}$ an integer, we have that $a(a-1) \equiv 0 \pmod{q}$ and hence either $a \equiv 0 \pmod{q}$ or $a \equiv 1 \pmod{q}$, since a and $a - 1$ are coprime and q is a prime power. □

Lemma 4. *If $a \equiv 0 \pmod{q}$, then:*

1. $q = 2$ and $k \in \{25, 33, 49\}$; furthermore:
   - If $k = 25$, then K is a set of type $(5, 9)_2$ of $PG(3, 4)$;
   - If $k \in \{33, 49\}$, then K is a set of type $(5, 9, 13)_2$ of $PG(3, 4)$;
   - If $k = 49$, then a line meets K in at most 4 points and therefore K contains no line;
2. $q = 3$ and $k = 244$; furthermore:
   - K is a set of type $(10, 28, 37)_2$ of $PG(3, 9)$;
   - A line meets K in at most 8 points and therefore K contains no line;
3. $q = 5$ and $k = 3126$; furthermore:
   - K is a set of type $(26, 126, 151)_2$ of $PG(3, 25)$;
   - A line meets K in at most 12 points and therefore K contains no line;
4. $k = q^5 + q^3 + 1$ for any $q \geq 2$; furthermore:
   - K is a set of type $(q^2 + 1, q^3 + 1, q^3 + q^2 + 1)_2$;
   - If α is a $(q^2 + 1)$-plane, then $\alpha \cap K$ is not a line.

Proof. Putting $a = bq$ into Equations (9)–(11) we obtain

$$(q-1)t_{q^2+1} = q(q^2+1)b^2 - (2\theta_5 - q^4 - 1)b + \theta_7 + 2q^4 + q^3 + q \tag{12}$$

$$(q-1)t_{q^3+1} = -q^2(q^2+1)b^2 + (\theta_5 + q^4 + 2q^2 + 1)qb - (\theta_7 + q^5 + 2q^4 + q^2) \tag{13}$$

$$t_{q^3+q^2+1} = q(q^2+1)b^2 - (\theta_5 + q^2)b + \theta_6 - q^5 + q^4 \tag{14}$$

where $\theta_d := \sum_{i=0}^{d} q^i$.

If $q = 2$, then we obtain $t_5 = 10b^2 - 109b + 297$, $t_9 = -20b^2 + 176b - 323$ and $t_{13} = 10b^2 - 67b + 111$. Since $t_5 \geq 1$, $t_9 \geq 0$ and $t_{13} \geq 0$, it is easy to prove that $b \in \{3, 4, 5, 6\}$ necessarily.

If $b = 3$, then $a = bq = 6$ and $k = aq^2 + 1 = 25$. Furthermore, we obtain that $(t_5, t_9, t_{13}) = (60, 25, 0)$. K is a 25-set of type $(5, 9)_2$ in $PG(3, 4)$.

If $b = 4$, then $a = bq = 8$ and $k = aq^2 + 1 = 33$. Furthermore, we obtain $(t_5, t_9, t_{13}) = (21, 61, 3)$. K is therefore a 33-set of type $(5, 9, 13)_2$ in $PG(3, 4)$.

If $b = 5$, then $a = bq = 10$ and $k = aq^2 + 1 = 41$. Let us note that in such a case, $k = 41 = 2^5 + 2^3 + 1 = q^5 + q^3 + 1$. This case is therefore included in item 4 in the statement of the lemma.

If $b = 6$, then $a = bq = 12$ and $k = aq^2 + 1 = 49$. Furthermore, we obtain that $(t_5, t_9, t_{13}) = (3, 13, 69)$. K is therefore a 49-set of type $(5, 9, 13)_2$ in $PG(3, 4)$. Finally, by point (2) of Lemma 1, we obtain that $h \leq 4$.

Now let us study the case $q \geq 3$.
Since $t_{q^2+1} \geq 1$, by Equation (12), we obtain that:

$$f(b) := q(q^2+1)b^2 - (2\theta_5 - q^4 - 1)b + \theta_7 + 2q^4 + q^3 + 1 \geq 0 \tag{15}$$

It is easy to see that:

- $f(q^2 + 1) = q^3 - 2q^2 + 1 > 0$ for any q;
- $f(q^2 + 2) = -q(q-1)(q^2 - 3q + 1) < 0$ for any $q \geq 3$;
- $f(q^2 + q - 1) = -2q^2 + 3q + 3 < 0$ for any $q \geq 3$;
- $f(q^2 + q) = (q-1)(q^3 - 2q - 2) > 0$ for any q.

For any $q \geq 3$, there are therefore two real numbers b_1 and b_2 such that $q^2 + 1 < b_1 < q^2 + 2$, $q^2 + q - 1 < b_2 < q^2 + q$ and $g(b_1) = g(b_2) = 0$. Thus, for any $q \geq 3$, we have $b \leq q^2 + 1$ or $b \geq q^2 + q$.

Since $t_{q^3+1} \geq 0$, by Equation (13), we obtain

$$g(b) := q^2(q^2+1)b^2 - (\theta_5 + q^4 + 2q^2 + 1)qb + \theta_7 + q^5 + 2q^4 + q^2 \leq 0 \quad (16)$$

It is easy to see that
- $g(q) = (q+1)(q^4+1) > 0$ for any q;
- $g(q+1) = -(q-1)(q^5 - q^3 + 1) < 0$ for any q;
- $g(q^2+q) = -2q^3 + q + 1 < 0$ for any q;
- $g(q^2+q+1) = (q-1)(q^5 + q^4 + 3q^3 - 1) > 0$ for any q.

Thus, for any q, there are two real numbers b_1 and b_2 such that $q < b_1 < q+1$, $q^2 + q < b_2 < q^2 + q + 1$ and $g(b_1) = g(b_2) = 0$. For any q, we therefore have $q + 1 \leq b \leq q^2 + q$.
Since $t_{q^3+q^2+1} \geq 0$, by Equation (14), we obtain

$$h(b) := q(q^2+1)b^2 - (\theta_5 + q^2)b + \theta_6 - q^5 + q^4 \geq 0 \quad (17)$$

It is easy to see that:
- $h(q) = q^4 + 1 > 0$ for any q;
- $h(q+1) = -(q-2)q^4 < 0$ for any $q \geq 3$;
- $h(q^2 - 1) = -(q-2)(q^4 + q^3 + 2q^2 + 2q + 1) < 0$ for any $q \geq 3$;
- $h(q^2) = q + 1 > 0$ for any q.

For any $q \geq 3$, there are therefore two real numbers b_1 and b_2 such that $q < b_1 < q+1$, $q^2 - 1 < b_2 < q^2$ and $g(b_1) = g(b_2) = 0$. Thus, for any $q \geq 3$, we have $b \leq q$ or $b \geq q^2$.
Finally, if $q \geq 3$, then $b \in \{q^2, q^2+1, q^2+q\}$ necessarily.
If $b = q^2$, then $k = q^5 + 1$, $t_{q^3+q^2+1} = q + 1$, $t_{q^3+1} = q^6 + q^4 - q^3 - 2q - 3 - \frac{4}{q-1}$, and $t_{q^2+1} = q^3 + q^2 + q + 3 + \frac{4}{q-1}$. Thus, $q-1$ must divide 4 with $q \geq 3$. Hence, $q \in \{3,5\}$ and K is a 244-set of type $(10, 28, 37)_2$ in $PG(3,9)$ or $q = 5$ and K is a 3126-set of type $(26, 126, 151)_2$ in $PG(3,25)$. Furthermore, since $t_{q^3+q^2+1} = q + 1$, by point (3) of Lemma 1, we have that $h \leq 2(q+1)$.
If $b = q^2 + 1$, then $k = q^5 + q^3 + 1$, $t_{q^2+1} = q^2 - q$, $t_{q^3+1} = q^6 - q^5 + 2q^4 - 2q^3 + 2q^2 + 1$, $t_{q^3+q^2+1} = q^5 - q^4 + 2q^3 - 2q^2 + q$; therefore, K is of type $(q^2+1, q^3+1, q^3+q^2+1)_2$. Now, let us suppose that there is a (q^2+1)-plane α such that $\alpha \cap K$ is a line r. Substituting $h = q^2 + 1$ into Equation (5), we obtain $u_{q^3+1} = 1 - qu_{q^2+1}$. Hence, $u_{q^2+1} = 0$ necessarily and no $(q^2 + 1)$-plane passes through line r, which is a contradiction.
If $b = q^2 + q$, then $k = q^5 + q^4 + 1$. Since $t_{q^2+1} \geq 2$, by point (4) of Lemma 1, we have that $q^5 + q^4 + 1 = k \leq q^5 + q^4 - q^3 + 2q^2 + 1$. Thus, $q \leq 2$, which is a contradiction. □

Lemma 5. *If $a \equiv 1 \pmod{q}$, then:*
1. $k = q^5 + q^2 + 1$ *for any $q \geq 2$; furthermore:*
   K is a set of type $(q^2+1, q^3+1, q^3+q^2+1)_2$.
2. $k = q^5 + q^4 - q^3 + q^2 + 1$ *for any $q \geq 3$; furthermore:*
   - *K is a set of type $(q^2+1, q^3+1, q^3+q^2+1)_2$;*
   - *A line meets K in at most $2q+1$ points and therefore K contains no line.*

Proof. Putting $a = cq + 1$ into Equations (9)–(11), we obtain

$$(q-1)t_{q^2+1} = q(q^2+1)c^2 - (2\theta_5 - q^4 - 2q^2 - 3)c + \theta_7 - 2q^2 - 2 \quad (18)$$

$$(q-1)t_{q^3+1} = -q^2(q^2+1)c^2 + (\theta_6 + q^5 - q - 1)c - (\theta_7 - \theta_3) \quad (19)$$

$$t_{q^3+q^2+1} = q(q^2+1)c^2 - (\theta_5 - q^2 - 2)c + q^6 + q^4 \quad (20)$$

where $\theta_d := \sum_{i=0}^{d} q^i$.

Since $t_{q^2+1} \geq 1$, by Equation (18) we obtain:

$$f(c) := q(q^2+1)c^2 - (2\theta_5 - q^4 - 2q^2 - 3)c + \theta_7 - 2q^2 - q - 1 \geq 0 \qquad (21)$$

It is easy to see that:
- $f(q^2) = (q-1)q^3 > 0$ for any q;
- $f(q^2+1) = -(q-1) < 0$ for any q;
- $f(q^2+q-2) = -q^4 + 4q^3 - 6q^2 + 9q - 2 < 0$ for any $q \geq 3$;
- $f(q^2+q-1) = (q-1)(q^2 - 3q + 1) > 0$ for any $q \geq 3$.

Thus, for any $q \geq 3$, there are two real numbers c_1 and c_2 such that $q^2 < c_1 < q^2 + 1$, $q^2 + q - 2 < c_2 < q^2 + q - 1$ and $g(c_1) = g(c_2) = 0$. For any $q \geq 3$, we therefore have $c \leq q^2$ or $c \geq q^2 + q - 1$.

If $q = 2$ and $c = q^2 + q - 2 = 4$, then $a = cq + 1 = 9$ and $k = aq^2 + 1 = 37 = (2^3+1)2^2 + 1 = (q^3+1)q^2 + 1$. Thus, this case is included in item 1 in the statement of the lemma.

If $q = 2$ and $c = q^2 + q - 1 = 5$, then $a = cq + 1 = 11$ and $k = aq^2 + 1 = 45 = 2^5 + 2^3 + 2^2 + 1 = q^5 + q^3 + q^2 + 1$. Therefore, $t_{q^2+1} = 0$, which is a contradiction.

Since $t_{q^3+1} \geq 0$, by Equation (19), we obtain

$$g(c) := q^2(q^2+1)c^2 - (\theta_6 + q^5 - q - 1)c + (\theta_7 - \theta_3) \leq 0 \qquad (22)$$

It is easy to see that:
- $g(q) = (q-1)q^3 > 0$ for any q;
- $g(q+1) = -(q-1)(q+1)q^4 < 0$ for any q;
- $g(q^2+q-1) = -(q-1)(q^3+2)q^2 < 0$ for any q;
- $g(q^2+q) = (q-1)(q+1)q^3 > 0$ for any q.

Therefore, for any q, there are two real numbers c_1 and c_2 such that $q < c_1 < q+1$, $q^2 + q - 1 < c_2 < q^2 + q$ and $g(c_1) = g(c_2) = 0$. For any q, we thus have $q + 1 \leq c \leq q^2 + q - 1$.

Since $t_{q^3+q^2+1} \geq 0$, by Equation (20), we obtain

$$h(b) := q(q^2+1)c^2 - (\theta_5 - q^2 - 2)c + q^6 + q^4 \geq 0 \qquad (23)$$

It is easy to see that:
- $h(q) = q(q^2 - q + 1) > 0$ for any q;
- $h(q+1) = -q^5 + q^4 + q^3 + q^2 + q + 1 < 0$ for any q;
- $h(q^2-1) = -q^5 + 2q^4 - q^3 + q^2 + 2q - 1 < 0$ for any q;
- $h(q^2) = q^2(q^2 - q + 1) > 0$ for any q.

Therefore, for any q there are two real numbers c_1 and c_2 such that $q < c_1 < q+1$, $q^2 - 1 < c_2 < q^2$ and $g(c_1) = g(c_2) = 0$. For any q, we thus have $c \leq q$ or $c \geq q^2$.

Finally, if $q \geq 3$, then $c \in \{q^2, q^2+q-1\}$ necessarily.

If $c = q^2$, then $k = q^5 + q^2 + 1$, $t_{q^2+1} = q^3 + 1$, $t_{q^3+1} = q^6$, $t_{q^3+q^2+1} = q^4 - q^3 + q^2$; K is therefore a set of type $(q^2+1, q^3+1, q^3+q^2+1)_2$.

If $c = q^2+q-1$, then $k = q^5 + q^4 - q^3 + q^2 + 1$, $t_{q^2+1} = (q-1)(q-2)$, $t_{q^3+1} = (q^3+2)q^2$, $t_{q^3+q^2+1} = q^6 - q^5 + q^4 - 2q^2 + 3q - 1$; K is therefore a set of type $(q^2+1, q^3+1, q^3+q^2+1)_2$.

Furthermore, by point (2) of Lemma 1, we have that $h \leq 2q + 1$. □

3. The Proof of the Main Result

In this section, K is a k-set of class $[q^2+1, q^3+1, q^3+q^2+1]_2$ in $PG(4, q^2)$. By Result 2, we immediately have that $k \equiv 1 \pmod{q^2}$.

We will prove that, apart from possible initial cases with $q \in \{2, 3, 5\}$ as in Corollary 1, K is the Hermitian variety $H(4, q^2)$.

As an immediate consequence of Theorem 2, we have the following

Corollary 1. *Apart from the following initial possible cases:*

1. *$q = 2$, K is a set of class $[25, 33, 37, 41, 45, 49]_3$, and there is at least one 25-solid or one 33-solid or one 49-solid (otherwise K is of class $[37, 41, 45]_3$ as in the next general case); furthermore:*
   - *If S is a 25-solid, then $K \cap S$ is a set of type $(5, 9)_2$ of $PG(3, 4)$;*
   - *If S is a 45-solid, then $K \cap S$ is a set of type $(9, 13)_2$ of $PG(3, 4)$;*
   - *If S is an n-solid with $n \in \{33, 37, 41, 49\}$, then $K \cap S$ is a set of type $(5, 9, 13)_2$ of $PG(3, 4)$.*

2. *$q = 3$, K is a set of class $[244, 253, 271, 280, 307]_3$ and there is at least one 244-solid (otherwise K is of class $[253, 271, 280, 307]_3$ as in the next general case); furthermore:*
   - *If S is a 280-solid, then $K \cap S$ is a set of type $(28, 37)_2$ of $PG(3, 9)$;*
   - *If S is a n-solid with $n \in \{244, 253, 271, 307\}$, then $K \cap S$ is a set of type $(10, 28, 37)_2$ of $PG(3, 9)$.*

3. *$q = 5$, K a set is of class $[3126, 3151, 3251, 3276, 3651]_3$ and there is at least one 3126-solid (otherwise K is of class $[3151, 3251, 3276, 3651]_3$ as in the next general case); furthermore:*
   - *If S is a 3276-solid, then $K \cap S$ is a set of type $(126, 151)_2$ of $PG(3, 25)$;*
   - *If S is a n-solid with $n \in \{3126, 3151, 3251, 3651\}$, then $K \cap S$ is a set of type $(26, 126, 151)_2$ of $PG(3, 25)$.*

K is of class $[q^5 + q^2 + 1, q^5 + q^3 + 1, q^5 + q^3 + q^2 + 1, q^5 + q^4 - q^3 + q^2 + 1]_3$; furthermore, if S is an n-solid, then

- *If $n = q^5 + q^3 + q^2 + 1$, then $K \cap S$ is a set of type $(q^3 + 1, q^3 + q^2 + 1)_2$ of $PG(3, q^2)$; otherwise, $K \cap S$ is of type $(q^2 + 1, q^3 + 1, q^3 + q^2 + 1)_2$;*
- *If $n \in \{q^5 + q^3 + 1, q^5 + q^4 - q^3 + q^2 + 1\}$, then for any $(q^2 + 1)$-plane α of S the set $\alpha \cap K$ is not a line.*

Remark 1. *If H is a set of type $(m)_d$ of $PG(r, q)$ with $1 \leq d \leq r$, then $m = 0$ or $m = \theta_d$. Furthermore, in the first case, H is the empty set, while in the second one, H is the whole space.*

Lemma 6. *K is of type $(q^2 + 1, q^3 + 1, q^3 + q^2 + 1)_2$.*

Proof. If there is no $(q^2 + 1)$-plane, then K is of type $(q^5 + q^3 + q^2 + 1)_3$, which is a contradiction since $q^5 + q^3 + q^2 + 1 \neq 0$ and $q^5 + q^3 + q^2 + 1 \neq \theta_3$. There is therefore at least one $(q^2 + 1)$-plane.

If there is no $(q^3 + 1)$-plane, then K has no type with respect to solids, which is a contradiction. There is therefore at least one $(q^3 + 1)$-plane.

If there is no $(q^3 + q^2 + 1)$-plane, then $q = 2$, and K is a set of type $(25)_3$, which is a contradiction since $25 \neq 0$ and $25 \neq 15 = \theta_3$. There is therefore at least one $(q^3 + q^2 + 1)$-plane. □

Lemma 7. *Apart from the possible initial cases as in Corollary 1, at least one $(q^5 + q^3 + q^2 + 1)$-solid passes through each $(q^3 + q^2 + 1)$-plane.*

Proof. Let α be a h-plane with $h \in \{q^2 + 1, q^3 + 1, q^3 + q^2 + 1\}$ such that no $(q^5 + q^3 + q^2 + 1)$-solid passes through α and let:

- *w be the number of $(q^5 + q^2 + 1)$-solids passing through α;*
- *x be the number of $(q^5 + q^3 + 1)$-solids passing through α;*
- *y be the number of $(q^5 + q^4 - q^3 + q^2 + 1)$-solids passing through α.*

Counting the point of K via the $q^2 + 1$ solids passing through α, we have

$$k = h + w(q^5 + q^2 + 1 - h) + x(q^5 + q^3 + 1 - h) + y(q^5 + q^4 - q^3 + q^2 + 1 - h) \quad (24)$$

Substituting $x = q^2 + 1 - w - y$ and $k = aq^2 + 1$ into (24), we obtain

$$w = q^4 + q^3 + 3q^2 + (y+3)q - y + 4 - \frac{a+h-5}{q-1} \qquad (25)$$

By Lemma 6, there is at least one (q^2+1)-plane α. By Corollary 1, no $(q^5+q^3+q^2+1)$-solid passes through α. Substituting $h = q^2 + 1$ into Equation (25), we obtain

$$w = q^4 + q^3 + 3q^2 + (y+2)q - y + 3 - \frac{a-3}{q-1} \qquad (26)$$

Thus, $q - 1$ divides $a - 3$. Now, let β a $(q^3 + q^2 + 1)$-plane and let us suppose that no $(q^5 + q^3 + q^2 + 1)$-solid passes through β. Substituting $h = q^3 + q^2 + 1$ into Equation (25), we obtain

$$w = q^4 + q^3 + 2q^2 + (y+1)q - y + 2 - \frac{a-2}{q-1} \qquad (27)$$

Thus, $q - 1$ divides $a - 2$, which is a contradiction. Thus, the statement is true. □

Lemma 8. *Apart from the possible initial cases as in Corollary 1, a $(q^3 + q^2 + 1)$-plane contains no external line.*

Proof. Let β be a $(q^3 + q^2 + 1)$-plane and r_h be a line of β meeting K in exactly h points. In view of the previous Lemma, at least one $(q^5 + q^3 + q^2 + 1)$-solid S passes through β. By Corollary 1, S contains no $(q^2 + 1)$-plane. Substituting $u^h_{q^2+1} = 0$ and $k = q^5 + q^3 + q^2 + 1$ into Equation (6), we obtain $h = u^h_{q^3+q^2+1}$. Since $u^h_{q^3+q^2+1} \geq 1$, we have the statement. □

Lemma 9. *Apart from the possible initial cases as in Corollary 1, only $(q^5 + q^3 + q^2 + 1)$-solids can pass through an external line.*

Proof. Let r_0 be an external line and let S an n-solid passing through r_0. By the previous Lemma, we have that no $(q^3 + q^2 + 1)$-plane passes through r_0. Substituting $h = 0$, $u^0_{q^3+q^2+1} = 0$ and $k = n$ into Equation (6), we obtain:

$$(q-1)u^0_{q^2+1} = q^3 + q + 1 - \frac{n-1}{q^2} \qquad (28)$$

- If $n = q^5 + q^2 + 1$, then we have that $u^0_{q^2+1} = 1 + \frac{1}{q-1}$;
- If $n = q^5 + q^3 + 1$, then we have that $u^0_{q^2+1} = \frac{1}{q-1}$;
- If $n = q^5 + q^3 + q^2 + 1$, then we have that $u^0_{q^2+1} = 0$;
- If $n = q^5 + q^4 - q^3 + q^2 + 1$, then we have that $u^0_{q^2+1} = 1 - q + \frac{1}{q-1}$.

Since $q > 2$ and $u^0_{q^2+1}$ are integers, we necessarily obtain $n = q^5 + q^3 + q^2 + 1$. □

Lemma 10. *Apart from the possible initial cases as in Corollary 1, if α is a $(q^2 + 1)$-plane, then $K \cap \alpha$ is a line.*

Proof. By Lemma 6, there is at least one $(q^2 + 1)$-plane α. Let S be a solid passing through α. If $K \cap \alpha$ is not a line, then $K \cap \alpha$ is not a blocking-set with respect to the lines of α. Hence, in α (and hence in S), there is at least one line r_0 external to K. By the previous Lemma, S is a $(q^5 + q^3 + q^2 + 1)$-solid. Finally, by Corollary 1, S contains no $(q^2 + 1)$-plane, which is a contradiction. □

Lemma 11. *Apart from the possible initial cases as in Corollary 1, K is a set of class $[q^5 + q^2 + 1, q^5 + q^3 + q^2 + 1]_3$. Furthermore, K has exactly $q^7 + q^5 + q^2 + 1$ points.*

Proof. By Corollary 1 and Lemma 10, we immediately have that K is of class $[q^5 + q^2 + 1, q^5 + q^3 + q^2 + 1]_3$.

Now, let α be a $(q^2 + 1)$-plane. Again, by Corollary 1, we have that only $(q^5 + q^2 + 1)$-solids pass through the plane α. Counting the points of K via these solids, we obtain $k = (q^2 + 1)q^5 + q^2 + 1 = q^7 + q^5 + q^2 + 1$. □

Finally, Theorem 1 follows either by Result 1 or, as can be seen in [11], by the following:

Result 3. *In $PG(4, q^2)$ with $q > 2$, let K be a $(q^7 + q^5 + q^2 + 1)$-set having two solid intersection numbers and three plane intersection numbers. If the minimum plane intersection number is $q^2 + 1$, then K is $H(4, q^2)$.*

4. Conclusions

The principal aim of this paper was to prove that the lower dimensional incidence assumption is stronger that the higher one. Therefore, applications and future developments are improvements on other combined characterizations that are obtained through different dimensional assumptions that remove the higher dimensional one.

Author Contributions: Investigation, S.I. and F.Z.; Writing—review, S.I. and F.Z.; Editing, F.Z. All authors have read and agreed to the published version of the manuscript.

Funding: This research received no external funding.

Conflicts of Interest: The authors declare no conflict of interest.

References

1. Hirschfeld, J.W.P.; Thas, J.A. *General Galois Geometries*; Clarendon Press: Oxford, UK, 1991.
2. Napolitano, V. A Characterization of the Hermitian Variety in Finite 3-Dimensional Projective Spaces. *Electron. J. Comb.* **2015**, *22*, #P1.22. [PubMed]
3. Tallini Scafati, M. Caratterizzazione grafica delle forme Hermitiane di un $S_{r,q}$. *Rend. Mat.* **1967**, *26*, 273–303.
4. Beelen, P.; Datta, M.; Homma, M. A proof of Sorensen's conjecture on Hermitian surfaces. *Proc. Am. Math. Soc.* **2021**, *149*, 1431–1441. [CrossRef]
5. Pradhan, P.; Sahu, B. A characterization of the family of secant lines to a hyperbolic quadric in $PG(3, q)$, q odd. *Discret. Math.* **2020**, *343*, 112044. [CrossRef]
6. Yahya, N.Y.K. Applications geometry of space in $PG(3, P)$. *J. Interdiscip. Math.* **2021**, *25*, 285–297. [CrossRef]
7. Aguglia, A. Quasi-Hermitian varieties in $PG(r, q^2)$, q even. *Contrib. Discret. Math.* **2013**, *8*, 31–37.
8. Aguglia, A.; Cossidente, A.; Korchmaros, G. On quasi-Hermitian varieties. *J. Comb. Des.* **2012**, *20*, 433–447. [CrossRef]
9. Schillewaert, J.; Thas, J.A. Characterizations of Hermitian varieties by intersection numbers. *Des. Codes Cryptogr.* **2009**, *50*, 41–60.
10. Innamorati, S.; Zuanni, F. Classifying sets of class $[1, q + 1, 2q + 1, q^2 + q + 1]_2$ in $PG(r, q)$, $r \geq 3$. *Des. Codes Cryptogr.* **2021**, *89*, 489–496. [CrossRef]
11. Innamorati, S.; Zannetti, M.; Zuanni, F. On two character $(q^7 + q^5 + q^2 + 1)$-sets in $PG(4, q^2)$. *J. Geom.* **2015**, *106*, 287–296. [CrossRef]

Article

Construction of an Infinite Cyclic Group Formed by Artificial Differential Neurons

Jan Chvalina [1], Bedřich Smetana [2,*] and Jana Vyroubalová [1]

[1] Department of Mathematics, Faculty of Electrical Engineeering and Comunication, Brno University of Technology, Technická 8, 616 00 Brno, Czech Republic; chvalina@vutbr.cz (J.C.); 143699@vut.cz (J.V.)
[2] Department of Quantitative Methods, University of Defence, Kounicova 65, 662 10 Brno, Czech Republic
* Correspondence: bedrich.smetana@unob.cz; Tel.: +420-973-443-429

Abstract: Infinite cyclic groups created by various objects belong to the class to the class basic algebraic structures. In this paper, we construct the infinite cyclic group of differential neurons which are modifications of artificial neurons in analogy to linear ordinary differential operators of the n-th order. We also describe some of their basic properties.

Keywords: time-varying artificial neuron; cyclic group; linear differential operator

MSC: 06F15; 68T07; 11F25

1. Introduction

In our paper, we study artificial, or formal, neurons. Recall that these are the building blocks of mathematically modeled neural networks, e.g., [1]. The design and functionality of artificial neurons are derived from observations of biological neural networks. Our investigation belongs to the theory which is developed and applied in various directions contained in many publications, cf. [2–6]. The bodies of artificial neurons compute the sum of the weighted inputs and bias and "process" this sum with a transfer function, cf. [1–10].

In the next step, the information is passed via outputs (output functions). Thus, artificial neural networks have the structure similar to that of weighted directed graphs with artificial neurons being their nodes and connections between neuron inputs and outputs being directed edges with weights. Recall that in the framework of artificial neural networks there are networks of simple neurons called perceptrons. The basic concept (perceptron) was introduced by Rosenblatt in 1958. Perceptrons compute single outputs (the output function) from multiple real-valued inputs by forming a linear combination according to input weights, and then possibly putting the output through some nonlinear activation functions. Mathematically, this can be written as

$$y(t) = \varphi\left(\sum_{i=1}^{n} w_i(t) x_i(t) + b\right) = \varphi\left(\vec{w}^T(t)\vec{x}(t) + b\right), \qquad (1)$$

where $\vec{w}(t) = (w_1(t), \ldots, w_n(t))$ denotes the vector of time dependent weight functions, $\vec{x}(t) = (x_1(t), \ldots, x_n(t))$ is the vector of time dependent (or time varying) input functions, b is the bias and φ is the activation function. The use of time varying functions as weights and inputs is a certain generalization of the classical concept of artificial neurons from the work of Warren McCulloch and Walter Pitts (1943); see also [1–10] and references mentioned therein.

2. Differential Neurons and Their Output Functions

In accordance with our previous papers [1,7–9], we regard the above mentioned artificial neurons such that inputs x_i and weights w_i will be functions of argument t belonging into a linearly ordered (tempus) set T with the least element 0. As the index set we use the interval of real numbers $[1, \infty) = \{x \in \mathbb{R}; 1 \leq x\}$, where $\mathbb{R}$ denotes the set of all real numbers. So, denote by W the set of all non-negative functions $w : T \to \mathbb{R}$ forming a subsemiring of the ring of all real functions of one real variable $x : \mathbb{R} \to \mathbb{R}$. Denote by $Ne(\vec{w}_r(t)) = Ne(w_{r1}(t), \ldots, w_{rn}(t))$ for $r \in [1, \infty)$, $n \in \mathbb{N}$ and the mapping

$$y_r(t) = \sum_{k=1}^{n} w_{r,k}(t) x_{r,k}(t) + b_r \qquad (2)$$

which will be called the artificial neuron with the bias $b_r \in \mathbb{R}$, in fact the output function of the corresponding neuron. By $\mathbb{AN}(T)$ we denote the collection of all such artificial neurons.

Neurons are usually denoted by capital letters X, Y or X_i, Y_i. However, we use also notation $Ne(\vec{w})$, where $\vec{w} = (w_1, \ldots, w_n)$ is the vector of weights.

We suppose, for the sake of simplicity, that transfer functions (activation functions) φ, σ (or f) are the same for all neurons from the collection $\mathbb{AN}(T)$ or that this function is the identity function $f(y) = y$.

Now, similarly as in the case of the collection of linear differential operators, we will construct a cyclic group of artificial neurons, extending their monoid, cf. [1].

Denote by δ_{ij} the so called Kronecker delta, $i, j \in \mathbb{N}$, i.e., $\delta_{ii} = \delta_{jj} = 1$ and $\delta_{ij} = 0$, whenever $i \neq j$.

Suppose $Ne(\vec{w}_r), Ne(\vec{w}_s) \in \mathbb{AN}(T)$, $r, s \in [1, \infty)$, $\vec{w}_r = (w_{r1}, \ldots, w_{r,n})$, $\vec{w}_s = (w_{s1}, \ldots, w_{s,n})$, $n \in \mathbb{N}$. Let $m \in \mathbb{N}$, $1 \leq m \leq n$ be a such an integer that $w_{r,m} > 0$. We define

$$Ne(\vec{w}_v(t)) = Ne(\vec{w}_r(t)) \cdot_m Ne(\vec{w}_s(t)), \qquad (3)$$

where

$$\vec{w}_v(t) = (w_{v,1}(t), \ldots, w_{v,n}(t)), \qquad (4)$$

$$w_{v,k}(t) = w_{r,m}(t) w_{s,k}(t) + (1 - \delta_{m,k}) w_{r,k}(t), \quad t \in T \qquad (5)$$

and, of course, the neuron $Ne(\vec{w}_v)$ is defined as mapping $y_v(t) = \sum_{k=1}^{n} w_k(t) x_k(t) + b_v$, $t \in T$, $b_v = b_r b_s$. Further, for a pair $Ne(\vec{w}_r(t))$, $Ne(\vec{w}_s(t))$ of neurons from $\mathbb{AN}(T)$ we put

$$Ne(\vec{w}_r(t)) \leq_m Ne(\vec{w}_s(t)),$$
$$\vec{w}_r(t) = (w_{r,1}(t), \ldots, w_{r,n}(t)), \quad \vec{w}_s(t) = (w_{s,1}(t), \ldots, w_{s,n}(t)) \qquad (6)$$

if $w_{r,k}(t) \leq w_{s,k}(t)$, $k \in \mathbb{N}$, $k \neq m$ and $w_{r,m}(t) = w_{s,m}(t)$, $t \in T$ and with the same bias.

Remark 1. *There exists a link between formal neurons and linear differential operators of the n-th order. This link is important for our future considerations. Recall the expression of formal neuron with inner potential $y_{-in} = \sum_{k=1}^{n} w_k(t) x_k(t)$, where $\vec{x}(t) = (x_1(t), \ldots, x_n(t))$ is the vector of inputs, $\vec{w}(t) = (w_1(t), \ldots, w_n(t))$ is the vector of weights. Using the bias b of the considered neuron and the transfer function σ we can expressed the output as $y(t) = \sigma\left(\sum_{k=1}^{n} w_k(t) x_k(t) + b\right)$.*

Now consider a fundamental function $u: J \to \mathbb{R}$, where $J \subseteq \mathbb{R}$ is an open interval; inputs are derived from the function $u \in \mathbb{C}^n(J)$ as follows:

$$x_1(t) = u(t), \quad x_2 = \frac{du(t)}{dt}, \quad \ldots, \quad x_n(t) = \frac{d^{n-1}(t)}{dt^{n-1}}, \quad n \in \mathbb{N}.$$

Further the bias $b = b_0 \frac{d^n u(t)}{dt^n}$. As weights we use continuous functions $w_k: J \to \mathbb{R}$, $k = 1, \ldots, n - 1$.

Then formula

$$y(t) = \sigma\left(\sum_{k=1}^{n} w_k(t) \frac{d^{k-1} u(t)}{dt^{k-1}} + b_0 \frac{d^n u(t)}{dt^n}\right) \qquad (7)$$

is a description of the action of the neuron Dn which will be called a *formal (artificial) differential neuron*. This approach allows to use solution spaces of corresponding linear differential equations.

3. Products and Powers of Differential Neurons

Suppose $\vec{w}(t) = (w_1(t), \ldots, w_n(t))$ are fixed vectors of continuous functions $w_k : \mathbb{R} \to \mathbb{R}$ and b_0 be the bias for any polynomial $p \in \mathbb{R}_s[t]$, $n \leq s$, $s \in \mathbb{N}_0$. We consider a differential neuron $DNe_p(\vec{w})$ by the action

$$y_1(t) = \sum_{k=1}^{n} w_{1,k}(t) \frac{d^{k-1} p(t)}{dt^{k-1}} + b_0 \frac{d^n p(t)}{dt^n} \tag{8}$$

with the identity activation function $\varphi(u) = u$. According to the formula, we can calculate the output function of the differential neuron $D^2 Ne_p(\vec{w}) = DNe_p(\vec{w}) \cdot DNe_p(\vec{w})$.

Firstly, we describe the product of neurons $Ne(\vec{w}_r) \cdot Ne(\vec{w}_s) = Ne(\vec{w}_u)$, i.e., outputs of neurons

$$y_r(t) = \sum_{k=1}^{n} w_{r,k}(t) x_k(t) + b_r, \quad y_s(t) = \sum_{k=1}^{n} w_{s,k}(t) x_k(t) + b_s. \tag{9}$$

The vector of weights of the neuron $Ne(\vec{w}_u)$ is of the form $\vec{w}_u(t) = (w_{u,1}, \ldots, w_{u,n})$, where

$$w_{u,k}(t) = w_{r,m}(t) w_{s,k}(t) + (1 - \delta_{m,k}) w_{r,k}(t), \quad t \in T \text{ and } 1 \leq m \leq n. \tag{10}$$

Then the neuron $Ne(\vec{w}_u)$ is defined using its output function $y_u(t) = \sum_{k=1}^{n} w_{u,k}(t) x_k(t) + b_r b_s$, $t \in T$.

In a greater detail:

$$w_{u,1}(t) = w_{r,m}(t) w_{s,1}(t) + w_{r,1}(t),$$
$$w_{u,2}(t) = w_{r,m}(t) w_{s,2}(t) + w_{r,2}(t),$$
$$\vdots$$
$$w_{u,m}(t) = w_{r,m}(t) w_{s,m}(t),$$
$$\vdots$$
$$w_{u,n}(t) = w_{r,m}(t) w_{s,n}(t) + w_{r,n}(t).$$

Application of the above product onto the case of differential neurons: Suppose $DNe_p(\vec{w}_r)$, $DNe_p(\vec{w}_s)$ are neurons with output functions

$$y_r(t) = \sum_{k=1}^{n} w_{r,k}(t) \frac{d^{k-1} p(t)}{dt^{k-1}} + b_r \frac{d^n p(t)}{dt^n},$$
$$y_s(t) = \sum_{k=1}^{n} w_{s,k}(t) \frac{d^{k-1} p(t)}{dt^{k-1}} + b_s \frac{d^n p(t)}{dt^n}, \tag{11}$$

where $p \in \mathbb{R}_l[t]$, $n \leq l$. Denote

$$DNe_p(\vec{w}_u) = DNe_p(\vec{w}_r) \cdot DNe_p(\vec{w}_s). \tag{12}$$

Then the output function of the neuron $DNe_p(\vec{w}_u)$ has the form

$$y_u(t) = \sum_{\substack{k=1 \\ k \neq m}}^{n} \left(w_{r,m}(t)w_{s,k}(t) + w_{r,k}(t)\right) \frac{d^{k-1}p(t)}{dt^{k-1}} +$$
$$+ w_{r,m}(t)w_{s,m}(t) \frac{d^{m-1}p(t)}{dt^{m-1}} + b_r b_s \left(\frac{d^n p(t)}{dt^n}\right)^2. \tag{13}$$

Now, using the above formula we can express output functions of powers $D^2 Ne_p(\vec{w}_r)$, $D^\alpha Ne_p(\vec{w}_r)$ (for $\alpha \in \mathbb{N}$) and $D^0 Ne_p(\vec{w}_r)$ (the neutral element-unit) of the infinite cyclic group $\{D^\alpha Ne_p(\vec{w}_r); \alpha \in \mathbb{Z}\}$. The output function $y_u^{[2]}(t)$ of the differential neuron is of the form

$$y_u^{[2]}(t) = \sum_{\substack{k=1 \\ k \neq m}}^{n} ((w_{r,m}(t) + 1)w_{r,k}(t)) \frac{d^{k-1}p(t)}{dt^{k-1}} + w_{r,m}^2(t) \frac{d^{m-1}p(t)}{dt^{m-1}} + b_r^2 \left(\frac{d^n p(t)}{dt^n}\right)^2 =$$
$$= (w_{r,m}(t) + 1) \sum_{\substack{k=1 \\ k \neq m}}^{n} w_{r,k}(t) \frac{d^{k-1}p(t)}{dt^{k-1}} + w_{r,m}^2(t) \frac{d^{m-1}p(t)}{dt^{m-1}} + b_r^2 \left(\frac{d^n p(t)}{dt^n}\right)^2. \tag{14}$$

In the paper [1] the following theorem is proved:

Theorem 1. *Consider a differential neuron $DNe_p(\vec{w})$ with the vector $\vec{w}(t) = (w_1(t), \ldots, w_n(t))$ of time variable weights and the vector of inputs $\vec{x}(t) = \left(p(t), \frac{dp(t)}{dt}, \ldots, \frac{d^n p(t)}{dt^n}\right)$ with polynomial $p \in \mathbb{R}_l[t]$, $n \leq l$, $t \in T$ and $1 \leq m \leq n$, $n \in \mathbb{N} = \{1, 2, \ldots\}$. The output function $y(t)$ of the above mentioned neuron is of the form*

$$y(t) = \sum_{k=1}^{n} w_k(t) \frac{d^{k-1}p(t)}{dt^{k-1}} + b \frac{d^n p(t)}{dt^n} \tag{15}$$

with the bias $b \frac{d^n p(t)}{dt^n}$. Suppose $\alpha \in \mathbb{N}$, $2 \leq \alpha$. Then the output function of the differential neuron $D^\alpha Ne_p(\vec{w})$ has the form

$$y^{[\alpha]}(t) = \sum_{k=0}^{\alpha-1} w_m^k(t) \sum_{\substack{k=1 \\ k \neq m}}^{n} w_k(t) \frac{d^{k-1}p(t)}{dt^{k-1}} + w_m^\alpha(t) \frac{d^{m-1}p(t)}{dt^{m-1}} + \left(b \frac{d^n p(t)}{dt^n}\right)^\alpha. \tag{16}$$

Now, we discuss a certain type of subgroup which appears in all groups. The following text up to Proposition 2 incl. contains well-known facts, which are overtaken from the monography [11] (Chapter 2, §2,4).

Take any group G and any element $a \in G$. Consider all powers of a: Define $a^0 = e$ (the neutral element), $a^1 = a$, and for $k > 1$, define a^k to be the product of k factors of a. (A little more properly, a^k is defined inductively by declaring $a^k = aa^{k-1}$.) For $k > 1$ define $a^{-k} = (a^{-1})^k$.

Recall briefly some well-known classical facts.

Definition 1. *Let a be an element of a group G. The set of powers of a $\langle a \rangle = \{a^k : k \in \mathbb{Z}\}$ is a subgroup of G, called the cyclic subgroup generated by a. If there is an element $a \in G$ such that $\langle a \rangle = G$, one says that G is a cyclic group. We say that a is a generator of the cyclic group.*

There are two possibilities for $\langle a \rangle$, one possibility is that all the powers a^k are distinct, in which case, of course, the subgroup $\langle a \rangle$ is infinite; if this is so, we say that a has infinite order.

The other possibility is that two powers of a coincide, but this is not our case.

Definition 2. *The order of the cyclic subgroup generated by a is called the order of a. If the order of a is finite, then it is the least positive integer n such that $a^n = e$.*

Proposition 1. *Let a be an element of a group G.*
(a) *If a has infinite order then $\langle a \rangle$ is isomorphic to $\mathbb{Z}$.*
(b) *If a has finite order n, then $\langle a \rangle$ is isomorphic to the group C_n of n-th roots of 1.*

Proposition 2.
(a) *Any non-trivial subgroup of $\mathbb{Z}$ is cyclic and isomorphic to $\mathbb{Z}$.*
(b) *Let $G = \langle a \rangle$ be a finite cyclic group. Any subgroup of G is also cyclic.*

For a construction of a cyclic group of artificial differential neurons we need to extend the cyclic monoid of differential neurons obtained in the paper [1] by negative powers of differential neurons, in particular to describe their output functions, so we need to construct negative powers $D^{-\alpha}Ne(\vec{w})$ of differential neurons which belong to the basic contribution of this paper. We suppose the existence of such inverse elements, i.e., negative powers of the generated element of the considered group.

In general, for the construction of the negative power $D^{-\alpha}Ne(\vec{w})$ with $\alpha \in \mathbb{N}$ it seems to be a suitable way of a using of this equality:

$$D^{\alpha+1}Ne(\vec{w}) \cdot_m D^{-\alpha}Ne_p(\vec{w}) = DNe_p(\vec{w}), \tag{17}$$

where on the right hand side is given an arbitrary general differential neuron with the vector $\vec{w}(t) = (w_1(t), \ldots, w_m(t), \ldots, w_n(t))$ of time variable weight functions, with the vector of inputs

$$\vec{x}(t) = \left(p(t), \frac{dp(t)}{d(t)}, \ldots, \frac{d^n p(t)}{dt^n}\right), \tag{18}$$

with a polynomial $p \in \mathbb{R}_l[t]$, $n \leq l$, $t \in T$ and $1 \leq m \leq n$, $n \in \mathbb{N} = \{1, 2, \ldots\}$. The neuron $DNe_p(\vec{w})$ has the output function

$$y(t) = \sum_{k=1}^{n} w_k(t) \frac{d^{k-1} p(t)}{dt^{k-1}} + b_0 \frac{d^n p(t)}{dt^n}, \tag{19}$$

with the bias $b = b_0 \frac{d^n p(t)}{dt^n}$. However, we will construct the proof using mathematical induction—similarly as in [1]—the proof of the Theorem 1, which seems to be a more convenient way. So we are going to prove the following theorem.

Theorem 2. *Suppose the existence of an inverse elements (i.e., negative powers of the generated element of the considered group). Let $DNe_p(\vec{w})$ be a differential neuron with the vector $\vec{w}(t) = (w_1(t), \ldots, w_m(t), \ldots, w_n(t))$ of time variable weights and with the vector of inputs $\vec{x}(t) = \left(p(t), \frac{dp(t)}{dt}, \ldots, \frac{d^n p(t)}{dt^n}\right)$, with a polynomial $p \in \mathbb{R}_l[t]$, $n \leq l$, $t \in T$ and $1 \leq m \leq n$, $n \in \mathbb{N} = \{1, 2, \ldots\}$, i.e., the output function $y(t)$ of the neuron $DNe_p(\vec{w})$ is of the form*

$$y(t) = \sum_{k=1}^{n} w_k(t) \frac{d^{k-1} p(t)}{dt^{k-1}} + b_0 \frac{d^n p(t)}{dt^n}, \tag{20}$$

with the bias $b = b_0 \frac{d^n p(t)}{dt^n}$. Suppose $\alpha \in \mathbb{N}$. Then the output function of the differential neuron $DNe_p^{-\alpha}(\vec{w})$ has the form

$$y^{[-\alpha]}(t) = -\frac{1}{w_m^\alpha(t)} \sum_{\zeta=0}^{\alpha-1} w_m^\zeta(t) \sum_{\substack{k=1 \\ k \neq m}}^{n} w_k(t) \frac{d^{k-1} p(t)}{dt^{k-1}} + \frac{1}{w_m^\alpha(t)} \cdot \frac{d^{m-1} p(t)}{dt^{m-1}} + \left(b_0 \frac{d^n p(t)}{dt^n}\right)^{-\alpha}$$

or

$$y^{[-\alpha]}(t) = -w_m^{-\alpha}(t) \sum_{k=0}^{\alpha-1} w_m^k(t) \sum_{\substack{k=1 \\ k \neq m}}^{n} w_k(t) \frac{d^{k-1}p(t)}{dt^{k-1}} + w_m^{-\alpha}(t) \frac{d^{m-1}p(t)}{dt^{m-1}} + \left(b_0 \frac{d^n p(t)}{dt^n}\right)^{-\alpha}.$$

Proof. Consider the equality

$$DNe_p(\vec{w}) \cdot_m D^{-1}Ne_p(\vec{w}) = N1(\vec{e})_m, \tag{21}$$

where the output function of the neuron $N1(\vec{e})_m$ (the identity element of the monoid $(S^1, \cdot_m)$ from [1]) is of the form $y_{N1}(t) = \frac{d^{m-1}p(t)}{dt^{m-1}} + 1$.

Let $y(t) = \sum_{k=1}^{n} w_k(t) \frac{d^{k-1}p(t)}{dt^{k-1}} + b_0 \frac{d^n p(t)}{dt^n}$ be the output function of the neuron $DNe_p(\vec{w})$ with the bias $b = b_0 \frac{d^n p(t)}{dt^n}$ and

$$y^{[-1]}(t) = \sum_{k=1}^{n} w_{s,k}(t) \frac{d^{k-1}p(t)}{dt^{k-1}} + b_s \tag{22}$$

be the output function of the neuron $D^{-1}Ne_p(\vec{w})$. Since $0 = w_{1,k} = w_m(t) \cdot w_{s,k}(t) + w_k(t)$ and $w_m(t) \cdot w_{s,m}(t) = 1$ for any $k \in \{1,2,\ldots,n\} \setminus \{m\}$, we have

$$w_{s,m}(t) = \frac{1}{w_m(t)} \quad \text{and} \quad w_{s,k}(t) = -\frac{w_k(t)}{w_m(t)}. \tag{23}$$

Moreover, $1 = b \cdot b_s = b_0 \frac{d^n p(t)}{dt^n} \cdot b_s$ which implies that the bias $b_s = \left(b_0 \frac{d^n p(t)}{dt^n}\right)^{-1}$. Thus, the output function is of the form

$$y^{[-1]}(t) = \sum_{\substack{k=1 \\ k \neq m}}^{n} \left(-\frac{w_k(t)}{w_m(t)}\right) \frac{d^{k-1}p(t)}{dt^{k-1}} + \frac{1}{w_m(t)} \cdot \frac{d^{m-1}p(t)}{dt^{m-1}} + b_s =$$

$$= \frac{-1}{w_m(t)} \sum_{\substack{k=1 \\ k \neq m}}^{n} w_k(t) \frac{d^{k-1}p(t)}{dt^{k-1}} + \frac{1}{w_m(t)} \cdot \frac{d^{m-1}p(t)}{dt^{m-1}} + \left(b_0 \frac{d^n p(t)}{dt^n}\right)^{-1}. \tag{24}$$

Using of Equation (16) we obtain after some simple calculation the expression:

$$y^{[-\alpha]}(t) = -\frac{1}{w_m^\alpha(t)} \sum_{\xi=0}^{\alpha-1} w_m^\xi(t) \sum_{\substack{k=1 \\ k \neq m}}^{n} w_k(t) \frac{d^{k-1}p(t)}{dt^{k-1}} +$$

$$+ \frac{1}{w_m^\alpha(t)} \cdot \frac{d^{m-1}p(t)}{dt^{m-1}} + \left(b_0 \frac{d^n p(t)}{dt^n}\right)^{-\alpha}. \tag{25}$$

This function is in a fact the output function of the neuron $D^{-\alpha}Ne_p(\vec{w})$.
Now, for $\alpha = 1$ we obtain

$$y^{[-1]}(t) = -\frac{1}{w_m(t)} \sum_{\substack{k=1 \\ k \neq m}}^{n} w_k(t) \frac{d^{k-1}p(t)}{dt^{k-1}} + \frac{1}{w_m(t)} \cdot \frac{d^{m-1}p(t)}{dt^{m-1}} + \left(b_0 \frac{d^n p(t)}{dt^n}\right)^{-1},$$

which is in fact the Expression (24).

We have

$$y^{[-\alpha-1]}(t) = -\frac{1}{w_m^{\alpha+1}(t)} \sum_{\xi=0}^{\alpha} w_m^\xi(t) \sum_{\substack{k=1 \\ k \neq m}}^{n} w_k(t) \frac{d^{k-1}p(t)}{dt^{k-1}} +$$

$$+ \frac{1}{w_m^{\alpha+1}(t)} \cdot \frac{d^{m-1}p(t)}{dt^{m-1}} + \left(b_0 \frac{d^n p(t)}{dt^n}\right)^{-\alpha-1},$$

which is the Equality (25) written for $-(\alpha+1)$ instead for $-\alpha$. The other negative powers can be also obtained from example we have. $\square$

Using output functions of corresponding differential neurons we verify a validity of equalities

$$D^{-\alpha}Ne_p(\vec{w}) \cdot_m N1(\vec{e})_m = D^{-\alpha}Ne_p(\vec{w}) = N1(\vec{e})_m \cdot_m D^{-\alpha}Ne_p(\vec{w}) \qquad (26)$$

certifying that the neuron $N1(\vec{e})_m$ is the neutral element also for negative powers of the neuron $DNe_p(\vec{w})$.

Denote by $y_u(t)$ the output function of the neuron

$$DNe_p(\vec{w}_u) = D^{-\alpha}Ne_p(\vec{w}) \cdot_m N1(\vec{e})_m. \qquad (27)$$

Since the output function of the neuron $N1(\vec{e})$ (the unit element) has the form

$$y_1(t) = w_{N1,m}(t) \frac{d^{m-1}p(t)}{dt^{m-1}} + 1 \text{ with } w_{N1,m}(t) = 1, \qquad (28)$$

we have

$$y_u(t) = \frac{1}{w_m^{\alpha}(t)} \left(-\sum_{\xi=0}^{\alpha-1} w_m^\xi(t)\right) \sum_{\substack{k=1 \\ k \neq m}}^{n} w_k(t) \frac{d^{k-1}p(t)}{dt^{k-1}} + 1 \cdot w^{-\alpha}(t) \frac{d^{m-1}p(t)}{dt^{m-1}} + 1 \cdot \left(b_0 \frac{d^n p(t)}{dt^n}\right)^{-\alpha},$$

which is in fact the output function $y^{[-\alpha]}(t)$ of the differential neuron $D^{-\alpha}Ne_p(\vec{w})$. In a similar way we can verify the second equality.

Remark 2. *In paper [12] there is defined a concept of a general n-hyperstructure as there follows:*

Let $n \in \mathbb{N}$ be an arbitrary positive integer and $\{X_k; k = 1, \ldots, n\}$ be a system of non-empty sets. By a general n-hyperstructure we mean the pair

$$(\{X_k; k = 1, \ldots, n\}, *_n),$$

where $*_n : \prod_{k=1}^{n} X_k \to \mathcal{P}^*\left(\bigcup_{k=1}^{n} X_k\right)$ *is a mapping assigning to any n-tuple $[x_1, \ldots, x_n] \in \prod_{k=1}^{n} X_k$ a non-empty subset $*_n(x_1, \ldots, x_n) \subseteq \bigcup_{k=1}^{n} X_k$. Here $\mathcal{P}^*(M)$ means the power set of M without the empty set $\emptyset$.*

Similarly as above, with this hyperoperation there is associated a mapping of power sets

$$\otimes_n : \prod_{k=1}^{n} \mathcal{P}^*(X_k) \to \mathcal{P}^*\left(\bigcup_{k=1}^{n} X_k\right) \qquad (29)$$

defined by

$$\otimes_n (A_1, \ldots, A_n) = \bigcup \left\{ *_n(x_1, \ldots, x_n); [x_1, \ldots, x_n] \in \prod_{k=1}^{n} A_k \right\}. \qquad (30)$$

This construction is also based on an idea of Nezhad and Hashemi for $N - 2$.

At the end of this section we give this example:

Let $J \subseteq \mathbb{R}$ be an open interval, $\mathbb{C}^n(J)$ be the ring (with respect to the usual addition and multiplication of functions) of all real functions $f: J \to \mathbb{R}$ with continuous derivatives up to the order $n \geq 0$ including. Now, as in suppositions of Theorems 1 and 2, we consider a differential neuron $DNe_p(\vec{w})$ with the vector $\vec{w}(t) = (w_1(t), \ldots, w_n(t))$ of time variable weights and the vector of inputs $\vec{x}(t) = \left(p(t), \frac{dp(t)}{dt}, \ldots, \frac{d^n p(t)}{dt^n}\right)$ with the polynomial $p \in \mathbb{R}_l[t]$, $n \leq l$, $t \in T$ and $1 \leq m \leq n$, $n \in \mathbb{N} = \{1, 2, \ldots\}$. The output function $y(t)$ of the mentioned neuron is of the form

$$y(t) = \sum_{k=1}^{n} w_k(t) \frac{d^{k-1} p(t)}{dt^{k-1}} + b \frac{d^n p(t)}{dt^n} \tag{31}$$

with the bias $b \frac{d^n p(t)}{dt^n}$ and $w_k : T \to \mathbb{R}$, $w_k \in \mathbb{C}^n(T)$. In accordance with [13], we put

$$\mathbb{DAN}_k(T) = \{DNe_p(\vec{w}_s); p \in \mathbb{R}_l[t], \vec{w}_s \in [\mathbb{C}^n(T)]^k\}.$$

As above, we put $DNe_p(\vec{w}_s) \leq DNe_p(\vec{w}_r)$ whenever $\vec{w}_s(t) = (w_{s,1}(t), \ldots, w_{s,n}(t))$, $\vec{w}_r(t) = (w_{r,1}(t), \ldots, w_{r,n}(t))$ and $w_{s,k}(t) \leq w_{r,k}(t)$, $t \in T$, $k = 1, 2, \ldots, n$. Defining

$$*_{n,p} \left(DNe_p(\vec{w}_1(t)), DNe_p(\vec{w}_2(t)), \ldots, DNe_p(\vec{w}_n(t))\right) =$$
$$= \bigcup_{k=1}^{n} \left\{DNe_p(\vec{w}(t) \in \mathbb{DAN}_k(T)_p; Ne_p(\vec{w}_k(t)) \leq Ne_p(\vec{w}(t))\right\} \tag{32}$$

for any n-tuple $[Ne_p(\vec{w}_1(t)), Ne_p(\vec{w}_2(t)), \ldots, Ne_p(\vec{w}_n(t))] \in \prod_{k=1}^{n} \mathbb{DAN}_k(T)_p$, we obtain that

$$\mathbb{D}_p(n) = (\{\mathbb{DAN}_k(T)_p; k = 1, 2, \ldots, n\}, *_{n,p}) \tag{33}$$

is a general n-hyperstructure for the polynomial $p \in \mathbb{R}_l[t]$.

It is to be noted, that the used concept of investigated neurons is in a certain sense motivated by ordinary differential operators forming of left-hand sides of corresponding differential equations, see, e.g., [13,14].

Therefore, the construction of differential neurons consists of a certain modification of the concept of an artificial neuron which is investigated in a certain formal analogy to linear differential operators as mentioned above. Using the obtained cyclic group of differential neurons, we will construct a certain other hyperstructure of differential neurons. The mentioned relationship is in [8] described by the construction of a homomorphism.

It is to be noted that a hypergroup is a multistructure $(H, *)$, where H is a non-empty set and $* : H \times H \to \mathcal{P}(H)$ is a mapping which is associative, i.e.,

$$(a * b) * c = a * (b * c)$$

for any triad $a, b, c \in H$, where $A * B = \bigcup_{(a,b) \in A \times B} a * b$ for $A \neq \emptyset \neq B$, $A, B \subseteq H$, and $b * A = \{b\} * A$. Further, the reproduction axiom

$$a * H = H = H * a$$

for any element $a \in H$ is satisfied.

The above definition of a hypergroup is in the sense of F. Marty.

Let $J \subseteq \mathbb{R}$ be an open interval (bounded or unbounded) of real numbers, $\mathbb{C}^k(J)$ be the ring (with respect to usual addition and multiplication of functions) of all real functions with continuous derivatives up to the order $k \geq 0$ including. We write $\mathbb{C}(J)$ instead of

$\mathbb{C}^0(J)$. For a positive integer $n \geq 2$ we denote by $\mathbb{A}_n$ the set of all linear homogeneous differential equations of the n-th order with continuous real coefficients on J, i.e.,

$$y^{(n)} + p_{n-1}(x)y^{(n-1)} + \cdots + p_0(x)y = 0, \tag{34}$$

(cf. [14–16]), where $p_k \in \mathbb{C}(J)$, $k = 0, 1, \ldots, n-1$, $p_0(x) > 0$ for any $x \in J$ (this is not an essential restriction). Denote $L(p_0, \ldots, p_{n-1}) : \mathbb{C}^n(J) \to \mathbb{C}^n(J)$ the above defined linear operator defined by

$$L(p_0, \ldots, p_{n-1})(y) = y^{(n)} + p_{n-1}(x)y^{(n-1)} + \cdots + p_0(x)y \tag{35}$$

and put

$$\mathbb{L}\mathbb{A}_n(J) = \{L(p_0, \ldots, p_{n-1}); p_k \in \mathbb{C}(J), p_0 > 0\}. \tag{36}$$

Further $\mathbb{N}_0(n) = \{0, 1, \ldots, n-1\}$ and δ_{ij} stands for the Kronecker δ, $\overline{\delta_{ij}} = 1 - \delta_{ij}$. For any $m \in \mathbb{N}_0(n)$ we denote by $\mathbb{L}\mathbb{A}_n(J)_m$ the set of all linear differential operators of the n-th order $L_0(p_0, \ldots, p_{n-1}) : \mathbb{C}^n(J) \to \mathbb{C}(J)$, where $p_k \in \mathbb{C}(J)$ for any $k \in \mathbb{N}_0(n)$, $p_m \in \mathbb{C}_1(J)$, (i.e., $p_m(x) > 0$ for each $x \in J$). Using the vector notation $\vec{p}(x) = (p_0(x), \ldots, p_{n-1}(x))$, $x \in J$ we can write $L_n(\vec{p}_0)y = y^{(n)} + (\vec{p}(x) \cdot (y, y', \ldots, y^{(n-1)}))$, i.e., a scalar product.

We define a binary operation $\circ_m$ and a binary relation $\leq_m$ on the set $\mathbb{L}\mathbb{A}_n(J)_m$ in this way: For arbitrary pair $L(\vec{p}), L(\vec{q}) \in \mathbb{L}\mathbb{A}_n(J)_m$, $\vec{p} = (p_0, \ldots, p_{n-1})$, $\vec{q} = (q_0, \ldots, q_{n-1})$ we put $L(\vec{p}) \circ_m L(\vec{q}) = L(\vec{u})$, $\vec{u} = (u_0, \ldots, u_{n-1})$, where

$$u_k(x) = p_m(x)q_k(x) + (1 - \delta_{km})p_k(x), \ x \in J \tag{37}$$

and $L(\vec{p}) \leq L(\vec{q})$ whenever $p_k(x) \leq q_k(x)$, $k \in \mathbb{N}_0(n)$, $p_m(x) = q_m(x)$, $x \in J$. Evidently, $(\mathbb{L}\mathbb{A}_n(J)_m, \leq_m)$ is an ordered set.

In paper [14] there is presented the sketch of the proof of the following lemma:

Lemma 1. *The triad* $(\mathbb{L}\mathbb{A}_n(J)_m, \circ_m, \leq_m)$ *is an ordered (non-commutative) group.*

4. Groups and Hypergroups of Artificial Neurons

As it is mentioned in the dissertation [2] neurons are the atoms of neural computation. Out of those simple computational units all neural networks are build up. For a pair $Ne(\vec{w}_r)$, $Ne(\vec{w}_s)$ of neurons from $\mathbb{AN}(T)$ we put $Ne(\vec{w}_r) \leq_m Ne(\vec{w}_s)$, $w_r = (w_{r,1}(t), \ldots, w_{r,n}(t))$, $w_s = (w_{s,1}(t), \ldots, w_{s,n}(t))$ if $w_{r,k}(t) \leq w_{s,k}(t)$, $k \in \mathbb{N}$, $k \neq m$ and $w_{r,m}(t) = w_{s,m}(t)$, $t \in T$ and with the same bias. Evidently $(\mathbb{AN}(T), \leq_m)$ is an ordered set. A relationship (compatibility) of the binary operation "·" and the ordering $\leq_m$ on $\mathbb{AN}(T)$ is given by this assertion analogical to the above one. In paper [1] there is established that the structure $(\mathbb{AN}(T), \cdot_m)$ is a non-commutative group.

Lemma 2. *The triad* $(\mathbb{AN}(T), \cdot_m, \leq_m)$ *(algebraic structure with an ordering) is a non-commutative ordered group.*

Sketch of the proof is presented in [8]. Denoting

$$\mathbb{AN}_1(T)_m = \{Ne(\vec{w}); \vec{w} = (w_1, \ldots, w_n), w_k \in \mathbb{C}(T), k = 1, \ldots, n, w_m(t) \equiv 1\}, \tag{38}$$

we get the following assertion, the proof of which with necessary concepts is contained in [1].

Proposition 3. *Let* $T = \langle 0, t_0 \rangle \subset \mathbb{R}$, $t_0 \in \mathbb{R} \cup \{\infty\}$. *Then for any positive integer* $n \in \mathbb{N}$, $n \geq 2$ *and for any integer m such that $1 \leq m \leq n$ the semigroup* $(\mathbb{AN}_1(T)_m, \cdot_m)$ *is an invariant subgroup of the group* $(\mathbb{AN}(T)_m, \cdot_m)$.

If $m, n \in \mathbb{N}$, $1 \leq m \leq n-1$, then a certain relationship between groups $(\mathbb{AN}_n(T)_m, \cdot_m)$, $(\mathbb{LA}_n(T)_{m+1}, \circ_{m+1})$ is contained in the following proposition:

Proposition 4. *Let $t_0 \in \mathbb{R}$, $t_0 > 0$, $T = \langle 0, t_0 \rangle \subset \mathbb{R}$ and m, $n \in \mathbb{N}$ are integers such that $1 \leq m \leq n-1$. Define a mapping $F: \mathbb{AN}_n(T)_m \to \mathbb{LA}_n(T)_{m+1}$ by this rule: For an arbitrary neuron $Ne(\vec{w}_r) \in \mathbb{AN}_n(T)_m$, where $\vec{w}_r = (w_{r,1}(t), \ldots, w_{r,n}(t)) \in [\mathbb{C}(T)]^n$ we put $F(Ne(\vec{w}_r)) = L(w_{r,1}, \ldots, w_{r,n}) \in \mathbb{LA}_n(T)_{m+1}$ with the action :*

$$L(w_{r,1}, \ldots, w_{r,n})y(t) = \frac{d^n y(t)}{dt^n} + \sum_{k=1}^{n} w_{r,k}(t) \frac{d^{k-1}(t)}{dt^{k-1}}, \quad y \in \mathbb{C}^n(T). \tag{39}$$

Then the mapping $F: \mathbb{AN}_n(T)_m \to \mathbb{LA}_n(T)_{m+1}$ is a homomorphism of the group $(\mathbb{AN}_n(T)_m, \cdot_m)$ into the group $(\mathbb{LA}_n(T)_{m+1}, \circ_{m+1})$.

Consider $Ne(\vec{w}_r), Ne(\vec{w}_s) \in \mathbb{AN}_n(T)_m$ and denote $F(Ne(\vec{w}_r)) = L(w_{r,1}, \ldots, w_{r,n})$, $F(Ne(\vec{w}_s) = L(w_{s,1}, \ldots, w_{s,n})$. Denote $Ne(\vec{w}_u) = Ne(\vec{w}_r) \cdot_m Ne(\vec{w}_s)$. There holds

$$F(Ne(\vec{w}_r) \cdot_m Ne(\vec{w}_s)) = F(Ne(\vec{w}_u)) = L(w_{u,1}, \ldots, w_{u,n}), \tag{40}$$

where

$$L(w_{u,1}, \ldots, w_{u,n})y(t) = y^{(n)}(t) + \sum_{k=1}^{n} w_{u,k}(t) y^{(k-1)}(t). \tag{41}$$

Here $w_{u,k}(t) = w_{r,m+1}(t)w_{s,k}(t) + w_{r,k}(t)$, $k \neq m$, and $w_{u,m+1}(t) = w_{r,m+1}(t)w_{s,m+1}(t)$. Then $L(w_{u,1}, \ldots, w_{u,n}) = L(w_{r,1}, \ldots, w_{r,n}) \cdot_m L(w_{s,1}, \ldots, w_{s,n}) = F(Ne(\vec{w}_r)) \cdot_m F(Ne(\vec{w}_s))$. The neutral element $Ne(\vec{w}) \in \mathbb{AN}_n(T)_m$ is also mapped onto the neutral element of the group $(\mathbb{L}_n\mathbb{A}(T)_{m+1}, \cdot_{m+1})$, thus the mapping $F: (\mathbb{AN}_n(T)_m, \cdot_m) \to (\mathbb{L}_n\mathbb{A}(T)_{m+1}, \circ_{m+1})$ is a group homomorphism.

Now, using the construction described in Lemma 2, we obtain the final transposition hypergroup (called also non-commutative join space). Denote by $\mathbb{P}(\mathbb{AN}(T)_m)^*$ the power set of $\mathbb{AN}(T)_m$ consisting of all nonempty subsets of the last set and define a binary hyperoperation

$$*_m : \mathbb{AN}(T)_m \times \mathbb{AN}(T)_m \to \mathbb{P}(\mathbb{AN}(T)_m)^* \tag{42}$$

by the rule

$$Ne(\vec{w}_r) *_m Ne(\vec{w}_s) = \{Ne(\vec{w}_u); Ne(\vec{w}_r) \cdot_m Ne(\vec{w}_s) \leq_m Ne(\vec{w}_u)\}$$

for all pairs $Ne(\vec{w}_r), Ne(\vec{w}_s) \in \mathbb{AN}(T)_m$. More in detail if $\vec{w}(u) = (w_{u,1}, \ldots, w_{u,n})$, $\vec{w}(r) = (w_{r,1}, \ldots, w_{r,n})$, $\vec{w}(s) = (w_{s,1}, \ldots, w_{s,n})$, then $w_{r,m}(t)w_{s,m}(t) = w_{u,m}(t)$, $w_{r,m}(t)w_{s,k}(t) + w_{r,k}(t) \leq w_{u,k}(t)$, if $k \neq m$, $t \in T$. Then we have that $(\mathbb{AN}(T)_m, *_m)$ is a non-commutative hypergroup. The above defined invariant (termed also normal) subgroup $(\mathbb{AN}_1(T)_m, \cdot_m)$ of the group $(\mathbb{AN}(T)_m, \cdot_m)$ is the carried set of a subhypergroup of the hypergroup $(\mathbb{AN}(T)_m, *_m)$ and it has certain significant properties.

Using certain generalization of methods from [8] we obtain after investigation of constructed structures this result:

Let $T = [0, t_0) \subset \mathbb{R}$, $t_0 \in \mathbb{R} \cup \{\infty\}$. Then for any positive integer $n \in \mathbb{N}$, $n \geq 2$ and for any integer m such that $1 \leq m \leq n$ the hypergroup $(\mathbb{AN}(T)_m, *_m)$, where

$$\mathbb{AN}(T)_m = \{Ne(\vec{w}_r); \vec{w}_r = (w_{r,1}(t), \ldots, w_{r,n}(t)) \in [\mathbb{C}(T)]^n, w_{r,m}(t) > 0, t \in T\},$$

is a transpozition hypergroup (i.e., a non-commutative join space) such that $(\mathbb{AN}(T)_m, *_m)$ is its subhypergroup, which is

- Invertible (i.e., $Ne(\vec{w}_r)/Ne(\vec{w}_s) \cap \mathbb{AN}_1(T)_m \neq \emptyset$ implies $Ne(\vec{w}_s)/Ne(\vec{w}_r) \cap \mathbb{AN}_1(T)_m \neq \emptyset$ and $Ne(\vec{w}_r) \backslash Ne(\vec{w}_s) \cap \mathbb{AN}_1(T)_m \neq \emptyset$ implies $Ne(\vec{w}_s) \backslash Ne(\vec{w}_r) \cap \mathbb{AN}_1(T)_m \neq \emptyset$ for all pairs of neurons $Ne(\vec{w}_r), Ne(\vec{w}_s) \in \mathbb{AN}_1(T)_m$,

- Closed (i.e., $Ne(\vec{w}_r)/Ne(\vec{w}_s) \subset \mathbb{AN}_1(T)_m$, $Ne(\vec{w}_r) \setminus Ne(\vec{w}_s) \subset \mathbb{AN}_1(T)_m$ for all pairs $Ne(\vec{w}_r), /, Ne(\vec{w}_s) \in \mathbb{AN}_1(T)_m$,
- Reflexive (i.e., $Ne(\vec{w}_r) \mathbb{AN}_1(T)_m = \mathbb{AN}_1(T)_m/Ne(\vec{w}_r)$ for any neuron $Ne(\vec{w}_r) \in \mathbb{AN}_1(T)_m$ and
- Normal (i.e., $Ne(\vec{w}_r) * \mathbb{AN}_1(T)_m = \mathbb{AN}_1(T)_m * Ne(\vec{w}_r)$ for any neuron $Ne(\vec{w}_r) \in \mathbb{AN}_1(T)_m$.

Remark 3. *We can define a certain transformation function which mappes the output function $y^{[\alpha]}(t)$ into the output function $y^{[\alpha+1]}(t)$. This function denoting by $\rho^{[\alpha]}$ also determines the transformation $S^{[\alpha]}$ of powers of corresponding differential neurons: $D^\alpha \xrightarrow{S^{[\alpha]}} D^{\alpha+1}$. In more detail, let us describe output functions $y^{[\alpha]}(t)$, $y^{[\alpha+1]}(t)$ and mentioned transformation function $\rho^{[\alpha]}$.*

$$y^{[\alpha]}(t) = (1 + w_m(t) + \cdots + w_m^{\alpha-1}(t))(w_1(t)p(t) + w_2(t)\frac{dp(t)}{dt} + \cdots + w_{m-1}(t)\frac{d^{m-2}p(t)}{dt^{m-2}} +$$

$$w_m^\alpha \frac{d^{m-1}p(t)}{dt^{m-1}} + w_{m+1}\frac{d^m p(t)}{dt^m} + \cdots + (b\frac{d^n p(t)}{dt^n})^\alpha),$$

$$y^{[\alpha+1]}(t) = (1 + w_m(t) + \cdots + w_m^{\alpha-1}(t) + w_m^\alpha(t))(w_1(t)p(t) + w_2(t)\frac{dp(t)}{dt} + \cdots$$

$$w_{m-1}(t)\frac{d^{m-2}p(t)}{dt^{m-2}} + w_m^{\alpha+1}\frac{d^{m-1}p(t)}{dt^{m-1}} + w_{m+1}\frac{d^m p(t)}{dt^m} + \cdots + (b\frac{d^n p(t)}{dt^n})^{\alpha+1}).$$

Transformation function $\rho^{[\alpha]}$ of the output function $y^{[\alpha]}(t)$ into the output function $y^{[\alpha+1]}(t)$ which determines the transformation $D^\alpha \xrightarrow{S^{[\alpha]}} D^{\alpha+1}$ of powers of corresponding differential neurons.

So,

$$\rho^{[\alpha]}\left[\left(\sum_{r=0}^{\alpha-1} w_m^r(t)\right) \cdot \left(\sum_{\substack{k=1 \\ k \neq m}}^{m+1} w_k(t)\frac{d^{k-1}p(t)}{dt^{k-1}} + w_m^\alpha(t)\frac{d^{m-1}p(t)}{dt^{m-1}} + \left(b\frac{d^n p(t)}{dt^n}\right)^\alpha\right)\right] =$$

$$= \left[\left(\sum_{r=0}^{\alpha} w_m^r(t)\right) \cdot \left(\sum_{\substack{k=1 \\ k \neq m}}^{m+1} w_k(t)\frac{d^{k-1}p(t)}{dt^{k-1}} + w_m^{\alpha+1}(t)\frac{d^{m-1}p(t)}{dt^{m-1}} + \left(b\frac{d^n p(t)}{dt^n}\right)^{\alpha+1}\right)\right].$$

Denoting

$$w_m^{[\alpha-1]} = \sum_{r=0}^{\alpha-1} w_m^r(t) \text{ and } v^{[\alpha]} = \sum_{\substack{k=1 \\ k \neq m}}^{m+1} w_k(t)\frac{d^{k-1}p(t)}{dt^{k-1}} + w_m^\alpha(t)\frac{d^{m-1}p(t)}{dt^{m-1}} + \left(b\frac{d^n p(t)}{dt^n}\right)^\alpha,$$

we can write

$$\rho^{[\alpha]}\left[\left(w_m^{[\alpha-1]}\right) \cdot w_m^{[\alpha]}\right] = w_m^{[\alpha]} \cdot v^{[\alpha+1]}.$$

5. Conclusions

We have constructed the infinite cyclic group $(\mathbb{G}Dn, \cdot_m)$ of differential neurons which is isomorphic to the cyclic group $(\mathbb{Z}, +)$, possessing the neuron $N1(\vec{e})_m$ as the identity element of $(\mathbb{G}Dn, \cdot_m)$. Thus,

$$(\{N1(\vec{e})_m\} \cup \{D^\alpha Ne_p(\vec{w}; \alpha \in \mathbb{Z}, \alpha \neq 0\}, \cdot_m) = (\mathbb{G}Dn, \cdot_m) \cong (\mathbb{Z}, +). \tag{43}$$

It is to be noted that the above constructed cyclic (infinite) group of artificial differential neurons can be also used for the construction of certain hyperstructures formed by such neurons [17–20]. So the above presented approach enables an additional elaboration of the hyperstructure theory ([8,9,11–32]) in connection with time varying weights and with vectors of differentiable input functions.

The construction of the considered infinite cyclic group of differential neurons can be onto other its isomorphic images under the using other weights and inputs. After those constructions there is possible to create abelian finitely or infinitely generated groups of artificial differential neurons and to investigate their direct products or sums. Using a suitable ordering these considerations involve to obtain neural networks with prescribed structures.

Author Contributions: Investigation, J.C., B.S.; writing—original draft preparation, J.C., B.S.; writing—review and editing, J.C., B.S., J.V. All authors have read and agreed to the published version of the manuscript.

Funding: The first author was supported by the FEKT-S-17-4225 grant of Brno University of Technology and the third author was supported by the FEKT-S-20-6225 grant of Brno University of Technology.

Institutional Review Board Statement: Not applicable.

Informed Consent Statement: Not applicable.

Data Availability Statement: Not applicable.

Acknowledgments: The authors would like to express their thanks to Dario Fasino and Domenico Freni.

Conflicts of Interest: The authors declare no conflict of interest. The funders had no role in the design of the study; in the collection, analyses, or interpretation of data; in the writing of the manuscript, or in the decision to publish the results.

References

1. Chvalina, J.; Novák, M.; Smetana, B. Construction on an Infinite Cyclic Monoid of Differential Neurons. In *Mathematics, Information Technologies and Applied Sciences 2021 Post-Conference Proceedings of Extended Versions of Selected Papers, Proceedings of the MITAV 2021, Brno, Czech Republic, 17–18 June 2021*; Baštinec, J., Hrubý, M., Eds.; lUniversity of Defence: Brno, Czech Republic, 2021; pp. 1–10.
2. Koskela, T. *Neural Network Methods in Analysing and Modelling Time Varying Processes*; Report B, 35; Helsinki University of Technology Laboratory of Computational Engineering Publications, Department of Electrical and Communications: Helsinki, Finland, 2003.
3. Buchholz, S. *A Theory of Neural Computation with Clifford-Algebras*; Technical Report Number 0504; Christian-Albrechts-Universität zu Kiel, Institut für Informatik und Praktische Mathematik: Kiel, Germany, 2005.
4. Tučková, J.; Šebesta, V. Data Mining Approach for Prosody Modelling by ANN in Text-to-Speech Synthesis. In Proceedings of the IASTED Inernational Conference on Artificial Intelligence and Applications—AIA 2001, Marbella, Spain, 4–7 September 2001; Hamza, M.H., Ed.; ACTA Press: Marbella, Spain, 2001; pp. 164–166.
5. Volná, E. *Neuronové Sítě 1. [Neural Networks]*, 2nd ed.; Ostravská Univerzita: Ostrava, Czech Republic, 2008.
6. Waldron, M.B. Time varying neural networks. In Proceedings of the Annual International Conference of the IEEE Engineering in Medicine and Biology Society, New Orleans, LA, USA, 4–7 November 1988.
7. Chvalina, J.; Smetana, B. Models of Iterated Artificial Neurons. In Proceedings of the 18th Conference on Aplied Mathematics Aplimat 2019, Bratislava, Slovakia, 5–7 February 2019; pp. 203–212.
8. Chvalina, J.; Smetana, B. Groups and Hypergroups of Artificial Neurons. In Proceedings of the 17th Conference on Aplied Mathematics Aplimat 2018, Bratislava, Slovakia, 6–8 February 2018; pp. 232–243.
9. Chvalina, J.; Smetana, B. Solvability of certain groups of time varying artificial neurons. *Ital. J. Pure Appl. Math.* **2021**, *45*, 80–94.
10. Pollock, D.; Waldron, M.B. Phase dependent output in a time varying neural net. In Proceedings of the Annual Conference on EMBS, lSeattle, WA, USA, 9–12 November 1989; pp. 2054–2055.
11. Goodman, F.M. *Algebra: Abstract and Concrete*; Prentice Hall: London, UK, 1998.
12. Chvalina, J.; Hošková-Mayerová, Š.; Dehghan Nezhad, A. General actions of hypergroups and some applications. *Analele Stiint. Univ. Ovidius Constanta* **2013**, *21*, 59–82.
13. Chvalina, J.; Chvalinová, L. Action of centralizer hypergroups of n-th order linear differential operators on rings on smooth functions. *J. Appl. Math.* **2008**, *1*, 45–53.
14. Chvalina, J.; Chvalinová, L. Modelling of join spaces by n-th order linear ordinary differential operators. In Proceedings of the 4th International Conference APLIMAT 2005, Bratislava, Slovakia, 1–4 February 2005; pp. 279–284.

15. Chvalina, J.; Novák, M.; Smetana, B.; Staněk, D. Sequences of Groups, Hypergroups and Automata of Linear Ordinary Differential Operators. *Mathematics* **2021**, *9*, 319. [CrossRef]
16. Chvalina, J.; Novák, M.; Staněk, D. Sequences of groups and hypergroups of linear ordinary differential operators. *Ital. J. Pure Appl. Math.* 2019, accepted. [CrossRef]
17. Novák, M. n-ary hyperstructures constructed from binary quasi-ordered semigroups. *Analele Stiint. Univ. Ovidius Constanta Ser. Mat.* **2014**, *22*, 147–168. [CrossRef]
18. Novák, M. On *EL*-semihypergroups. *Eur. J. Comb.* **2015**, *44*, 274–286. [CrossRef]
19. Novák, M. Some basic properties of *EL*-hyperstructures. *Eur. J. Comb.* **2013**, *34*, 446–459. [CrossRef]
20. Cristea, I.; Novák, M.; Křehlík, Š. A class of hyperlattices induced by quasi-ordered semigroups. In Proceedings of the 16th Conference on Aplied Mathematics Aplimat 2017, Bratislava, Slovakia, 31 January–2 February 2017; pp. 1124–1135.
21. Corsini, P. *Prolegomena of Hypergroup Theory*; Aviani Editore Tricesimo: Udine, Italy, 1993.
22. Corsini, P.; Leoreanu, V. *Applications of Hyperstructure Theory*; Kluwer: Dordrecht, The Netherlands; Boston, MA, USA; London, UK, 2003.
23. Cristea, I. Several aspects on the hypergroups associated with n-ary relations. *Analele Stiint. Univ. Ovidius Constanta* **2009**, *17*, 99–110.
24. Cristea, I.; Ştefănescu, M. Binary relations and reduced hypergroups. *Discrete Math.* **2008**, *308*, 3537–3544. [CrossRef]
25. Cristea, I.; Ştefănescu, M. Hypergroups and n-ary relations. *Eur. J. Combin.* **2010**, *31*, 780–789. [CrossRef]
26. Leoreanu-Fotea, V.; Ciurea, C.D. On a P-hypergroup. *J. Basic Sci.* **2008**, *4*, 75–79.
27. Račková, P. Hypergroups of symmetric matrices. In Proceedings of the 10th International Congress of Algebraic Hyperstructures and Applications, (AHA), Brno, Czech Republic, 3–9 September 2008; pp. 267–272.
28. Vougiouklis, T. *Hyperstructures and their Representations*; Hadronic Press: lPalm Harbor, FL, USA, 1994.
29. Novák, M.; Cristea, I. Composition in *EL*–hyperstructures. *Hacet. J. Math. Stat.* **2019**, *48*, 45–58. [CrossRef]
30. Vougiouklis, T. Cyclicity in a special class of hypergroups, *Acta Univ. Carol. Math. Phys.* **1981**, *22*, 3–6.
31. Chvalina, J.; Svoboda, Z. Sandwich semigroups of solutions of certain functional equations and hyperstructures determined by sandwiches of functions. *J. Appl. Math.* **2009**, *2*, 35–43.
32. Borzooei, R.A.; Varasteh, H.R.; Hasankhani, A. $\mathcal{F}$-Multiautomata on Join Spaces Induced by Differential Operators. *Appl. Math.* **2014**, *5*, 1386–1391. [CrossRef]

Article

Commutativity and Completeness Degrees of Weakly Complete Hypergroups

Mario De Salvo [1], Dario Fasino [2], Domenico Freni [2,*] and Giovanni Lo Faro [1]

[1] Dipartimento di Scienze Matematiche e Informatiche, Scienze Fisiche e della Terra, Università di Messina, 98122 Messina, Italy; desalvo@unime.it (M.D.S.); lofaro@unime.it (G.L.F.)
[2] Dipartimento di Scienze Matematiche, Informatiche e Fisiche, Università di Udine, 33100 Udine, Italy; dario.fasino@uniud.it
* Correspondence: domenico.freni@uniud.it

Abstract: We introduce a family of hypergroups, called weakly complete, generalizing the construction of complete hypergroups. Starting from a given group G, our construction prescribes the β-classes of the hypergroups and allows some hyperproducts not to be complete parts, based on a suitably defined relation over G. The commutativity degree of weakly complete hypergroups can be related to that of the underlying group. Furthermore, in analogy to the degree of commutativity, we introduce the degree of completeness of finite hypergroups and analyze this degree for weakly complete hypergroups in terms of their β-classes.

Keywords: hypergroups; complete hypergroup; fundamental relations

MSC: 20N20

Citation: De Salvo, M.; Fasino, D.; Freni, D.; Lo Faro, G. Commutativity and Completeness Degrees of Weakly Complete Hypergroups. *Mathematics* **2022**, *10*, 981. https://doi.org/10.3390/math10060981

Academic Editors: Irina Cristea, Emeritus Mario Gionfriddo and Takayuki Hibi

Received: 4 February 2022
Accepted: 16 March 2022
Published: 18 March 2022

Publisher's Note: MDPI stays neutral with regard to jurisdictional claims in published maps and institutional affiliations.

Copyright: © 2022 by the authors. Licensee MDPI, Basel, Switzerland. This article is an open access article distributed under the terms and conditions of the Creative Commons Attribution (CC BY) license (https://creativecommons.org/licenses/by/4.0/).

1. Introduction

We refer to hypercompositional algebra as the branch of algebra concerned with hypercompositional structures, that is, algebraic structures where the composition of two elements is a nonempty set rather than a single element [1]. Although hypercompositional algebra differs from classic algebra in its subjects, methods, and goals, the two fields are connected by certain equivalence relations, called fundamental relations [2,3]. Through the fundamental relations, hypercompositional algebra can make use of the wealth of tools typical of traditional algebra.

A fundamental relation is the smallest equivalence relation defined on a hypercompositional structure such that the corresponding quotient is a classic structure whose operational properties are analogous to those of the original structure [4,5]. For example, the quotient of a hypergroup modulo the equivalence β is isomorphic to a group [6–8]. On the other hand, given a group G and a family $\mathfrak{F} = \{A_k\}_{k \in G}$ of nonempty and pairwise disjoint sets, the set $H = \bigcup_{k \in G} A_k$ equipped with the hyperproduct $x \circ y = A_{ij}$, for all $x \in A_i$ and $y \in A_j$, is a hypergroup. Hypergroups built in this way are called complete [4] and have the property that the β-classes are the sets A_k. For any nonempty subset A of a hypergroup $(H, \circ)$, the set $\mathfrak{C}(A) = \bigcup_{a \in A} \beta(a)$ is the complete closure of A. Hence, a hypergroup $(H, \circ)$ is complete if and only if $x \circ y = \mathfrak{C}(x \circ y)$, for all $x, y \in H$. Complete hypergroups have been the subject of many studies, see, e.g., [9–12], because they have a variety of group-like properties. Notably, in [13], the authors define the commutativity degree of complete hypergroups and characterize it with an identity that is analogous to the class equation for groups. Recall that the commutativity degree of a finite group G was defined by W. Gustafson in [14] as the probability that two randomly chosen elements commute,

$$d(G) = \frac{|\{(x,y) \in G^2 \mid xy = yx\}|}{|G|^2}.$$

Inspired by this concept, in [13] the commutativity degree of a finite hypergroup $(H, \circ)$ is defined as
$$d(H) = \frac{|\{(x,y) \in H^2 \mid x \circ y = y \circ x\}|}{|H|^2}.$$

The probabilistic interpretation of this number is completely analogous to that for groups. In this paper, we define the completeness degree of a finite hypergroup $(H, \circ)$ as the number
$$\Delta(H) = \frac{|\{(x,y) \in H^2 \mid \mathfrak{C}(x \circ y) = x \circ y\}|}{|H|^2},$$
and determine some formulas which allow us to compute the previous numbers for a special class of hypergroups, called weakly complete, that include complete hypergroups.

The plan of this paper is the following: In Section 2, we introduce definitions, notations, and fundamental facts to be used throughout the paper. In Section 3, we give the definition of product-free relations on a group G and study their main characteristics. In particular, we characterize product-free relations that are maximal with respect to inclusion. In Section 4, we present a new construction of hypergroups. These hypergroups are called weakly complete and are defined using a product-free relation I on a group G, a family $\{A_k : k \in G\}$ of nonempty and pairwise disjoint sets and a special family of functions $\{\varphi_{i,j} : (i,j) \in I\}$. The main features of these hypergroups are discussed in this section. The completeness degree $\Delta(H)$ of finite weakly complete hypergroups is defined and analyzed in Section 5. There, we prove lower bounds for $\Delta(H)$ that depend only on the size of the β-classes of H. Finally, in Section 6, we discuss the commutativity degree $d(H)$ of finite weakly complete hypergroups, and establish relations between $d(H)$ and $\Delta(H)$. In particular, in our last theorem we prove that, if the cardinality of A_k does not depend on k, then $|d(H) - \Delta(H)| \leq \frac{1}{4}$.

2. Basic Definitions and Notations

We adopt from known textbooks [1,4,5] standard definitions of basic concepts in hypercompositional algebra, such as semihypergroups and hypergroups. For the reader's convenience, we present below a few concepts that are needed in this work.

Given a semihypergroup $(H, \circ)$, the relation $\beta \subseteq H \times H$ is defined as $\beta = \cup_{n \geq 1} \beta_n$, where β_1 is the diagonal relation in H and, for every integer $n > 1$, β_n is defined as follows:
$$x \beta_n y \iff \exists z_1, \ldots, z_n \in H : \{x, y\} \subseteq z_1 \circ z_2 \circ \cdots \circ z_n, \tag{1}$$

see, e.g., [2,3]. This relation is one of the main fundamental relations alluded to in the Introduction. For some special families of semihypergroups, β is transitive; see, e.g., [15,16]. In particular, if $(H, \circ)$, is a hypergroup then β is an equivalence relation, see [7,8], and we have the chain of inclusions
$$\beta_1 \subseteq \beta_2 \subseteq \beta_3 \subseteq \cdots \subseteq \beta_n \cdots. \tag{2}$$

Moreover, the quotient set H/β equipped with the operation $\beta(x) \otimes \beta(y) = \beta(z)$ for all $x, y \in H$ and $z \in x \circ y$, is a group. More precisely, β is the smallest strongly regular equivalence on H such that the quotient H/β is a group [2]. The canonical epimorphism $\pi : H \mapsto H/\beta$ fulfills the identity $\pi(x \circ y) = \pi(x) \otimes \pi(y)$ for all $x, y \in H$, and the kernel $\omega_H = \pi^{-1}(1_{H/\beta})$ of π is the *heart* of $(H, \circ)$.

Let $(H, \circ)$ be a hypergroup. We say that a nonempty subset $A \subseteq H$ is a complete part if for every $n \geq 1$ and $x_1, x_2, \ldots, x_n \in H$,
$$(x_1 \circ x_2 \circ \cdots \circ x_n) \cap A \neq \emptyset \implies x_1 \circ x_2 \circ \cdots \circ x_n \subseteq A.$$

The complete closure of A is the intersection of all complete parts containing A and is denoted with $\mathfrak{C}(A)$. Using the canonical projection $\pi : H \mapsto H/\beta^*$, the complete closure of A can be characterized as follows:

$$\mathfrak{C}(A) = \pi^{-1}(\pi(A)) = A \circ \omega_H = \omega_H \circ A.$$

A hypergroup $(H, \circ)$ is complete if $x \circ y = \mathfrak{C}(x \circ y)$ for all $x, y \in H$. In other words, $(H, \circ)$ is a complete hypergroup if $x \circ y = \mathfrak{C}(a) = \beta(a)$ for every $(x, y) \in H^2$ and $a \in x \circ y$.

Finally, let G be a group and let $I \subseteq G \times G$ be a binary relation on G. We denote I^T the transpose relation of $I \subseteq G \times G$, that is, $(a, b) \in I \Leftrightarrow (b, a) \in I^T$. Furthermore, we associate with I the span and support sets defined below:

$$\text{Span}(I) = \{ij : (i, j) \in I\},$$
$$\text{Supp}(I) = \{i \in G^* : \exists j \in G^* : (i, j) \in I \text{ or } (j, i) \in I\}.$$

Here and in the following, G^* denotes the set $G \setminus \{1_G\}$.

3. Product-Free Relations on a Group

The class of complete hypergroups is among the best known in hypergroup theory, and is characterized by the fact that the hyperproduct of any two elements is a β-class. These hypergroups were introduced by P. Corsini in [4] and can be built by considering a group G and a family $\mathfrak{F} = \{A_k\}_{k \in G}$ of nonempty and pairwise disjoint sets. The set $H = \bigcup_{k \in G} A_k$ is endowed with the product $x \circ y = A_{ij}$ for $x \in A_i$ and $y \in A_j$. Then, $(H, \circ)$ is a complete hypergroup and the β-classes of $(H, \circ)$ are the sets A_k. In this section, we introduce a special family of binary relations in a group G. These relations will allow us to define in the next section the class of hypergroups that generalize that of complete hypergroups and is the main subject of this work.

Definition 1. *Let G be a group. A binary relation $I \subseteq G \times G$ is called product-free or PF-relation if, for all $i, j, k \in G$,*

$$(i, j) \in I \implies (ij, k) \notin I \text{ and } (k, ij) \notin I.$$

PF_G denotes the family of all PF-relations in the group G. If $I \in \text{PF}_G$ and $(i, j) \in I$, then the elements i, j are different from 1_G. Otherwise, if, for example, $i = 1_G$, then we have the contradiction $(i, j) = (i, 1_G j) = (i, ij) \notin I$. As a consequence, if $|G| = 1$, then PF_G reduces to the empty relation. Hence, if $I \neq \emptyset$, then $|G| \geq 2$ and $I \subset G^* \times G^*$.

Our first result provides a characterization of PF-relations in terms of support and span sets. Subsequently, we analyze the structure of PF_G and provide some examples.

Lemma 1. *Let $I \subseteq G \times G$. Then, $I \in \text{PF}_G$ if and only if $\text{Supp}(I) \cap \text{Span}(I) = \emptyset$.*

Proof. If $x \in \text{Supp}(I) \cap \text{Span}(I)$, then $(x, y) \in I$ or $(y, x) \in I$ for some $y \in G^*$ and there exists $(i, j) \in I$ such that $ij = x$. We obtain $(ij, y) \in I$ or $(y, ij) \in I$, a contradiction. Conversely, if $I \notin \text{PF}_G$, then there exists $(i, j) \in I$ and $y \in G$ such that $(ij, y) \in I$ or $(y, ij) \in I$. However, then we have $ij \in \text{Supp}(I) \cap \text{Span}(I)$. □

Below we provide a couple of examples of how PF-relations can be built.

Example 1. *Let G be a group. For any subset $S \subseteq G^*$, let I_S be the relation*

$$I_S = \{(i, j) : i, j \in S, ij \notin S\}.$$

It is can be seen that $\text{Supp}(I_S) \subseteq S$ and $\text{Span}(I_S) \subseteq G \setminus S$. Hence, $I_S \in \text{PF}_G$ by Lemma 1. For example, I_{G^*} is the relation consisting of the pairs (x, x^{-1}) for $x \in G^*$. On the other hand, I_{1_G} is the empty relation.

Example 2. Let G and G' be groups. Moreover, let $I \in \operatorname{PF}_G$ and $I' \in \operatorname{PF}_{G'}$. Then, the direct product relation
$$I \otimes I' = \{((a,a'),(b,b')) \mid (a,b) \in I, (a',b') \in I'\}$$
is a PF-relation on the direct product $G \times G'$. Indeed, $\operatorname{Supp}(I \otimes I') = \operatorname{Supp}(I) \times \operatorname{Supp}(I')$ and $\operatorname{Span}(I \otimes I') \subseteq \operatorname{Span}(I) \times \operatorname{Span}(I')$, so the claim follows from Lemma 1.

The following features of PF-relations are self-evident, so we refrain from including a proof.
- Every subset of a PF-relation is a PF-relation.
- If $I_1, I_2 \in \operatorname{PF}_G$, then $I_1 \cap I_2 \in \operatorname{PF}_G$.
- Let G be abelian. Then, $I \in \operatorname{PF}_G$ if and only if $I^T \in \operatorname{PF}_G$.

Hereafter, we show that no PF-relation can contain more than a quarter of all possible pairs of elements in the group. This result will play an important role in the forthcoming sections.

Theorem 1. *Let G be a finite group and $I \in \operatorname{PF}_G$. Then, $|I| \leq |G|^2/4$.*

Proof. For notational simplicity, let $S = \operatorname{Supp}(I)$. For any element $i \in S$, let $\mathcal{S}(i) = \{j \in G : (i,j) \in I\}$ and $\mathcal{R}(i) = \{ij : j \in \mathcal{S}(i)\}$. Obviously, $\mathcal{S}(i)$ and $\mathcal{R}(i)$ have the same cardinality, since the application $f_i : \mathcal{S}(i) \to \mathcal{R}(i)$ such that $f_i(j) = ij$ is bijective. Since $\mathcal{R}(i) \subseteq G \setminus S$, we have
$$|\mathcal{R}(i)| \leq |G \setminus S| = |G| - |S|.$$
Moreover,
$$|I| = \left| \bigcup_{i \in S} \mathcal{S}(i) \right| \leq \sum_{i \in S} |\mathcal{S}(i)| = \sum_{i \in S} |\mathcal{R}(i)| \leq |S|(|G| - |S|).$$
To maximize the rightmost quantity, we set $|S| = |G|/2$, and we have the claim. □

The following example shows that the inequality in the preceding theorem is the best possible, since it can hold as an equality.

Example 3. Let $G = (\mathbb{Z}_m, +)$, where $m \geq 2$ is even. Consider the following relation $I \subset G \times G$:
$$(i,j) \in I \iff i \equiv j \equiv 1 \pmod{2}.$$
It is easy to see that $\operatorname{Span}(I) = \{i \in \mathbb{Z}_m : i \equiv 0 \pmod{2}\}$ and $\operatorname{Supp}(I) = \{i \in \mathbb{Z}_m : i \equiv 1 \pmod{2}\}$. Hence, $I \in \operatorname{PF}_G$ by Lemma 1. Finally, $|\operatorname{Span}(I)| = |\operatorname{Supp}(I)| = m/2$ and $|I| = |G|^2/4$.

Maximal PF-Relations

PF-relations can be semi-ordered by inclusion; hence, it is worth considering maximal elements in PF_G, with regard to their existence and characterization. The existence of maximal relations is shown in the forthcoming result.

Proposition 1. *The family PF_G of PF-relations on G has at least one maximal element.*

Proof. The family PF_G is nonempty because it contains the empty relation. Moreover, for each chain $\{R_j\}_{j \in J}$ in the partially ordered set $(\operatorname{PF}_G, \subseteq)$, the relation $\widehat{R} = \bigcup_{j \in J} R_j$ is product free. Indeed, if $(x,y) \in \widehat{R}$ and by chance there exists $z \in G$ such that $(xy, z) \in \widehat{R}$, then there exist $j_1, j_2 \in J$ such that $(x,y) \in R_{j_1}$ and $(xy, z) \in R_{j_2}$. Since $\{R_j\}_{j \in J}$ is a chain, we can assume that $R_{j_1} \subset R_{j_2}$, and so $\{(x,y),(xy,z)\} \subseteq R_{j_2}$, which is impossible because $R_{j_2} \in \operatorname{PF}_G$. Hence, $\widehat{R}$ is an upper bound of $\{R_j\}_{j \in J}$. By Zorn's Lemma, in PF_G there exists a maximal element. □

Using an argument similar to the previous one, we also have that every PF-relation I on a group G is contained in a maximal PF-relation M. It suffices to apply Zorn's lemma to the family of PF-relations that contain I. Hence, we have the following result:

Proposition 2. *Let $I \in \text{PF}_G$. Then, there exists a maximal PF-relation $M \in \text{PF}_G$ such that $I \subseteq M$.*

Remark 1. *Every maximal PF-relation M in an abelian group G is symmetric. Indeed, if $(x,y) \in M$ and $(y,x) \notin M$, then $M \cup \{(y,x)\}$ is a PF-relation and $M \subset M \cup \{(y,x)\}$. The same fact is not true if the group is not abelian, as shown in the following example. Let G be a noncommutative group with two elements $a, b \in G - \{1_G\}$ such that $ab \neq 1_G$, $a^2 = 1_G$ and $ab \neq ba$, e.g., the symmetric group S_3. In these hypotheses, $a \neq b$ and the relation $I = \{(a,b), (a,ba)\}$ are product free. If $M \in \text{PF}_G$ is maximal and $I \subseteq M$, then we have $(b,a) \notin M$ since $(a,ba) \in M$.*

The empty relation is maximal if and only if G is trivial. In the next result, we give a necessary and sufficient condition for a PF-relation to be maximal.

Theorem 2. *Let G be a group and let $I \in \text{PF}_G$. Moreover, let*

$$\bar{I} = \{(x,y) : xy \in \text{Supp}(I) \text{ or } \{x,y\} \cap \text{Span}(I) \neq \varnothing\}.$$

Then, we have
1. *$I \cap \bar{I} = \varnothing$;*
2. *I is maximal if and only if $I \cup \bar{I} = G^* \times G^*$.*

Proof. If $I = \varnothing$ then the claim is trivial, so suppose $I \neq \varnothing$. Note that $\bar{I}$ admits the alternative definition

$$\bar{I} = \{(x,y) \in G^* \times G^* \mid \exists (i,j) \in I : xy \in \{i,j\} \text{ or } ij \in \{x,y\}\}.$$

1. Let $(x,y) \in I \cap \bar{I}$. By hypotesis, there exists $(i,j) \in I$ such that $xy \in \{i,j\}$ or $ij \in \{x,y\}$. If $xy = i$ (resp., $xy = j$), then $(xy,j) \in I$ (resp., $(i,xy) \in I$), which contradicts $(x,y) \in I$. Similarly, if $ij = x$ (resp., $ij = y$) then $(ij,y) \in I$ (resp., $(x,ij) \in I$), which contradicts $(i,j) \in I$.

2. By point 1, if $I \cup \bar{I} = G^* \times G^*$, then I is maximal. On the other hand, let I be maximal and $(x,y) \in G^* \times G^*$ with $(x,y) \notin I$. Since $I \cup \{(x,y)\}$ is not a PF-relation, two cases are possible:

(a) There exist $(i,j) \in I$ and $k \in G^*$ such that $(x,y) = (ij,k)$ or $(x,y) = (k,ij)$.
(b) There exists $k \in G^*$ such that $(xy,k) \in I$ or $(k,xy) \in I$.

In the first case, we obtain $x = ij$ or $y = ij$; hence, $(x,y) \in \bar{I}$. In the second case, we have $(x,y) \in \bar{I}$ because $xy \in \{xy,k\}$. In both cases, we obtain $I \cup \bar{I} = G^* \times G^*$. □

Remark 2. *We observe that if I and I' are maximal PF-relations, then the tensor product relation $I \otimes I'$ is not necessarily maximal. For example, let $G = \{1_G, a\}$ and $G' = \{1_{G'}, a', b'\}$ be groups isomorphic to $(\mathbb{Z}_2, +)$ and $(\mathbb{Z}_3, +)$, respectively. Moreover, let $I = \{(a,a)\} \subset G \times G$ and $I' = \{(a',b'), (b',a')\} \subset G' \times G'$. The relations I and I' are maximal PF-relations. However, the tensor product relation $I \otimes I' = \{((a,a'), (a,b')), ((a,b'), (a,a'))\}$ is not maximal because it is contained in the following PF-relation on $G \times G'$:*

$$T = I \otimes I' \cup \{((1_G, a'), (1_G, a'))\}.$$

4. Weakly Complete Hypergroups

In this section, we introduce a new class of hypergroups, whose construction is fundamentally based on PF-relations. We introduce a few auxiliary concepts and notations for background information. In what follows, we denote $\mathcal{P}^*(X)$ the collection of nonempty subsets of the set X.

Definition 2. *Let A, B, C be nonempty sets. A function $\varphi : A \times B \mapsto \mathcal{P}^*(C)$ is a double covering, or* bi-covering *for short, if for all $a \in A$ and $b \in B$ we have*

$$\bigcup_{x \in B} \varphi(a, x) = \bigcup_{x \in A} \varphi(x, b) = C. \tag{3}$$

A bi-covering $\varphi : A \times B \mapsto \mathcal{P}^*(C)$ is called trivial *if $\varphi(a, b) = C$ for all $a \in A$ and $b \in B$, and* proper *if $\varphi(a, b) \subset C$ for all $a \in A$ and $b \in B$.*

Example 4. *Bi-covering functions can be constructed by considering a group G and three nonempty sets A, B, C of size $\geq |G|$. If $\alpha : A \to G$, $\beta : B \to G$ and $\gamma : C \to G$ are three surjective functions; then, the function $\varphi : A \times B \to \mathcal{P}^*(C)$ such that $\varphi(a, b) = \gamma^{-1}(\alpha(a)\beta(b))$, for all $(a, b) \in A \times B$, is bi-covering. Indeed, we trivially have $\bigcup_{x \in B} \varphi(a, x) \subseteq C$, for all $a \in A$. Moreover, if $c \in C$, then, taking $b \in \beta^{-1}(\alpha(a)^{-1}\gamma(c))$, we have $\beta(b) = \alpha(a)^{-1}\gamma(c)$ and we obtain*

$$c \in \gamma^{-1}(\gamma(c)) = \gamma^{-1}(\alpha(a)\beta(b)) = \varphi(a, b) \subseteq \bigcup_{x \in B} \varphi(a, x).$$

Hence, $\bigcup_{x \in B} \varphi(a, x) = C$ for all $a \in A$. Analogous arguments prove that $\bigcup_{x \in A} \varphi(x, b) = C$, for all $b \in B$. Thus, φ is a bi-covering. We note in passing that in the previous construction the role of the group G can be played by an arbitrary hypergroup.

Let G be a group and let I be a relation on G. Consider a family $\mathfrak{F} = \{A_k\}_{k \in G}$ of nonempty and pairwise disjoint sets, and let $\mathfrak{J} = \{\varphi_{i,j}\}_{(i,j) \in I}$ be a family of bi-coverings $\varphi_{i,j} : A_i \times A_j \mapsto \mathcal{P}^*(A_{ij})$. In particular, if $I = \emptyset$, then $\mathfrak{J} = \emptyset$. In the set, $H = \bigcup_{k \in G} A_k$ introduce the hyperproduct $\circ : H \times H \mapsto \mathcal{P}^*(H)$, defined as follows:

$$x \circ y = \begin{cases} A_{ij} & \text{if } x \in A_i, y \in A_j \text{ and } (i,j) \notin I \\ \varphi_{i,j}(x, y) & \text{if } x \in A_i, y \in A_j \text{ and } (i,j) \in I \end{cases} \tag{4}$$

for all $x, y \in H$. This hyperproduct is well defined because the sets in the family $\mathfrak{F} = \{A_k\}_{k \in G}$ are nonempty and pairwise disjointed. The hyperproduct is naturally extended to nonempty subsets of H as usual: For $X, Y \in \mathcal{P}^*(H)$ let

$$x \circ Y = \bigcup_{y \in Y} x \circ y, \qquad X \circ y = \bigcup_{x \in X} x \circ y, \qquad X \circ Y = \bigcup_{x \in X, y \in Y} x \circ y.$$

In particular, for every $i, j \in G$ and $x \in A_j$, we have

$$A_i \circ x = A_{ij}, \qquad x \circ A_i = A_{ji}. \tag{5}$$

Indeed, if $(i, j) \notin I$ then $A_i \circ x = \bigcup_{y \in A_i} y \circ x = A_{ij}$. Otherwise, if $(i, j) \in I$, then from (3) we obtain $A_i \circ x = \bigcup_{y \in A_i} \varphi_{i,j}(y, x) = A_{ij}$. Analogously we can deduce that $x \circ A_i = A_{ji}$. From this observation, it is not difficult to derive that if $I = \emptyset$ or all functions $\varphi_{i,j}$ are trivial; for every $(i, j) \in I$, then $(H, \circ)$ is a complete hypergroup. The following result shows that $(H, \circ)$ is always a hypergroup under the sole condition that $I \in \text{PF}_G$.

Theorem 3. *Let $I \in \text{PF}_G$. Then, in the previous notations,*
(a) *for every $i, j, k \in G$, $x \in A_i$, $y \in A_j$ and $z \in A_k$, we have*

$$(x \circ y) \circ z = A_{ijk} = x \circ (y \circ z);$$

(b) *for every integer $n \geq 3$ and for every $z_1, z_2, \ldots, z_n \in H$ there exists $i \in G$ such that $z_1 \circ z_2 \circ \cdots \circ z_n = A_i$;*

(c) $(H, \circ)$ is a hypergroup such that $\beta = \beta_2$;

Proof. (a) Let $i, j, k \in G$, $x \in A_i$, $y \in A_j$ and $z \in A_k$. If $(i, j) \notin I$ and $(j, k) \notin I$, then we have $x \circ y = A_{ij}$, $y \circ z = A_{jk}$. Consequently, by (5) we obtain

$$(x \circ y) \circ z = A_{ij} \circ z = A_{(ij)k} = A_{i(jk)} = x \circ A_{jk} = x \circ (y \circ z).$$

If $(i, j) \in I$ and $(j, k) \notin I$, we have $(ij, k) \notin I$, $x \circ y = \varphi_{i,j}(x, y) \subseteq A_{ij}$ and $y \circ z = A_{jk}$. Moreover, for every $a \in A_{ij}$ we have $a \circ z = A_{(ij)k}$. Hence,

$$(x \circ y) \circ z = \bigcup_{a \in \varphi_{i,j}(x,y)} a \circ z = A_{(ij)k}.$$

Moreover, by (5), we obtain $x \circ (y \circ z) = x \circ A_{jk} = A_{i(jk)}$. Therefore, $(x \circ y) \circ z = x \circ (y \circ z)$. We obtain same result also when $(i, j) \notin I$ and $(j, k) \in I$. Finally, if $(i, j) \in I$ and $(j, k) \in I$, we have $x \circ y = \varphi_{i,j}(x, y) \subseteq A_{ij}$ and $y \circ z = \varphi_{j,k}(y, z) \subseteq A_{jk}$. Since I is product free, we have $(ij, k) \notin I$ and $(i, jk) \notin I$. Thus,

$$(x \circ y) \circ z = \bigcup_{a \in \varphi_{i,j}(x,y)} a \circ z = A_{(ij)k},$$

$$x \circ (y \circ z) = \bigcup_{b \in \varphi_{j,k}(y,z)} x \circ b = A_{i(jk)}.$$

Hence, also in this case $(x \circ y) \circ z = x \circ (y \circ z) = A_{ijk}$.

(b) It suffices to apply (5) and the previous part (a) and proceed by induction on n.

(c) To prove that $(H, \circ)$ is a hypergroup, we only need to show that $\circ$ is reproducible. Let $x \in H$ and $x \in A_i$. Clearly, $iG = G$ for all $i \in G$ and, by Equation (5), we obtain

$$x \circ H = x \circ \left(\bigcup_{j \in G} A_j \right) = \bigcup_{j \in G} x \circ A_j = \bigcup_{j \in G} A_{ij} = H.$$

The identity $H \circ x = H$ follows analogously for every $x \in H$, so $(H, \circ)$ is a hypergroup. Finally, let $x \beta y$. By (2), there exists $n \geq 3$ such that $x \beta_n y$. By point b), there exists $i \in G$ such that $\{x, y\} \subseteq A_i$. Now, let $a \in A_{1_G}$. Since $(i, 1_G) \notin I$, by (4) we have $\{x, y\} \subseteq A_i = x \circ a$ and we deduce $x \beta_2 y$. □

Example 5. *Let G be a group and let $I \subset G \times G$ be a relation on G. Consider a family $\mathfrak{F} = \{A_k\}_{k \in G}$ of nonempty and pairwise disjoint sets such that $|A_k| \geq |G|$, for all $k \in G$. Moreover, let $\{f_k : A_k \to G\}_{k \in G}$ be a family of surjective functions. Proceeding as in Example 4, we obtain a family of bi-covering functions $\mathfrak{I} = \{\varphi_{i,j} : A_i \times A_j \to \mathcal{P}^*(A_{ij})\}_{(i,j) \in I}$. If $I \in PF_G$, then Theorem 3 provides a hypergroup $(H, \circ)$.*

Remark 3. *Product-free relations have a kind of optimality with respect to the rule (4). As shown in Theorem 3, every hyperproduct defined in terms of a PF-relation is associative and reproducible, independent of families $\mathfrak{F} = \{A_k\}_{k \in G}$ and $\mathfrak{I} = \{\varphi_{i,j}\}_{(i,j) \in I}$. The same property does not hold in general if the relation I is not a PF-relation. For example, consider the group $(\mathbb{Z}_3, +)$, the relation $I = \{(1, 1), (2, 2)\}$, the sets $A_0 = \{a, b\}$, $A_1 = \{c, d, e\}$, $A_2 = \{f, g, h\}$ and the bi-coverings $\varphi_{1,1} : A_1 \times A_1 \mapsto \mathcal{P}^*(A_2)$, $\varphi_{2,2}, \varphi'_{2,2} : A_2 \times A_2 \mapsto \mathcal{P}^*(A_1)$ defined as follows:*

$\varphi_{1,1}$	c	d	e
c	A_2	A_2	f, g
d	A_2	f, g	A_2
e	A_2	A_2	f, g

$\varphi_{2,2}$	f	g	h
f	A_1	A_1	c, d
g	d, e	A_1	A_1
h	d, e	A_1	d, e

$\varphi'_{2,2}$	f	g	h
f	A_1	A_1	c
g	d, e	A_1	A_1
h	d, e	A_1	d, e

Considering the functions $\varphi_{1,1}$ and $\varphi_{2,2}$, definition (4) returns the following hypergroup:

$\circ_1$	a	b	c	d	e	f	g	h
a	A_0	A_0	A_1	A_1	A_1	A_2	A_2	A_2
b	A_0	A_0	A_1	A_1	A_1	A_2	A_2	A_2
c	A_1	A_1	A_2	A_2	f,g	A_0	A_0	A_0
d	A_1	A_1	A_2	f,g	A_2	A_0	A_0	A_0
e	A_1	A_1	A_2	A_2	f,g	A_0	A_0	A_0
f	A_2	A_2	A_0	A_0	A_0	A_1	A_1	c,d
g	A_2	A_2	A_0	A_0	A_0	d,e	A_1	A_1
h	A_2	A_2	A_0	A_0	A_0	d,e	A_1	d,e

On the other hand, considering the functions $\varphi_{1,1}$ and $\varphi'_{2,2}$, we have the hyperproduct

$\circ_1$	a	b	c	d	e	f	g	h
a	A_0	A_0	A_1	A_1	A_1	A_2	A_2	A_2
b	A_0	A_0	A_1	A_1	A_1	A_2	A_2	A_2
c	A_1	A_1	A_2	A_2	f,g	A_0	A_0	A_0
d	A_1	A_1	A_2	f,g	A_2	A_0	A_0	A_0
e	A_1	A_1	A_2	A_2	f,g	A_0	A_0	A_0
f	A_2	A_2	A_0	A_0	A_0	A_1	A_1	c
g	A_2	A_2	A_0	A_0	A_0	d,e	A_1	A_1
h	A_2	A_2	A_0	A_0	A_0	d,e	A_1	d,e

which is not associative since $(f \circ_2 h) \circ_2 e \neq f \circ_2 (h \circ_2 e)$. This example reveals a specific quality of PF-relations: If a hyperproduct defined as in (4) is associative and reproducible, independent of families $\mathfrak{F} = \{A_k\}_{k \in G}$ and $\mathfrak{I} = \{\varphi_{i,j}\}_{(i,j) \in I}$, then the relation I is product free. This fact is formalized in the following result.

Theorem 4. Let G be a group and suppose that $I \subset G \times G$ is not product free. Then, there exists a family $\mathfrak{F} = \{A_k\}_{k \in G}$ of nonempty and pairwise disjoint sets and there exists a family of bi-coverings $\mathfrak{I} = \{\varphi_{i,j}\}_{(i,j) \in I}$ such that the hyperproduct defined in (4) is not associative.

Proof. Firstly, note that we have $I \neq \emptyset$ as $I \notin \text{PF}_G$. The proof can be reduced to the analysis of two cases: (a) there exists $(i,j) \in I$ such that $1_G \in \{i,j\}$; and (b) there exists $i,j,k \in G^*$ such that $(i,j) \in I$ and $(ij,k) \in I$ (or, equivalently, $(k,ij) \in I$).

(a) If $i = j = 1_G$ then it suffices to consider arbitrary families $\mathfrak{F}$ and $\mathfrak{I}$ where $A_{1_G} = \{a,b\}$ and the function $\varphi_{1_G,1_G}$ is described by the following table:

$\varphi_{1_G,1_G}$	a	b
a	b	a
b	a,b	a,b

Then, associativity fails because $(a \circ a) \circ a = \{a,b\} \neq \{a\} = a \circ (a \circ a)$. Otherwise, without loss of generality, assume $j = 1_G$ and $(1_G, 1_G) \notin I$. Let $\mathfrak{F}$ and $\mathfrak{I}$ verify the following conditions: $|A_\ell| = 2$ for every $\ell \in G$ and $|\varphi_{p,q}(x,y)| = 1$ for all $(p,q) \in I$. Let $x \in A_i$ and $y, z \in A_{1_G}$. Then,

$$(x \circ y) \circ z = \varphi_{i,1_G}(x,y) \circ z = \varphi_{i,1_G}(\varphi_{i,1_G}(x,y),z).$$

Hence, $|(x \circ y) \circ z| = 1$. On the other hand, $x \circ (y \circ z) = x \circ A_{1_G} = A_i$; hence $(x \circ y) \circ z \neq x \circ (y \circ z)$.

(b) Let $\mathfrak{F} = \{A_\ell : \ell \in G\}$ and $\mathfrak{I} = \{\varphi_{p,q} : (p,q) \in I\}$ be arbitrary families verifying the following conditions: (b1) $|A_\ell| = 2$ for every $\ell \in G$; (b2) if $(i,jk) \in I$ then $\varphi_{i,jk}(x,y) = A_{ijk}$ for every $x \in A_i$ and $y \in A_{jk}$; $|\varphi_{p,q}(x,y)| = 1$ in all remaining cases. Let $x \in A_i$, $y \in A_j$, and $z \in A_k$. Then,

$$(x \circ y) \circ z = \varphi_{i,j}(x,y) \circ z = \varphi_{ij,k}(\varphi_{i,j}(x,y),z).$$

Since $1_G \notin \{i, j, k\}$, then $(i, j) \neq (i, jk)$ and $(ij, k) \neq (i, jk)$. Hence, $|(x \circ y) \circ z| = 1$ by (b2). On the other hand, for some $w \in y \circ z \subseteq A_{jk}$, we have

$$A_{ijk} = x \circ w \subseteq x \circ (y \circ z).$$

By (b1) we can conclude that $(x \circ y) \circ z \neq x \circ (y \circ z)$. (The proof proceeds in a similar way if $(k, ij) \in I$.) □

Definition 3. *The hypergroups $(H, \circ)$ defined as in (4) with a PF-relation I are called* weakly complete. *A weakly complete hypergroup is n-uniform if $|A_i| = n$ for all $i \in G$; if the size n is not relevant, then we simply call it* uniform.

The term "weakly complete" originates from the following observations: Let $(H, \circ)$ be a weakly complete hypergroup built from families $\mathfrak{F} = \{A_k\}_{k \in G}$ and $\mathfrak{I} = \{\varphi_{i,j}\}_{(i,j) \in I}$, and let $\diamond$ be the hyperproduct obtained from the same set family $\mathfrak{F}$ using only trivial bi-coverings. Then, $(H, \diamond)$ is a complete hypergroup and $x \circ y \subseteq x \diamond y$ for all $x, y \in H$. We also obtain the same conclusion by replacing the given relation I with the empty relation. Furthermore, both in complete hypergroups and weakly complete hypergroups, the fundamental relation β coincides with β_2, as shown in Theorem 3.

In the following, we use the notation $(H, \circ) = \mathcal{W}(G, I, \mathfrak{F}, \mathfrak{I})$ to indicate a weakly complete hypergroup whose hyperproduct $\circ$ is defined as (4) from $I \in PF_G$ and the families $\mathfrak{F} = \{A_k\}_{k \in G}$ and $\mathfrak{I} = \{\varphi_{i,j}\}_{(i,j) \in I}$. We call $\mathcal{W}(G, I, \mathfrak{F}, \mathfrak{I})$ a representation of $(H, \circ)$. It is worth noting that a weakly complete hypergroup may have multiple representations. Indeed, let $(H, \circ) = \mathcal{W}(G, I, \mathfrak{F}, \mathfrak{I})$ and let $(i, j) \notin I$. If the relation $\tilde{I} = I \cup \{(i, j)\}$ is product free, then the same hypergroup admits the representation $\mathcal{W}(G, \tilde{I}, \mathfrak{F}, \tilde{\mathfrak{I}})$ where $\tilde{\mathfrak{I}} = \mathfrak{I} \cup \{\varphi_{i,j}\}$ and $\varphi_{i,j}(x, y) = A_{ij}$ for every $x \in A_i$ and $y \in A_j$. However, all possible representations of a given weakly complete hypergoup share the same group G and family $\mathfrak{F}$. This fact should be evident from the following proposition, where we explain the algebraic role of the parameters of a representation of a weakly complete hypergroup.

Proposition 3. *Let $(H, \circ) = \mathcal{W}(G, I, \mathfrak{F}, \mathfrak{I})$. Then, we have:*

1. *The sets $A_i \in \mathfrak{F}$ are the β-classes of H, i.e, for every $x \in H$, $x \in A_i \Leftrightarrow \beta(x) = A_i$.*
2. *$H/\beta \simeq G$ and $\omega_H = A_{1_G}$.*
3. *Every subhypergroup K of $(H, \circ)$ is a complete part of H, that is, $\mathfrak{C}(K) = K$.*
4. *A subset $K \subseteq H$ is a subhypergroup of $(H, \circ)$ if and only if there exists a subgroup G' of G such that $K = \bigcup_{i \in G'} A_i$.*

Proof. 1. Let $x \in A_k$ and $a \in A_{1_G}$. Then, $A_k = x \circ a$, and so $y \in A_k$ implies $y\beta_2 x$. Conversely, if $y\beta_2 x$, then there exist $a, b \in H$ such that $\{x, y\} \subseteq a \circ b$. By construction, there exists $r \in G$ such that $a \circ b \subseteq A_r$. Therefore, since $x \in A_k \cap A_r$ and the sets of the family $\mathfrak{F}$ are pairwise disjoint, we obtain $y \in a \circ b \subseteq A_r = A_k$. Hence, $y \in A_k$ if and only if $y\beta_2 x$. By Theorem 3, we conclude $A_k = \beta(x)$.

2. The map $f : G \mapsto H/\beta$ such that $f(k) = A_k$, for every $k \in G$, is a group isomorphism. Moreover, we have $\omega_H = A_{1_G}$ since $1_{H/\beta} = f(1_G) = A_{1_G}$.

3. We must prove that $\beta(x) \subseteq K$, for all $x \in K$. By reproducibility of K, if $x \in K$ then there exists $u \in K$ such that $x \in x \circ u$. Considering the canonical epimorphism $\pi : H \mapsto H/\beta$, we obtain $\pi(x) = \pi(x) \otimes \pi(u)$ and so $\pi(u) = 1_{H/\beta}$. Hence, from point 2., we have $u \in \pi^{-1}(1_{H/\beta}) = \omega_H = A_{1_G}$. Consequently, $\omega_H = A_{1_G} = u \circ u \subseteq K$ and $\beta(x) = x \circ \omega_H \subseteq K \circ K = K$, for all $x \in K$.

4. Since $iG' = G' = G'i$, for all $i \in G'$, the proof of the implication $\Leftarrow$ is similar to the one used in point 3. of Theorem 3 to prove that $(H, \circ)$ is a hypergroup. Now, we prove the implication $\Rightarrow$. By point 1, the β-classes of $(H, \circ)$ are the sets A_i, for all $i \in G$. Let $\pi : H \mapsto H/\beta$ be the canonical epimorphism and $f : H/\beta \mapsto G$ be the isomorphism such that $f(A_i) = i$, for all $i \in G$. If K is a subhypergroup of $(H, \circ)$, then $G' = (f \circ \pi)(K)$ is a subgroup of G. Moreover, if $x \in K$ then there exists $i \in G$ such that $x \in A_i$. By point 1,

we have $A_i = \beta(x)$ and $i = f(A_i) = f(\pi(x)) = (f \circ \pi)(x) \in G'$. Hence, $K \subseteq \bigcup_{i \in G'} A_i$. On the other hand, if $x \in \bigcup_{i \in G'} A_i$, there exists $i \in G'$ such that $x \in A_i$. Clearly, there exists $y \in K$ such that $(f \circ \pi)(y) = i$. If we suppose that $y \in A_j$, then we have $A_j = \beta(y)$ and $i = (f \circ \pi)(y) = f(\pi(y)) = f(A_j) = j$. Finally, by point 3, $x \in A_i = A_j = \beta(y) \subseteq K$. Therefore, $\bigcup_{i \in G'} A_i \subseteq K$. □

The following result, which follows from the definition of hyperproduct in (4) and point 1 in Proposition 3, describes all cases where a weakly complete hypergroup is complete.

Corollary 1. *Let $(H, \circ) = W(G, I, \mathfrak{F}, \mathfrak{J})$.*
1. *If $I = \varnothing$, then $(H, \circ)$ is complete;*
2. *if $I \neq \varnothing$, then $(H, \circ)$ is complete $\iff \varphi_{i,j}$ is trivial, for every $(i, j) \in I$.*

Example 6. *Let $(H, \circ) = W(G, I, \mathfrak{F}, \mathfrak{J})$ such that $|A_k| > 1$ for some $k \in G$ and $|A_i| = 1$ for $i \neq k$. Then, $(H, \circ)$ is complete, as a consequence of the previous corollary. Indeed, if $(i, j) \in I$ and $ij \neq k$, then $|A_{ij}| = 1$ and $\varphi_{i,j}$ is trivial. On the other hand, if $k = ij$ then $k \notin \{i, j\}$ because I is product free. Thus, $|A_i| = |A_j| = 1$ and $\varphi_{i,j}$ are trivial since it is a bi-covering.*

The next example shows a weakly complete hypergroup that contains both complete and noncomplete subhypergroups.

Example 7. *Let $G = \{1, 2, 3, 4\}$ be a group isomorphic to the Klein group $\mathbb{Z}_2 \times \mathbb{Z}_2$ where $1 = 1_G$. Consider $I = \{(2, 2), (3, 3)\}$, $A_1 = \{a, b\}$, $A_2 = \{c, d\}$, $A_3 = \{e, f\}$ and $A_4 = \{g\}$. In the set $H = \{a, b, c, d, e, f, g, h\}$, define the hyperproduct represented in the following table:*

∘	a	b	c	d	e	f	g
a	A_1	A_1	A_2	A_2	A_3	A_3	A_4
b	A_1	A_1	A_2	A_2	A_3	A_3	A_4
c	A_2	A_2	a	A_1	A_4	A_4	A_3
d	A_2	A_2	b	a	A_4	A_4	A_3
e	A_3	A_3	A_4	A_4	b	a	A_2
f	A_3	A_3	A_4	A_4	a	b	A_2
g	A_4	A_4	A_3	A_3	A_2	A_2	A_1

Then, $(H, \circ)$ ia a weakly complete hypergroup. The subsets $K_1 = A_1 \cup A_2$, $K_2 = A_1 \cup A_3$, $K_3 = A_1 \cup A_4$ are a subhypergroup of $(H, \circ)$. Moreover, K_3 is complete and K_1 and K_2 are not complete.

The next theorem characterizes weakly complete hypergroups, in that it yields a necessary and sufficient condition for a given hypergroup to be weakly complete, based on the structure of its quotient group.

Theorem 5. *Let $(H, \circ)$ be a hypergroup, and let $\pi : H \mapsto H/\beta$ be the canonical projection. Consider the following relation $J \subseteq H/\beta \times H/\beta$:*

$$J = \{(i, j) : \exists x \in \pi^{-1}(i),\ \exists y \in \pi^{-1}(j) : x \circ y \neq \mathfrak{C}(x \circ y)\}.$$

The following conditions are equivalent:
1. *J is product free;*
2. *$(H, \circ)$ is a weakly complete hypergroup.*

Proof. Suppose that J is product free. For every $i \in H/\beta$, let $A_i = \pi^{-1}(i)$, and note that $\bigcup_i A_i = H$. For every $(i, j) \in J$ introduce the function $f_{i,j} : A_i \times A_j \mapsto A_{ij}$ such that

$f_{i,j}(x,y) = x \circ y$. It is not difficult to see that $f_{i,j}$ is a bi-covering. Indeed, for any fixed $x \in A_i$ we have by construction

$$\bigcup_{y \in A_j} f_{i,j}(x,y) = \bigcup_{y \in A_j} x \circ y = x \circ A_j$$
$$= x \circ (y \circ \omega_H)$$
$$= (x \circ y) \circ \omega_H = \mathfrak{C}(x \circ y) = A_{ij}.$$

The identity $\bigcup_{x \in A_i} f_{i,j}(x,y) = A_{ij}$ can be derived analogously, so $f_{i,j}$ is a bi-covering. It remains to observe that $(H, \circ) = \mathcal{W}(H/\beta, J, \{A_i\}, \{f_{i,j}\})$, and we have the first part of the claim.

Conversely, suppose that $(H, \circ)$ is a weakly complete hypergroup, $(H, \circ) = \mathcal{W}(G, I, \mathfrak{F}, \mathfrak{I})$. Identifying G with H/β modulo an isomorphism, we have $J \subseteq I$. Indeed, let $(i,j) \in J$. By hypotesis, there exist $x, y \in H$ such that $\pi(x) = i$, $\pi(y) = j$ and $x \circ y \neq \mathfrak{C}(x \circ y)$. Hence, $(i,j) \in I$ by (4). This conclusion follows immediately from the fact that a subset of a PF-relation is a PF-relation. □

5. Completeness Degree of Finite Hypergroups

In this section, we introduce the notion of completeness degree of finite hypergroups and analyze the completeness degree of finite weakly complete hypergroups.

Definition 4. *Let $(H, \circ)$ be a finite hypergroup. Define the set $\mathcal{C}_H \subseteq H \times H$,*

$$\mathcal{C}_H = \{(x,y) \in H \times H \mid \mathfrak{C}(x \circ y) = x \circ y\}.$$

The rational number

$$\Delta(H) = \frac{|\mathcal{C}_H|}{|H|^2}$$

is the completeness degree *of $(H, \circ)$.*

Thus, the completeness degree of a hypergroup is the probability that the hyperproduct of two randomly chosen elements is a β-class. Clearly, $\Delta(H) \in [0,1]$ and $\Delta(H) = 1$ if and only if $(H, \circ)$ is complete. In the next lemma, we deduce an explicit formula for the completeness degree of finite weakly complete hypergroups. For this purpose, we make use of the following auxiliary notation. Let $(H, \circ) = \mathcal{W}(G, I, \mathfrak{F}, \mathfrak{I})$. For every $i, j \in G$, let

$$\mathcal{C}_{i,j} = \{(x,y) \in A_i \times A_j \mid x \circ y = A_{ij}\}.$$

Lemma 2. *Let $(H, \circ) = \mathcal{W}(G, I, \mathfrak{F}, \mathfrak{I})$. Then,*

$$\Delta(H) = \frac{\sum_{(i,j) \notin I} |A_i||A_j| + \sum_{(i,j) \in I} |\mathcal{C}_{i,j}|}{|H|^2}. \qquad (6)$$

Moreover, if $(H, \circ)$ is uniform, then

$$\Delta(H) = 1 - \frac{|I|}{|G|^2} + \frac{\sum_{(i,j) \in I} |\mathcal{C}_{i,j}|}{|H|^2}. \qquad (7)$$

Proof. Firstly, note that $\mathcal{C}_H = \bigcup_{i,j} \mathcal{C}_{i,j}$. From the definition of the hyperproduct $\circ$ in (4), we deduce the alternative formula

$$\mathcal{C}_{i,j} = \begin{cases} A_i \times A_j & \text{if } (i,j) \notin I \\ \{(x,y) \in A_i \times A_j \mid \varphi_{i,j}(x,y) = A_{ij}\} & \text{if } (i,j) \in I; \end{cases}$$

hence,
$$\mathcal{C}_H = \left(\bigcup_{(i,j)\notin I} A_i \times A_j\right) \cup \left(\bigcup_{(i,j)\in I} \mathcal{C}_{i,j}\right).$$

Recalling that the sets of the family $\mathfrak{F}$ are pairwise disjoint, we obtain
$$|\mathcal{C}_H| = |\bigcup_{(i,j)\notin I} A_i \times A_j| + |\bigcup_{(i,j)\in I} \mathcal{C}_{i,j}|$$
$$= \sum_{(i,j)\notin I} |A_i||A_j| + \sum_{(i,j)\in I} |\mathcal{C}_{i,j}|,$$

and Equation (6) follows. Moreover, if $(H, \circ)$ is n-uniform, then $|H| = n|G|$ and
$$\frac{\sum_{(i,j)\notin I} |A_i||A_j|}{|H|^2} = \frac{\sum_{(i,j)\notin I} n^2}{|H|^2} = \frac{(|G|^2 - |I|)n^2}{|G|^2 n^2} = 1 - \frac{|I|}{|G|^2},$$

and we also obtain (7). □

Our next result provides two lower bounds on $\Delta(H)$ that depend only on the size of the β-classes of H.

Theorem 6. *Let $(H, \circ)$ be a finite weakly complete hypergroup. Then,*
$$\Delta(H) \geq \frac{|\omega_H|}{|H|}\left(2 - \frac{|\omega_H|}{|H|}\right),$$

where ω_H is the heart of H. Moreover, if $(H, \circ)$ is uniform, then $\Delta(H) \geq \frac{3}{4}$.

Proof. Let $\{A_i\}_{i\in G}$ be the family of disjoint sets in the representation of $(H, \circ)$. Then,
$$|\mathcal{C}_H| \geq \sum_{(i,j)\notin I} |A_i||A_j|$$
$$= \sum_{i,j=1}^{|G|} |A_i||A_j| - \sum_{(i,j)\in I} |A_i||A_j|$$
$$= |H|^2 - \sum_{(i,j)\in I} |A_i||A_j|$$
$$\geq |H|^2 - (\sum_{i\in G^*} |A_i|)^2$$
$$= |H|^2 - (|H| - |A_{1_G}|)^2 = |A_{1_G}|(2|H| - |A_{1_G}|).$$

Recalling that $A_{1_G} = \omega_H$ and using (6), we obtain the first inequality. Moreover, from (7) we have $\Delta(H) \geq 1 - |I|/|G|^2$; hence, the second part of the claim is an immediate consequence of Theorem 1. □

The next example shows that the inequalities in Theorem 6 can hold as equalities.

Example 8. *Let $m \geq 2$ be an even number, and let G and I be the same as in Example 3. Let $(H, \circ) = \mathcal{W}(G, I, \mathfrak{F}, \mathfrak{I})$ be any uniform weakly complete hypergroup such that $\mathcal{C}_{i,j} = \emptyset$ for all $(i,j) \in I$; i.e., all bi-coverings are proper. A straightforward application of Lemma 2 proves that $\Delta(H) = 3/4$. Moreover, if $m = 2$, then $|H| = 2n$ and $|\omega_H| = n$. Thus, also the first inequality in Theorem 6 holds as an equality.*

In the forthcoming example, we construct uniform weakly complete hypergroups where all bi-coverings are proper, that is, $\mathcal{C}_{i,j} = \emptyset$, for all $(i,j) \in I$. According to Lemma 2, these hypergroups achieve the smallest $\Delta(H)$ possible for a given PF-relation.

Example 9. *Let G be a group and $I \in \mathrm{PF}_G$. Let $\mathfrak{F} = \{A_k\}_{k \in G}$ be a family of finite, pairwise disjoint sets such that $|A_k| = n \geq 2$ for all $k \in G$. We assume $A_k = B_k \cup C_k$, with B_k, C_k nonempty disjoint sets. For every $(i,j) \in I$, let $\varphi_{i,j} : A_i \times A_j \mapsto A_{ij}$ be defined as follows:*

$$\varphi_{i,j}(x,y) = \begin{cases} B_{ij} & \text{if } (x \in B_i \text{ and } y \in B_j) \text{ or } (x \in C_i \text{ and } y \in C_j) \\ C_{ij} & \text{else.} \end{cases}$$

It is not difficult to verify that $\varphi_{i,j}$ is a proper bi-covering. Moreover, the hypergroup $(H, \circ) = \mathcal{W}(G, I, \{A_i\}, \{\varphi_{i,j}\})$ is n-uniform. Owing to (7) and the finiteness of G, the completeness degree of $(H, \circ)$ is

$$\Delta(H) = 1 - \frac{|I|}{|G|^2},$$

i.e., the smallest possible value for the given relation I.

6. Commutativity Degree of Weakly Complete Hypergroups

In a nonabelian group and, more generally, in any nonabelian algebraic structure, it makes sense to compute the probability that two randomly chosen elements commute. This problem was popularized by Gustafson in [14], who defined the commutativity degree $d(G)$ of a group G as the probability that two arbitrary elements commute,

$$d(G) = \frac{|\{(x,y) \in G^2 \,:\, xy = yx\}|}{|G|^2}, \tag{8}$$

and proved that if $d(G) > \frac{5}{8}$ then G is abelian. Moreover, we have $d(G) = \frac{5}{8}$ if and only if $G/Z(G) \simeq \mathbb{Z}_2 \times \mathbb{Z}_2$, where $Z(G)$ is the center of G. The basic technique adopted for the proof relies on the relationship between $d(G)$ and the number of conjugacy classes of G, and can be traced back to a paper by Erdős and Turán [17]. Later on, there has been considerable interest in the use of probabilistic techniques in group theory, and this concept has had significant developments.

Recently, the concept of commutativity degree has been introduced also in hypergroup theory [13,18]. In particular, in [13] the authors defined the commutativity degree of a finite hypergroup $(H, \circ)$ as

$$d(H) = \frac{|\{(x,y) \in H^2 \,:\, x \circ y = y \circ x\}|}{|H|^2} \tag{9}$$

and characterized this index when $(H, \circ)$ is complete by considering a partitioning of H into suitably defined conjugacy classes. In this section, we study the commutativity degree of weakly complete hypergroups. Our main tool is the partitioning of H into β-classes. To begin with, we point out an important observation. For any $i,j \in G$ and for any $x \in A_i$ and $y \in A_j$, a necessary condition for the identity $x \circ y = y \circ x$ to be valid is $ij = ji$, because $x \circ y \subseteq A_{ij}$, $y \circ x \subseteq A_{ji}$ and $A_{ij} \cap A_{ji} = \emptyset$ if $ij \neq ji$. Hence, we can restrict our attention to pairs (i,j) belonging to the set

$$c(G) = \{(i,j) \in G \times G \,:\, ij = ji\}.$$

This set is directly related to the commutativity degree of G, since $d(G) = |c(G)|/|G|^2$.

Definition 5. *We say that a relation $I \in \mathrm{PF}_G$ is G-symmetric if its restriction to $c(G)$ is symmetric; that is, for every $i,j \in G$, if $ij = ji$ then $(i,j) \in I \iff (j,i) \in I$.*

Equivalently, $I \in \mathrm{PF}_G$ is G-symmetric if and only if $I \cap c(G) = I^T \cap c(G)$. It can be observed that if G is abelian then a relation in PF_G is G-symmetric if and only if it is symmetric. The relevance of the previous definition lies in the fact that every weakly complete hypergroup admits a representation with a G-symmetric relation, as shown in the following lemma.

Lemma 3. *Let $(H, \circ)$ be a weakly complete hypergroup. Then, there exists a representation $(H, \circ) = \mathcal{W}(G, I, \mathfrak{F}, \mathfrak{J})$ where I is G-symmetric.*

Proof. Let $(H, \circ) = \mathcal{W}(G, \hat{I}, \hat{\mathfrak{F}}, \hat{\mathfrak{J}})$ be any representation of $(H, \circ)$. If $\hat{I}$ is G-symmetric, then it is complete. Otherwise, $\hat{I} \cap c(G) \neq \hat{I}^T \cap c(G)$ and we define the relation

$$I = \hat{I} \cup (\hat{I}^T \cap c(G)).$$

We have $I \cap c(G) = I^T \cap c(G)$, so I is G-symmetric, and I properly extends $\hat{I}$. Moreover, from Lemma 1 we can deduce that $I \in \text{PF}_G$, because both the support and the span of I coincide with those of $\hat{I}$.

For every $(i, j) \in I \setminus \hat{I}$ let $\psi_{i,j} : A_i \times A_j \mapsto A_{ij}$ be the trivial bi-covering, and define $\mathfrak{J} = \hat{\mathfrak{J}} \cup \{\psi_{i,j}\}$. To conclude the proof, it suffices to show that the hypergroup $(H, \diamond) = \mathcal{W}(G, I, \mathfrak{F}, \mathfrak{J})$ coincides with $(H, \circ)$. Indeed, for arbitrary $x \in A_i$ and $y \in A_j$, if $(i, j) \in I \setminus \hat{I}$, then

$$x \diamond y = \psi_{i,j}(x, y) = A_{ij} = x \circ y.$$

Otherwise, if either $(i, j) \in \hat{I}$ or $(i, j) \notin I$ then the identity $x \diamond y = x \circ y$ follows trivially from the construction (4). We can conclude that $(H, \circ) = \mathcal{W}(G, I, \mathfrak{F}, \mathfrak{J})$. □

In what follows, we obtain different characterizations of the commutativity degree of a weakly complete hypergroup $(H, \circ) = \mathcal{W}(G, I, \mathfrak{F}, \mathfrak{J})$ in terms of the parameters of its representation. By virtue of Lemma 3, we can safely assume that I is G-symmetric. In this case, for every pair $(i, j) \in c(G) \cap I$ the sets

$$\mathcal{D}_{i,j} = \{(x, y) \in A_i \times A_j : \varphi_{i,j}(x, y) = \varphi_{j,i}(y, x)\} \tag{10}$$

$$\mathcal{E}_{i,j} = \{(x, y) \in A_i \times A_j : \varphi_{i,j}(x, y) \neq \varphi_{j,i}(y, x)\} \tag{11}$$

are well defined.

Theorem 7. *Let $(H, \circ) = \mathcal{W}(G, I, \mathfrak{F}, \mathfrak{J})$ where I is G-symmetric. Then,*

$$d(H) = \frac{\sum_{(i,j) \in c(G) \setminus I} |A_i||A_j| + \sum_{(i,j) \in c(G) \cap I} |\mathcal{D}_{i,j}|}{|H|^2}. \tag{12}$$

Moreover, if $(H, \circ)$ is uniform then

$$d(H) = d(G) - \frac{|c(G) \cap I|}{|G|^2} + \frac{\sum_{(i,j) \in c(G) \cap I} |\mathcal{D}_{i,j}|}{|H|^2} \tag{13}$$

$$= d(G) - \frac{\sum_{(i,j) \in c(G) \cap I} |\mathcal{E}_{i,j}|}{|H|^2}. \tag{14}$$

Proof. Let $i, j \in c(G)$, $x \in A_i$ and $y \in A_j$. Two cases are possible:

(a) $(i, j) \in c(G) \setminus I$. In this case, $x \circ y = A_{ij} = A_{ji} = y \circ x$; hence

$$\{(x, y) \in A_i \times A_j : x \circ y = y \circ x\} = A_i \times A_j.$$

(b) $(i, j) \in I \cap c(G)$. Owing to the G-symmetry of I, we have both $x \circ y = \varphi_{ij}(x, y)$ and $y \circ x = \varphi_{ji}(y, x)$. By (10),

$$\{(x, y) \in A_i \times A_j : x \circ y = y \circ x\} = \mathcal{D}_{i,j}.$$

The first claim follows from the fact that the set $c(G)$ is the disjoint union of $c(G) \setminus I$ and $I \cap c(G)$. Moreover, if $|A_i| = n$ for all $i \in G$, then

$$\sum_{(i,j) \in c(G) \setminus I} |A_i||A_j| = n^2 |c(G) \setminus I| = n^2 (|c(G)| - |c(G) \cap I|).$$

Since $|H| = n|G|$, we also have

$$\frac{\sum_{(h,k) \in c(G) \setminus I} |A_h||A_k|}{|H|^2} = \frac{|c(G)| - |c(G) \cap I|}{|G|^2} = d(G) - \frac{|c(G) \cap I|}{|G|^2},$$

and (13) follows. Finally, using (11) we obtain

$$\frac{|c(G) \cap I|}{|G|^2} - \frac{\sum_{(i,j) \in c(G) \cap I} |\mathcal{D}_{i,j}|}{|H|^2} = \frac{\sum_{(i,j) \in c(G) \cap I} (n^2 - |\mathcal{D}_{i,j}|)}{|H|^2}$$

$$= \frac{\sum_{(i,j) \in c(G) \cap I} |\mathcal{E}_{i,j}|}{|H|^2},$$

which yields (14), and the proof is complete. □

The previous theorem yields a few notable consequences. For example, taking $I = \emptyset$ we conclude that if $(H, \circ)$ is complete and

$$d(H) = \frac{\sum_{(i,j) \in c(G)} |A_i||A_j|}{|H|^2}.$$

In particular, if $(H, \circ)$ is also uniform, then $d(H) = d(G)$. More generally, $d(H) \leq d(G)$ for any uniform weakly complete hypergroup, and the equality holds if and only if $\varphi_{i,j}(x,y) = \varphi_{j,i}(y,x)$ for every $(i,j) \in c(G) \cap I$.

Finally, the similarity between formulas (6) and (12) suggests that we should study the relationship between the degrees of commutativity and completeness, at least in the commutative case. We propose our result below. Before doing so, we recall that if G is abelian, then G-symmetric relations are symmetric. Hence, by Lemma 3, every weakly complete hypergroup built from an abelian group admits a representation whose PF-relation is symmetric.

Theorem 8. *Let G be abelian and let $(H, \circ) = \mathcal{W}(G, I, \mathfrak{F}, \mathfrak{J})$, where I is symmetric. Then,*

$$d(H) = \Delta(H) + \frac{\sum_{(i,j) \in I} (|\mathcal{D}_{i,j}| - |\mathcal{C}_{i,j}|)}{|H|^2}. \tag{15}$$

Moreover, if $(H, \circ)$ is uniform then $|d(H) - \Delta(H)| \leq |I|/|G|^2 \leq \frac{1}{4}$.

Proof. Since G is abelian, we have $c(G) \cap I = I$ and the condition $(i,j) \in c(G) \setminus I$ reduces to $(i,j) \notin I$. Therefore, subtracting (13) from (6) we obtain (15). Furthermore, for every $(i,j) \in I$ we have $\mathcal{D}_{i,j} \cup \mathcal{C}_{i,j} \subseteq A_i \times A_j$. If H is n-uniform, then $|H| = n|G|$ and $|A_i \times A_j| = n^2$. Hence,

$$-n^2 \leq |\mathcal{D}_{i,j}| - |\mathcal{C}_{i,j}| \leq n^2.$$

Thus, $|d(H) - \Delta(H)| \leq n^2 |I|/|H|^2 = |I|/|G|^2$. The rightmost inequality in the claim comes from Theorem 1. □

7. Conclusions

The class of complete hypergroups is among the best known in hypergroup theory. Complete hypergroups have a variety of group-like properties and are characterized by the fact that the composition of two elements is a β-class [9–12]. In this paper, we introduce a class of hypergroups $(H, \circ)$ that includes complete hypergroups as a particular case. The construction of these hypergroups, called weakly complete, is crucially based on particular binary relations defined on the quotient group H/β. We call these relations product free because no group element is in relation with the product of two elements that are related to each other. Product-free relations are interesting by themselves, and we show a number of their main properties on generic groups in Section 2. For example, we prove an attainable upper bound on the cardinality of product-free relations in finite groups.

The main motivation of introducing weakly complete hypergroups lies in the possibility of measuring their "closeness" to complete hypergroups. Indeed, to every finite hypergroup, we can associate a completeness degree, which quantifies how close to completion the hypergroup is. We introduce and analyze this concept in Section 5. More precisely, the completeness degree of a hypergroup is the probability that the composition of two randomly chosen elements is a β-class. For a weakly complete hypergroup whose β-classes have the same cardinality, this probability is bounded from below by $\frac{3}{4}$. Indeed, the completeness degree of weakly complete hypergroups admits simple closed formulas. Furthermore, it can be related to the commutativity degree, which has been recently brought into hypercompositional algebra from group theory [13,18].

Completeness concepts and probabilistic methods are relevant topics nowadays not only in classical algebra but also in hypercompositional algebra, and this discipline is continually expanding with the introduction of structures with distinctive properties [19]. It would be interesting to discover more hypergroup classes, and more general hypercompositional structures, for which useful results can be found along these directions.

Author Contributions: Conceptualization and investigation: D.F. (Dario Fasino), D.F. (Domenico Freni), and G.L.F.; writing—original draft: D.F. (Domenico Freni); writing—review and editing: M.D.S. and D.F. (Dario Fasino). All authors have read and agreed to the published version of the manuscript.

Funding: The research work of Mario De Salvo was funded by Università di Messina, Italy, grant FFABR Unime 2019. Giovanni Lo Faro was supported by INdAM-GNSAGA, Italy, and by Università di Messina, Italy, grant FFABR Unime 2020. The work of Dario Fasino was partially supported by INdAM-GNCS, Italy.

Institutional Review Board Statement: Not applicable.

Informed Consent Statement: Not applicable.

Data Availability Statement: Not applicable.

Conflicts of Interest: The authors declare no conflict of interest.

References

1. Massouros, G.; Massouros, C. Hypercompositional algebra, computer science and geometry. *Mathematics* **2020**, *8*, 1338. [CrossRef]
2. Koskas, H. Groupoïdes, demi-hypergroupes et hypergroupes. *J. Math. Pures Appl.* **1970**, *49*, 155–192.
3. Vougiouklis, T. Fundamental relations in hyperstructures. *Bull. Greek Math. Soc.* **1999**, *42*, 113–118.
4. Corsini, P. *Prolegomena of Hypergroup Theory*; Aviani Editore: Tricesimo, Italy, 1993.
5. Davvaz, B.; Leoreanu-Fotea, V. *Hyperring Theory and Applications*; International Academic Press: Palm Harbor, FL, USA, 2007.
6. De Salvo, M.; Fasino, D.; Freni, D.; Lo Faro, G. G-Hypergroups: Hypergroups with a group-isomorphic heart. *Mathematics* **2022**, *10*, 240. [CrossRef]
7. Freni, D. Strongly transitive geometric spaces: Applications to hypergroups and semigroups theory. *Commun. Algebra* **2004**, *32*, 969–988. [CrossRef]
8. Gutan, M. On the transitivity of the relation β in semihypergroups. *Rend. Circ. Mat. Palermo* **1996**, *45*, 189–200. [CrossRef]
9. Cristea, I.; Davvaz, B.; Hassani, S.E. Special intuitionistic fuzzy subhypergroups of complete hypergroups. *J. Intell. Fuzzy Syst.* **2015**, *28*, 237–245. [CrossRef]
10. Singha, M.; Das, K.; Davvaz, B. On topological complete hypergroups. *Filomat* **2017**, *31*, 5045–5056. [CrossRef]
11. Sadeghi, M.M.; Hassankhani, A.; Davvaz, B. n-abelian and μ_m complete n-abelian hypergroups. *Int. J. Appl. Math. Stat.* **2017**, *56*, 130–141.
12. De Salvo, M.; Fasino, D.; Freni, D.; Lo Faro, G. On hypergroups with a β-class of finite height. *Symmetry* **2020**, *12*, 168. [CrossRef]
13. Sonea, A.; Cristea, I. The class equation and the commutativity degree for complete hypergroups. *Mathematics* **2020**, *8*, 2253. [CrossRef]

14. Gustafson, W.H. What is the probability that two group elements commute? *Am. Math. Mon.* **1973**, *80*, 1031–1034. [CrossRef]
15. De Salvo, M.; Freni, D.; Lo Faro, G. Fully simple semihypergroups. *J. Algebra* **2014**, *399*, 358–377. [CrossRef]
16. De Salvo, M.; Fasino, D.; Freni, D.; Lo Faro, G. Fully simple semihypergroups, transitive digraphs, and Sequence A000712. *J. Algebra* **2014**, *415*, 65–87. [CrossRef]
17. Erdős, P.; Turán, P. On Some Problems of a Statistical Group Theory, IV. *Acta Math. Acad. Hung.* **1968**, *19*, 413–435. [CrossRef]
18. Sonea, A.C. New aspects in polygroup theory. *Analele Științ. Univ. Ovidius Constanța Ser. Mat.* **2020**, *28*, 241–254. [CrossRef]
19. Massouros, C.G. *Hypercompositional Algebra and Applications*; MDPI: Basel, Switzerland, 2021.

Article

Uniform (C_k, P_{k+1})-Factorizations of $K_n - I$ When k Is Even

Giovanni Lo Faro [1,*], Salvatore Milici [2] and Antoinette Tripodi [1]

[1] Dipartimento di Scienze Matematiche e Informatiche, Scienze Fisiche e Scienze della Terra, Università di Messina, 98166 Messina, Italy; atripodi@unime.it
[2] Dipartimento di Matematica e Informatica, Università di Catania, 95125 Catania, Italy; milici@dmi.unict.it
* Correspondence: lofaro@unime.it

Abstract: Let H be a connected subgraph of a graph G. An H-factor of G is a spanning subgraph of G whose components are isomorphic to H. Given a set $\mathcal{H}$ of mutually non-isomorphic graphs, a uniform $\mathcal{H}$-factorization of G is a partition of the edges of G into H-factors for some $H \in \mathcal{H}$. In this article, we give a complete solution to the existence problem for uniform (C_k, P_{k+1})-factorizations of $K_n - I$ in the case when k is even.

Keywords: graph factorization; complete graph; block design

MSC: 05B30

1. Introduction

Let $V(G)$ and $E(G)$ denote the vertex set and the edge set of a graph G, respectively. As per standard notations, K_n denotes the complete graph on n vertices, C_k is the k-cycle (i.e., the cycle of length k) and P_{k+1} is the path on $k+1$ vertices. For missing notions and terms that are not explicitly defined in this paper, we point the reader to [1] and its online updates. If $\mathcal{H}$ is a set of mutually non-isomorphic connected graphs, an $\mathcal{H}$-decomposition of a graph G is a partition of $E(G)$ into subgraphs (*blocks*) that are isomorphic to some element of $\mathcal{H}$. An $\mathcal{H}$-factor of G is a spanning subgraph of G, i.e., a subgraph of G with the same vertex set as G, whose connected components are isomorphic to some element of $\mathcal{H}$. An $\mathcal{H}$-factorization of G is an $\mathcal{H}$-decomposition of G whose set of blocks admits a partition into $\mathcal{H}$-factors. An $\mathcal{H}$-factorization of G is also known as a *resolvable $\mathcal{H}$-decomposition* of G and an $\mathcal{H}$-factor of G can be called a *parallel class* of G. When $\mathcal{H} = \{H\}$, then we simply write H-factor and H-factorization. A K_2-factorization of G is better known as a *1-factorization* and its factors are said *1-factors*; a 1-factor of K_n is a set of $\frac{n}{2}$ mutually vertex disjoint edges of K_n and a 1-factorization of K_n exists if and only if n is even [2]. A C_k-factorization of K_n exists if and only if $3 \leq k \leq n$, n and k are odd and $n \equiv 0 \pmod{k}$ [3]. An $\mathcal{H}$-factorization of a graph G is said to be *uniform* if each factor is an H-factor for some $H \in \mathcal{H}$ (sometimes it is referred to as a uniformly resolvable $\mathcal{H}$-decomposition of G).

In the context of graph factorizations, and in particular of cycle factorizations, the most famous problems are the *Oberwolfach Problem* and the *Hamilton–Waterloo Problem*. The first one was first posed in 1967 by G. Ringel and asks whether it is possible to seat n mathematicians at m round tables in $(n-1)/2$ dinners so that every two mathematicians sit next to each other exactly once. This puzzle can be formalized in terms of graph factorizations as follows. If integers $p_1, p_2, \ldots, p_m$ denote the sizes of the m round tables, then the solution of the Oberwolfach Problem is a factorization of K_n where each factor has m components which are isomorphic to cycles of length $p_1, p_2, \ldots, p_m$, $\sum_{i=1}^m p_i = n$. It is well known that such a factorization can exist only if n is odd. If the number n is even, then an analogous problem is reformulated in terms of decomposition of $K_n - I$, that is the graph obtained by removing a 1-factor from K_n. The version where all cycles of a factor have the same size is called the uniform Oberwolfach Problem, which has been completely

solved by Alspach and Häggkvist [4] and Alspach, Schellenberg, Stinson and Wagner [3].
The Hamilton–Waterloo Problem is a variation of the Oberwolfach Problem and requires
that the dining mathematicians have their dinners in two different venues. In this case,
the factors of the sought decomposition of K_n (when n is odd) or $K_n - I$ (when n is even)
can have either s components that are isomorphic to cycles of length $p_1, p_2, \ldots, p_s$ or t
components that are isomorphic to cycles of length $q_1, q_2, \ldots, q_t$, $\sum_{i=1}^{s} p_i = \sum_{i=1}^{t} q_i = n$. If
the tables in one venue sit p mathematicians and those in the other venue sit q each, then the
problem is called the uniform Hamilton–Waterloo Problem, which asks for a decomposition
of K_n or $K_n - I$ into C_p-factors and C_q-factors. For both problems, the Hamilton–Waterloo
Problem and the non-uniform case of the Oberwolfach Problem, many partial results are
known, but a complete solution is far to be achieved.

Existence problems for $\mathcal{H}$-factorizations are usually considered for the complete graph
K_n or the graph $K_n - I$. For these graph families, many results have been obtained especially
in the uniform case; just to give some examples, when $\mathcal{H}$ contains two complete graphs
of order $k \leq 5$ [5–8], when $\mathcal{H}$ contains two or three paths of order $2 \leq k \leq 4$ [9,10], for
$\mathcal{H} = \{K_2, K_{1,3}\}$ [11,12], for $\mathcal{H} = \{K_2, K_{1,4}\}$ [13], and for $\mathcal{H} = \{C_{2k}, K_{1,2k}\}$ [14].

A uniform $\{H_1, H_2\}$-factorization of G with r_i H_i-factors, $i = 1, 2$, is denoted by
URD$(G; H_1^{r_1}, H_2^{r_2})$. When $G = K_n$ we simply write URD$(n; H_1^{r_1}, H_2^{r_2})$. In this paper, we deal
with uniform $\mathcal{H}$-factorizations of K_n or $K_n - I$ in the case when $\mathcal{H} = \{C_k, P_{k+1}\}$. In [9],
a solution to the existence problem of a URD$(n; C_k^{r_1}, P_{k+1}^{r_2})$ is given for $k = 2$ (note that
$C_2 = K_2$). Here, we are interested in the case when $k \equiv 0 \pmod{2}$ and $k \geq 4$. As for the
k even case, it is known that a URD$(n; C_k^0, P_{k+1}^{r_2})$ exists if and only if $n \equiv 0 \pmod{k+1}$
and $(k+1)(n-1)n \equiv 0 \pmod{2k}$, see [15,16], while no URD$(n; C_k^{r_1}, P_{k+1}^0)$ exists because n
must be odd and divisible by k; a URD$(K_n - I; C_k^{r_1}, P_{k+1}^0)$ exists if and only if $n \equiv 0 \pmod{2}$
and k divides n, see [17]. When n is even, no URD$(n; C_k^{r_1}, P_{k+1}^{r_2})$ exists with $r_1 > 0$ because,
otherwise, the resolvability implies $2(k+1)r_1 + 2kr_2 = (k+1)(n-1)$ and, clearly, this is
impossible. Therefore, it would be interesting to prove whether or not there exist uniformly
resolvable decompositions of $K_n - I$ in terms of factors belonging to $\mathcal{H} = \{C_k, P_{k+1}\}$. For
brevity, we introduce the notation URD$^*(n; C_k^{r_1}, P_{k+1}^{r_2})$ for a decomposition of this kind.
Moreover, since k and $k+1$ must divide n, we assume $n \equiv 0 \pmod{k(k+1)}$. Finally, we
must have $r_1 > 0$ because $(n-2)(k+1)/(2k)$ is not an integer.

The goal of this paper is to characterize the existence of URD$^*(n; C_k^{r_1}, P_{k+1}^{r_2})$ in the
previously defined cases, namely, $k \equiv 0 \pmod{2}$, $k \geq 4$, and $n \equiv 0 \pmod{k(k+1)}$. Our
main result, shown in the last section, proves that such decompositions exist if and only if
the (ordered) pair (r_1, r_2) belongs to the set $I(n)$ defined in Table 1.

Table 1. The set $I(n)$.

n	$I(n)$
$0 \pmod{2k(k+1)}$	$(\frac{n-2}{2} - kx, (k+1)x)$, $x = 0, 1, \ldots, \frac{n-2k}{2k}$
$k(k+1) \pmod{2k(k+1)}$	$(\frac{n-2}{2} - kx, (k+1)x)$, $x = 0, 1, \ldots, \frac{n-k}{2k}$

We remark that, since n is an even positive integer, $I(n)$ is a set of ordered pairs
of integers.

To this goal, we firstly recall in Section 2 known decompositions of simple cases and
two basic constructions, the so-called GDD Construction and Filling Construction, which
allow to derive decompositions of more general cases from the knowledge of simpler
cases. Our main theorem is crucially based on a clever use of these two constructions. In
Section 3, we derive the necessary conditions for the existence of URD$^*(n; C_k^{r_1}, P_{k+1}^{r_2})$, see
Lemma 2. Moreover, we set up preliminary results for later use, consisting of particular
decompositions of certain simple graphs. On the basis of these results, in Section 4, we
prove that the necessary conditions derived in Lemma 2 are also sufficient. Section 5

essentially contains the statement of our main theorem, the proof of which boils down to a recall of the partial results of the previous sections.

2. Two Constructions

In what follows, $K_{u(g)}$ denotes the complete multipartite graph with u partite sets of size g. An $\mathcal{H}$-decomposition of $K_{u(g)}$ is known as a *group divisible decomposition* (briefly, $\mathcal{H}$-GDD) of type g^u; the partite sets are called *groups*. An $\mathcal{H}$-decomposition of K_n can be regarded as an $\mathcal{H}$-GDD of type 1^n. When $\mathcal{H} = \{H\}$, we simply write H-GDD. In what follows, a (uniformly) resolvable $\mathcal{H}$-GDD is denoted by $\mathcal{H}$-(U)RGDD. More precisely, a $\{H_1, H_2\}$-URGDD with r_i H_i-factors is denoted by URGDD($H_1^{r_1}, H_2^{r_2}$). It is not hard to see that the number of H-factors of an H-RGDD is

$$\alpha = \frac{g(u-1)|V(H)|}{2|E(H)|}.$$

Let H be a given graph. For any positive integer t, $H_{(t)}$ denotes the graph with vertex set $V(H) \times \mathbb{Z}_t$ and edge set $\{\{x_i, y_j\} : \{x,y\} \in E(H), i,j \in \mathbb{Z}_t\}$, where the subscript notation a_i denotes the pair (a,i). We say that the graph $H_{(t)}$ is obtained from H by expanding each vertex t times. When $H = K_m$, the graph $H_{(t)}$ is the complete equipartite graph

$$\underbrace{K_{t,t,\ldots,t}}_{m \text{ times}}$$

with m partite sets of size t and is denoted by $K_{m(t)}$. Analogously, $C_{m(t)}$ denotes the graph $H_{(t)}$ where H is an m-cycle.

Remark 1. *The graph $H_{(t)}$ admits t 1-factors for each 1-factor of G. Therefore, since a $2k$-cycle has two 1-factors, then $C_{2k(t)}$ admits $2t$ 1-factors.*

Given two pairs (r_1, r_2) and (r'_1, r'_2) of non-negative integers, define $(r_1, r_2) + (r'_1, r'_2) = (r_1 + r'_1, r_2 + r'_2)$. Given two sets I and I' of pairs of non-negative integers and a positive integer α, then $I + I'$ denotes the set

$$\{(r_1, r_2) + (r'_1, r'_2) : (r_1, r_2) \in I, (r'_1, r'_2) \in I'\}.$$

Moreover, we denote $\alpha * I$ the set whose elements are all pairs of non-negative integers obtained by adding any α elements of I (repetitions of elements of I are allowed).

To obtain our main result we firstly construct RGDDs with appropriate parameters by means of the GDD-Construction defined here below, see Theorem 1. Subsequently, we fill their groups using the Filling Construction, stated in Theorem 2. The GDD-Construction can be derived from the more general construction described in [14]. The Filling Construction is a minor variation of the corresponding construction in [14].

Theorem 1 (GDD-Construction). *Let t be a positive integer and suppose there exists an $\mathcal{H}$-RGDD of type g^u, whose blocks are graphs of order at least 2 and whose factors are F_i, $i = 1, 2, \ldots, \alpha$. If for any $i = 1, 2, \ldots, \alpha$ there exists a URD($B_{(t)}; C_k^{\bar{r}_1}, P_{k+1}^{\bar{r}_2}$) for each $B \in F_i$ and for each $(\bar{r}_1, \bar{r}_2) \in I_i$, then so does URGDD($C_k^{r_1}, P_{k+1}^{r_2}$) of type $(gt)^u$ for each $(r_1, r_2) \in I_1 + I_2 + \cdots + I_\alpha$.*

Theorem 2 (Filling Construction). *Suppose there exists a URGDD($C_k^{r_1}, P_{k+1}^{r_2}$) of type g^u for each $(r_1, r_2) \in I$. If there exists a URD*$(g; C_k^{r'_1}, P_{k+1}^{r'_2})$, for each $(r'_1, r'_2) \in I'$, then so does a URD*$(ug; C_k^{\bar{r}_1}, P_{k+1}^{\bar{r}_2})$, for each $(\bar{r}_1, \bar{r}_2) \in I' + I$.*

Proof. Fixed any pairs $(r_1, r_2) \in I$ and $(r'_1, r'_2) \in I'$, for every group G_i, $i = 1, 2, \ldots, u$, of a URGDD$(C_k^{r_1}, P_{k+1}^{r_2})$ of type g^u, place a copy of a URD*$(g; C_k^{r'_1}, P_{k+1}^{r'_2})$ on G_i so to obtain a URD*$(gu; C_k^{r'_1+r_1}, P_{k+1}^{r'_2+r_2})$. □

We conclude this section by quoting from [18] the following result for a later use.

Lemma 1. *Let $m \geq 3$ and $u \geq 2$. There exists a C_m-RGDD of type g^u if and only if $g(u-1) \equiv 0$ (mod 2), $gu \equiv 0$ (mod m), $m \equiv 0$ (mod 2) if $u = 2$, and $(g, u, m) \neq (2, 3, 3)$, $(2, 6, 3)$, $(6, 2, 6)$, or $(6, 3, 3)$.*

3. Necessary Conditions and Basic Decompositions

Let $k \equiv 0$ (mod 2), $k \geq 4$. In this section, we start by giving necessary conditions for the existence of a URD*$(n; C_k^{r_1}, P_{k+1}^{r_2})$, and then, we construct the basic decompositions which will be used as ingredients in the GDD and Filling Constructions. From now on, throughout the paper, we set $p = k(k+1)$. Recall that the set $I(n)$ is defined in Table 1.

Lemma 2. *Let $n \equiv 0$ (mod p). If there exists a URD*$(n; C_k^{r_1}, P_{k+1}^{r_2})$ then $(r_1, r_2) \in I(n)$.*

Proof. The resolvability implies

$$\frac{r_1 k n}{k} + \frac{r_2 k n}{k+1} = \frac{n(n-2)}{2}$$

and so

$$2(k+1)r_1 + 2kr_2 = (k+1)(n-2). \tag{1}$$

Since $k+1$ cannot divide $2k$, Equation (1) implies $r_2 = (k+1)x$. Replacing $r_2 = (k+1)x$ in the above equation gives $r_1 = \frac{n-2}{2} - kx$, where $x < \frac{n-2}{2k}$ (because r_1 is a positive integer) and so $0 \leq x \leq \lfloor \frac{n-2}{2k} \rfloor$. □

From now on, we denote by $(a_1, a_2, \ldots, a_k)$ the k-cycle on $\{a_1, a_2, \ldots, a_k\}$ with edge set $\{\{a_1, a_2\}, \{a_2, a_3\}, \ldots, \{a_{k-1}, a_k\}, \{a_k, a_1\}\}$, and by $[a_1, a_2, \ldots, a_{k+1}]$ the path P_{k+1} on the vertex set $\{a_1, a_2, \ldots, a_{k+1}\}$ with edge set $\{\{a_1, a_2\}, \{a_2, a_3\}, \ldots, \{a_k, a_{k+1}\}\}$.

Lemma 3. *There exists a C_{2l-2}-decomposition of $P_{l(2)}$ for any integer $l \geq 3$.*

Proof. Let $P_{l(2)}$ be the graph obtained from the path $[1, 2, \ldots, l]$ by expanding each vertex twice. Consider the $(2l-2)$-cycles

$$C = (1_0, 2_0, \ldots, (l-1)_0, l_1, (l-1)_1, (l-2)_1, \ldots, 2_1),$$

and

$$\tilde{C} = (1_1, 2_0, 3_1, 4_0, \ldots, (l-1)_1, l_0, (l-1)_0, (l-2)_1, (l-3)_0, (l-4)_1, \ldots, 3_0, 2_1)$$

if l is even, or

$$\tilde{C} = (1_1, 2_0, 3_1, 4_0, \ldots, (l-2)_1, (l-1)_0, l_0, (l-1)_1, (l-2)_0, (l-3)_1, (l-4)_1, \ldots, 3_0, 2_1),$$

if l is odd. It is easy to see that C and $\tilde{C}$ decompose the graph $P_{l(2)}$. □

Lemma 4. *Let $q = \frac{k}{2}(1+k)$. There exists a C_k-factorization of $C_{q(2)}$.*

Proof. Start from the cycle $C_q = (1, 2, \ldots, q)$ and decompose it into the following copies of P_l, with $l = 1 + \frac{k}{2}$,

$$P^{(i)} = \left[1 + \frac{k}{2}i, 2 + \frac{k}{2}i, \ldots, 1 + \frac{k}{2}(1+i)\right], \; i = 0, 1, \ldots, k.$$

Expand twice each vertex of C_q and for every $i = 0, 1, \ldots, k$, decompose the graph $P_{l(2)}$ on $V(P^{(i)}) \times \mathbb{Z}_2$ into the k-cycles C_i and $\tilde{C}_i$ by using Lemma 3. The set of k-cycles $\{C_i, \tilde{C}_i\}_{i=0,1,\ldots,k}$ is a decomposition of $C_{q(2)}$ whose cycles can be partitioned into the factors $\{C_i\}_{i=0,1,\ldots,k}$ and $\{\tilde{C}_i\}_{i=0,1,\ldots,k}$. □

Lemma 5. *There exists a 1-factorization of $C_{l(2)}$ for any integer $l \geq 3$.*

Proof. If l is even, since C_l can be decomposed into two 1-factors, then $C_{l(2)}$ can be decomposed into four 1-factors (see Remark 1). If l is odd, $C_{l(2)}$ can be decomposed into the $2l$-cycles

$$C_1 = (1_0, 2_1, 3_0, 4_1, \ldots, l_0, 1_1, l_1, (l-1)_0, (l-2)_1 (l-3)_0, \ldots, 2_0)$$

and

$$C_2 = (1_0, l_0, (l-1)_0, (l-2)_0, (l-3)_0, \ldots, 2_0, 1_1, 2_1, 3_1, 4_1, \ldots, l_1),$$

each of which provides two 1-factors and so a 1-factorization of $C_{l(2)}$ is given. □

The following lemma follows by a result first proved by R. Laskar in [19]. For the ease of the reader, here, we propose an alternative proof which uses Graeco-Latin squares.

Lemma 6. *Let $k \neq 4, 12$ and $m = k + 1$. Then, there exists a Hamiltonian cycle decomposition of $C_{m(\frac{k}{2})}$.*

Proof. Consider the graph $C_{m(\frac{k}{2})}$ obtained from the cycle $(1, 2, \ldots, m)$ by expanding each vertex $\frac{k}{2}$ times. Let Q be a Graeco-Latin square of order $\frac{k}{2}$ on the sets $X_1 = \{1\} \times \mathbb{Z}_{\frac{k}{2}}$ and $X_2 = \{2\} \times \mathbb{Z}_{k/2}$, which exists for any $\frac{k}{2} \neq 2, 6$, see [20]. The columns of Q give a 1-factorization F_j, $j \in \mathbb{Z}_{k/2}$, of the complete bipartite graph with partite sets X_1 and X_2. For $i = 1, 2, \ldots, m$ and $j \in \mathbb{Z}_{k/2}$, consider the $\frac{k}{2} \times 1$ matrices $A_i^j = [i_j \; i_j \; \cdots \; i_j]^t$ and

$$A_{(i,j)} = [i_j \; i_{j+1} \; \cdots \; i_{j+\frac{k}{2}-1}]^t.$$

Now, for each $j \in \mathbb{Z}_{\frac{k}{2}}$, construct the $\frac{k}{2} \times m$ matrix

$$A_j = [F_j \; A_3^j \; A_{(4,j)} \; A_5^j \; A_{(6,j)} \; \cdots \; A_{(m-1,j)} \; A_m^j].$$

The rows of the $\frac{k}{2} \times \frac{km}{2}$ matrix $A = [A_0 \; A_1 \; \cdots \; A_{\frac{k}{2}-1}]$ give a Hamiltonian cycle decomposition of $C_{m(k/2)}$. □

Lemma 7. *A URD$(C_{m(k)}; C_k^{r_1}, P_{k+1}^{r_2})$ with $m = k+1$ exists for every $(r_1, r_2) \in \{(k, 0), (0, k+1)\}$.*

Proof. The proof is divided into two parts, which respectively cover the case $(r_1, r_2) = (k, 0)$ and $(r_1, r_2) = (0, k+1)$.

1. Case $(r_1, r_2) = (k, 0)$. If $k \neq 4, 12$, start from a Hamiltonian cycle decomposition of $C_{m(\frac{k}{2})}$ (which exists by Lemma 6 and has $\frac{k}{2}$ cycles) and, after expanding each vertex twice, for each cycle C on $V(C) \times \mathbb{Z}_2$, place a C_k-factorization of $C_{q(2)}$, $q = \frac{k}{2}(1+k)$ (which exists by Lemma 4 and has two factors) so to obtain a C_k-factorization of $C_{m(k)}$ with k factors, i.e., a URD$(C_{m(k)}; C_k^k, P_{k+1}^0)$. For $k = 4, 12$, start from a 1-factorization of

$C_{m(2)}$ (by Lemma 5, it exists and has four factors) and after expanding each vertex $\frac{k}{2}$ times, for each 1-factor F and each edge $e \in F$ on $e \times \mathbb{Z}_{\frac{k}{2}}$, place a C_k-RGDD of type $(\frac{k}{2})^2$, which is known to exist and have $\frac{k}{4}$ C_k-factors [21], so obtain a URD$(C_{m(k)}; C_k^k, P_{k+1}^0)$.

2. Case $(r_1, r_2) = (0, k+1)$. Starting from $C = (0, 1, \ldots, k)$ on $\mathbb{Z}_{k+1}$, expand each vertex k times and take the factors

$$F_j = \{[j_i, (1+j)_i, (2+j)_{i+1}, (3+j)_i, (4+j)_{i+2}, \ldots, (k-1+j)_i, (k+j)_{i+\frac{k}{2}}] : i \in \mathbb{Z}_k\},$$

for $j \in \mathbb{Z}_{k+1}$.

□

4. Sufficient Conditions

Lemma 8. *If $n \equiv 0 \pmod{2p}$, then a URD$^*(n; C_k^{r_1}, P_{k+1}^{r_2})$ exists for every $(r_1, r_2) \in I(n)$.*

Proof. Let $n = 2ph$, $h \geq 1$. Apply the GDD-Construction with $t = k$ to a C_{k+1}-RGDD of type $2^{(k+1)h}$ (which exists by Lemma 1 and has $\alpha = (k+1)h - 1$ factors) to obtain a URGDD$(C_k^{\bar{r}_1}, P_{k+1}^{\bar{r}_2})$ of type $(2k)^{(k+1)h}$ for each

$$(\bar{r}_1, \bar{r}_2) \in [(k+1)h - 1] * \{(k, 0), (0, k+1)\}$$

using as ingredients designs from Lemma 7. Finally, apply the Filling Construction by using copies of a URD$^*(2k; C_k^{k-1}, P_{k+1}^0)$ (see [17]) to get a URD$^*(2ph; C_k^{r_1}, P_{k+1}^{r_2})$ for every

$$\begin{aligned}(r_1, r_2) &\in \{(k-1, 0)\} + [(k+1)h - 1] * \{(k, 0), (0, k+1)\} \\ &= \{(ph - 1 - kx, (k+1)x) : x = 0, 1, \ldots, (k+1)h - 1\} \\ &= I(2ph) = I(n).\end{aligned}$$

□

Lemma 9. *If $n \equiv p \pmod{2p}$, then a URD$^*(n; C_k^{r_1}, P_{k+1}^{r_2})$ exists for every $(r_1, r_2) \in I(n)$.*

Proof. Let $n = p(1 + 2h)$, $h \geq 0$. Apply the GDD-Construction with $t = k$ to a C_{k+1}-RGDD of type $1^{(k+1)(1+2h)}$ (which exists by Lemma 1 and has $\alpha = \frac{(k+1)(1+2h)-1}{2}$ factors) to obtain a URGDD$(C_k^{\bar{r}_1}, P_{k+1}^{\bar{r}_2})$ of type $k^{(k+1)(1+2h)}$ for each

$$(\bar{r}_1, \bar{r}_2) \in \frac{(k+1)(1+2h) - 1}{2} * \{(k, 0), (0, k+1)\}$$

using as ingredients the designs from Lemma 7. Finally, apply the Filling Construction by using copies of a URD$^*(k; C_k^{\frac{k-2}{2}}, P_{k+1}^0)$ (see [17]) to get a URD$^*(p(1+2h); C_k^{r_1}, P_{k+1}^{r_2})$ for every

$$\begin{aligned}(r_1, r_2) &\in \left\{\left(\frac{k-2}{2}, 0\right)\right\} + \frac{(k+1)(1+2h) - 1}{2} * \{(k, 0), (0, k+1)\} \\ &= \left\{\left(\frac{p(1+2h) - 2}{2} - kx, (k+1)x\right) : x = 0, 1, \ldots, \frac{(k+1)(1+2h) - 1}{2}\right\} \\ &= I(p(1+2h)) = I(n).\end{aligned}$$

□

5. Conclusions

Combining together Lemmas 2, 8 and 9 gives our main result.

Theorem 3. *Let* $n \equiv 0 \pmod{k(k+1)}$. *There exists a* $URD^*(n; C_k^{r_1}, P_{k+1}^{r_2})$ *if and only if* $(r_1, r_2) \in I(n)$.

We emphasize that our main result fits in the context of a series of papers, where the authors investigated the existence of $\mathcal{H}$-factorizations of K_n or $K_n - I$ in the case that $\mathcal{H}$ contains at least one cycle. As a final note, we stress that determining necessary and sufficient conditions for the existence of analogous decompositions for odd values of k is still an open problem of definite interest for further research.

Author Contributions: Conceptualization, G.L.F., S.M. and A.T.; formal analysis, G.L.F., S.M. and A.T.; writing-original draft preparation, G.L.F., S.M. and A.T.; writing-review and editing, G.L.F., S.M. and A.T. All authors have read and agreed to the published version of the manuscript.

Funding: This research was funded by GNSAGA INDAM (Giovanni Lo Faro, Antoinette Tripodi).

Institutional Review Board Statement: Not applicable.

Informed Consent Statement: Not applicable.

Data Availability Statement: Not applicable.

Conflicts of Interest: The authors declare no conflict of interest.

References

1. Colbourn, C.J.; Dinitz, J.H. (Eds.) *Handbook of Combinatorial Designs*, 2nd ed.; Chapman and Hall/CRC: Boca Raton, FL, USA, 2007. Available online: https://site.uvm.edu/jdinitz/?page_id=312 (accessed on 7 March 2022).
2. Lucas, E. *Récréations Mathématiques*; Gauthier-Villars: Paris, France, 1883; Volume 2.
3. Alspach, B.; Schellenberg, P.; Stinson, D.R.; Wagner, D. The Oberwolfach problem and factors of uniform length. *J. Combin. Theory Ser. A* **1989**, *52*, 20–43. [CrossRef]
4. Alspach, B.; Häggkvist, R. Some observations on the Oberwolfach problem. *J. Graph Theory* **1985**, *9*, 177–187. [CrossRef]
5. Rees, R. Uniformly resolvable pairwise balanced designs with block sizes two and three. *J. Combin. Theory Ser. A* **1987**, *45*, 207–225. [CrossRef]
6. Dinitz, J.H.; Ling, A.C.H.; Danziger, P. Maximum Uniformly resolvable designs with block sizes 2 and 4. *Discrete Math.* **2009**, *309*, 4716–4721. [CrossRef]
7. Schuster, E.; Ge, G. On uniformly resolvable designs with block sizes 3 and 4. *Des. Codes Cryptogr.* **2010**, *57*, 57–69. [CrossRef]
8. Wei, H.; Ge, G. Uniformly resolvable designs with block sizes 3 and 4. *Discret. Math.* **2016**, *339*, 1069–1085. [CrossRef]
9. Gionfriddo, M.; Milici, S. Uniformly resolvable $\{K_2, P_k\}$-designs with $k = \{3, 4\}$. *Contrib. Discret. Math.* **2015**, *10*, 126–133.
10. Lo Faro, G.; Milici, S.; Tripodi, A. Uniformly resolvable decompositions of K_v into paths on two, three and four vertices. *Discret. Math.* **2015**, *338*, 2212–2219. [CrossRef]
11. Küçükçifçi, S.; Lo Faro, G.; Milici, S.; Tripodi, A. Resolvable 3-star designs. *Discret. Math.* **2015**, *338*, 608–614. [CrossRef]
12. Chen, F.; Cao, H. Uniformly resolvable decompositions of K_v into K_2 and $K_{1,3}$ graphs. *Discret. Math.* **2016**, *339*, 2056–2062. [CrossRef]
13. Keranen, M.S.; Kreher, D.L.; Milici, S.; Tripodi, A. Uniformly resolvable decompositions of K_v into 1-factors and 4-stars. *Australas. J. Combin.* **2020**, *76*, 55–72.
14. Lo Faro, G.; Milici, S.; Tripodi, A. Uniformly Resolvable Decompositions of $K_v - I$ into n-Cycles and n-Stars, for Even n. *Mathematics* **2020**, *8*, 1755. [CrossRef]
15. Horton, D.G. Resolvable paths designs. *J. Combin. Theory Ser. A* **1985**, *39*, 117–131. [CrossRef]
16. Bermond, J.C.; Heinrich, K.; Yu, M.L. Existence of resolvable paths designs. *Europ. J. Combin.* **1990**, *11*, 205–211. [CrossRef]
17. Hoffman, D.G.; Schellenberg, P.J. The existence of C_k-factorizations of $K_{2n} - I$. *Discret. Math.* **1991**, *97*, 243–250. [CrossRef]
18. Cao, H.; Niu, M.; Tang, C. On the existence of cycle frames and almost resolvable cycle systems. *Discret. Math.* **2011**, *311*, 2220–2232. [CrossRef]
19. Laskar, R. Decomposition of some composite graphs into Hamilton cycles. In *Combinatorics, Proceedings of the Fifth Colloquium of the János Bolyai Mathematical Society, Keszthely, Hungary, 28 June–3 July 1976*; Hajnal, A., Sós, V.T., Eds.; North-Holland: New York, NY, USA, 1978; pp. 705–716.
20. Bose, R.C.; Shrikhande, S.S.; Parker, E.T. Further results on the construction of mutually orthogonal Latin squares and the falsity of Euler's conjecture. *Canad. J. Math.* **1960**, *12*, 189–203. [CrossRef]
21. Enomoto, H.; Miyamoto, T.; Ushio, K. C_k-factorization of complete bipartite graphs. *Graphs Comb.* **1988**, *4*, 111–113. [CrossRef]

Article

G-Hypergroups: Hypergroups with a Group-Isomorphic Heart

Mario De Salvo [1], Dario Fasino [2], Domenico Freni [2,*] and Giovanni Lo Faro [1]

[1] Dipartimento di Scienze Matematiche e Informatiche, Scienze Fisiche e Scienze della Terra, Università di Messina, 98122 Messina, Italy; desalvo@unime.it (M.D.S.); lofaro@unime.it (G.L.F.)
[2] Dipartimento di Scienze Matematiche, Informatiche e Fisiche, Università di Udine, 33100 Udine, Italy; dario.fasino@uniud.it
* Correspondence: domenico.freni@uniud.it

Abstract: Hypergroups can be subdivided into two large classes: those whose heart coincide with the entire hypergroup and those in which the heart is a proper sub-hypergroup. The latter class includes the family of 1-hypergroups, whose heart reduces to a singleton, and therefore is the trivial group. However, very little is known about hypergroups that are neither 1-hypergroups nor belong to the first class. The goal of this work is to take a first step in classifying G-hypergroups, that is, hypergroups whose heart is a nontrivial group. We introduce their main properties, with an emphasis on G-hypergroups whose the heart is a torsion group. We analyze the main properties of the stabilizers of group actions of the heart, which play an important role in the construction of multiplicative tables of G-hypergroups. Based on these results, we characterize the G-hypergroups that are of type U on the right or cogroups on the right. Finally, we present the hyperproduct tables of all G-hypergroups of size not larger than 5, apart of isomorphisms.

Keywords: hypergroups; heart; group action; 1-hypergroups; cogroups

Citation: De Salvo, M.; Fasino, D.; Freni, D.; Lo Faro, G. G-Hypergroups: Hypergroups with a Group-Isomorphic Heart. *Mathematics* **2022**, *10*, 240. https://doi.org/10.3390/math10020240

Academic Editors: Elena Guardo and Domenico Freni

Received: 17 December 2021
Accepted: 10 January 2022
Published: 13 January 2022

Publisher's Note: MDPI stays neutral with regard to jurisdictional claims in published maps and institutional affiliations.

Copyright: © 2022 by the authors. Licensee MDPI, Basel, Switzerland. This article is an open access article distributed under the terms and conditions of the Creative Commons Attribution (CC BY) license (https://creativecommons.org/licenses/by/4.0/).

1. Introduction

Hypercompositional algebra is a branch of Algebra that falls under the many generalizations of group theory [1]. Therefore, it is not surprising that there is a great deal of overlap between the tools and problems of group theory and those of hypergroup theory. In fact, one of the best developed research areas in hypergroup theory is that of their classification. Although a complete classification of hypergroups is well beyond any current research horizon, several important results have been obtained in characterizing classes of hypergroups having certain properties. For example, the class of D-hypergroups consists of those hypergroups that are isomorphic to the quotient set of a group with respect to a non-normal subgroup, and is a subclass of cogroups [2–4], and cogroups appear as generalizations of C-hypergroups, that were introduced as hyperstructures having an identity element and a weak form of the cancellation law [5,6].

A strong link between group theory and hypergroup theory is established by the relation β, which is the smallest equivalence relation defined on a hypergroup H such that the corresponding quotient set H/β is a group [7–9]. This relation is a very expressive tool for classifying significant families of hypergroups. In particular, the β-class of the identity of the quotient group H/β is called heart [10–12]. The heart is a special sub-hypergroup of H that gives detailed informations on the partition of H determined by β. Notably, a 1-hypergroup is a hypergroup whose heart consists of only one element [13,14]. In this case, that element is also the identity of the hypergroup. In [15,16], the authors characterized 1-hypergroups in terms of the height of their heart and provided a classification of the 1-hypergroups with $|H| \leq 6$ based on the partition of H induced by β. By means of this technique, the authors were able to enumerate all 1-hypergroups of size up to 6 and construct explicitly all non-isomorphic 1-hypergroups of size up to 5.

Motivated by these studies, in this paper we consider the class of hypergroups whose heart is isomorphic to a group. These hypergroups are called G-hypergroups. Clearly, this class contains that of 1-hypergroups as the heart of a 1-hypergroup is the trivial group. The plan of this paper is the following. In the next section, we introduce basic definitions and notations to be used throughout the paper. In Section 3, we introduce G-hypergroups and their main properties, and give a flexible construction of G-hypergroups that allows to prescribe arbitrarily both the heart and the quotient group H/β. Moreover, we analyze G-hypergroups whose heart is isomorphic to a torsion group. We denote this subclass of G-hypergroups with $\mathfrak{T}(H)$. If $(H, \circ) \in \mathfrak{T}(H)$ then the identity ε of ω_H is also identity of $(H, \circ)$, that is $x \in x \circ \varepsilon \cap \varepsilon \circ x$ for all $x \in H$. Consequently, we prove that the singleton $\{\varepsilon\}$ is an invertible sub-hypergroup of $(H, \circ)$ and the family of right (or left) cosets $\varepsilon \circ x$ (or $x \circ \varepsilon$, respectively) is a partition of H. Moreover, all β-classes are a disjoint union of right (left) cosets of $\{\varepsilon\}$. In Section 4, we analyze the main properties of the stabilizers of special actions of ω_H on the set families $\mathfrak{L} = \{x \circ g \mid x \in H - \omega_H, g \in \omega_H\}$ and $\mathfrak{R} = \{g \circ x \mid x \in H - \omega_H, g \in \omega_H\}$. These stabilizers play an important role in the construction of multiplicative tables of G-hypergroups, as they fix the hyperproducts $g \circ x$ and $x \circ g$ for all $g \in \omega_H$ and $x \in H$. The results of Section 5 concern products of elements $x, y \in H - \omega_H$ such that $x \circ y \subseteq \omega_H$. In Section 6, we characterize the G-hypergroups in $\mathfrak{T}(H)$ that are of type U on the right. Moreover, we find a sufficient condition for a G-hypergroup of type U on the right to be a cogroup. Finally, in Section 7, we classify the G-hypergroups of size ≤ 5 and $|\omega_H| \in \{2, 3, 4\}$. Apart of isomorphisms, all the multiplicative tables of these hypergroups are listed and, using the results on 1-hypergroups found in [16], we conclude that there are 48 non-isomorphic G-hypergroups of size ≤ 5.

2. Fundamentals of Hypergroup Theory

Throughout this paper, we will use standard definitions of fundamental concepts in hyperstructure theory, such as hyperproduct, semi-hypergroup, hypergroup, and sub-hypergroup, see, e.g., in [17–19]. To keep the exposition self-contained, we recall below some auxiliary definitions and results that will be needed in the sequel.

A sub-hypergroup K of a hypergroup $(H, \circ)$ is invertible on the right (resp., on the left) if for all $x, y \in H$, $x \in y \circ K \Rightarrow y \in x \circ K$ (resp., $x \in K \circ y \Rightarrow y \in K \circ x$). Moreover, if K is invertible both on the right and on the left then it is called invertible.

A sub-hypergroup K of a hypergroup $(H, \circ)$ is said to be conjugable if for all $x \in H$ there exists $x' \in H$ such that $xx' \subseteq K$.

An element ε of a semihypergroup $(H, \circ)$ is an identity if $x \in x \circ \varepsilon \cap \varepsilon \circ x$, for all $x \in H$. Moreover, if $\{x\} = x \circ \varepsilon = \varepsilon \circ x$ then ε is a scalar identity.

Given a semihypergroup $(H, \circ)$, the relation β^* of H is the transitive closure of the relation $\beta = \cup_{n \geq 1} \beta_n$, where β_1 is the diagonal relation in H and, for every integer $n > 1$, β_n is defined as follows:

$$x \beta_n y \iff \exists (z_1, \ldots, z_n) \in H^n \;:\; \{x, y\} \subseteq z_1 \circ z_2 \circ \cdots \circ z_n.$$

The relations β and β^* are among the so-called fundamental relations [7,9,11,20]. Their relevance in hyperstructure theory stems from the following facts. If $(H, \circ)$ is a semihypergroup (resp., a hypergroup), then the quotient set H/β^* endowed with the operation $\beta^*(x) \otimes \beta^*(y) = \beta^*(z)$ for $x, y \in H$ and $z \in x \circ y$ is a semigroup (resp., a group) [21,22]. The canonical projection $\varphi : H \to H/\beta^*$ verifies the identity $\varphi(x \circ y) = \varphi(x) \otimes \varphi(y)$ for all $x, y \in H$, that is, φ is said to be a good homomorphism. Moreover, if $(H, \circ)$ is a hypergroup then β is transitive [8], H/β is a group and the kernel $\omega_H = \varphi^{-1}(1_{H/\beta})$ of φ is the heart of $(H, \circ)$.

If A is a non-empty set of a semihypergroup $(H, \circ)$, then we say that A is a complete part if for every $n \geq 1$ and $(x_1, x_2, \ldots, x_n) \in H^n$,

$$(x_1 \circ x_2 \circ \ldots \circ x_n) \cap A \neq \emptyset \implies x_1 \circ x_2 \circ \cdots \circ x_n \subseteq A.$$

The transposed hypergroup of a hypergroup $(H, \circ)$ is the hypergroup $(H, \star)$ where $x \star y = y \circ x$ for all $x, y \in H$.

For later reference, we collect in the following theorem some classic results of hypergroup theory, see in [8,17].

Theorem 1. *Let $(H, \circ)$ be a hypergroup. Then,*

1. *the relation β is transitive;*
2. *if K is a subhypergroup invertible on the right (resp., on the left) of $(H, \circ)$, then the family $\{x \circ K\}_{x \in H}$ (resp., $\{K \circ x\}_{x \in H}$) is a partition of H;*
3. *a subhypergroup K of $(H, \circ)$ is a complete part if and only if it is conjugable;*
4. *the heart ω_H is the intersection of all conjugable subhypergroups (or complete parts) of $(H, \circ)$;*
5. *the heart ω_H is a reflexive subhypergroup of $(H, \circ)$, that is, $x \circ y \cap \omega_H \neq \emptyset \Rightarrow y \circ x \cap \omega_H \neq \emptyset$.*

3. G-Hypergroups

The heart of a hypergroup $(H, \circ)$ allows us to explicitly compute the partition determined by β, as $\beta(x) = \omega_H \circ x = x \circ \omega_H$ for all $x \in H$. For this reason, the heart of hypergroups has been the subject of much research, in particular, to characterize it as the union of particular hyperproducts [12]. A special class of hypergroups is that of 1-hypergroups, where the heart is a singleton. Clearly, the heart of a 1-hypergroup is isomorphic to a trivial group and if $\omega_H = \{\varepsilon\}$ then the element ε is an identity since $x \in \beta(x) = \varepsilon \circ x = x \circ \varepsilon$. Other relevant results on 1-hypergroups can be found, e.g., in [13–16]. In this section, we will study the main properties of hypergroups whose heart is isomorphic to a group G, which we call G-hypergroups.

Notably, the class of G-hypergroups is closed under direct product. Indeed, if $(H, \circ)$ and $(H', \star)$ are G-hypergroups then the direct product $H \times H'$ is a G-hypergroup as $\omega_{H \times H'} = \omega_H \times \omega'_H$. Indeed, for all $(x,y) \in H \times H'$, we have $\beta_{H \times H'}(x,y) = \beta_H(x) \times \beta_{H'}(y)$. Non-trivial examples of G-hypergroups can be built by means of the construction shown in Example 2 of [15], which we recall hereafter. Let $Aut(H)$ be the automorphism group of a hypergroup $(H, \circ)$. For $f \in Aut(H)$, let $\langle f \rangle$ denote the subgroup of $Aut(H)$ generated by f. In $H \times \langle f \rangle$, define the following hyperproduct: for $(a, f^m), (b, f^n) \in H \times \langle f \rangle$, let

$$(a, f^m) \star (b, f^n) = \{(c, f^{m+n}) \mid c \in a \circ f^m(b)\} = (a \circ f^m(b)) \times \{f^{m+n}\}.$$

with respect to this hyperproduct $(H \times \langle f \rangle, \star)$ is a hypergroup whose heart is $\omega_H \times \{f^0\}$. Clearly, if $(H, \circ)$ is a G-hypergroup then also $(H \times \langle f \rangle, \star)$ is a G-hypergroup.

3.1. A Construction of G-Hypergroups

Let T and G be groups with $|T| \geq 2$. Consider a family $\mathfrak{F} = \{A_k\}_{k \in T}$ of non-empty and pairwise disjoint sets such that $A_{1_T} = G$ and $|A_i| = |G|$, for all $i \in T$. In these hypotheses we pose $A_i = \{a_{i,h}\}_{h \in G}$, for all $i \neq 1_T$. In the set $H = \bigcup_{k \in T} A_k$ we consider the hyperproduct $\circ : H \times H \to \mathcal{P}^*(H)$ defined as follows: for all $x, y \in H$,

$$x \circ y = \begin{cases} \{xy\} & \text{if } x, y \in A_{1_T}; \\ \{a_{i,hy}\} & \text{if } x = a_{i,h}, y \in A_{1_T} \text{ and } i \neq 1_T; \\ A_{ij} & \text{if } x \in A_i, y \in A_j \text{ and } j \neq 1_T; \end{cases} \quad (1)$$

We note that, by definition of hyperproduct $\circ$, we have $x \circ 1_G = \{x\}$ and $x \in 1_G \circ x$ for all $x \in H$. Moreover, for every $i, j \in T$ and $x \in A_j$ we obtain

$$A_i \circ x = A_{ij}, \quad x \circ A_i = A_{ji}. \quad (2)$$

Indeed, if $j \neq 1_T$ then $A_i \circ x = \bigcup_{y \in A_i} y \circ x = A_{ij}$. Otherwise, if $i = j = 1_T$ then $A_{1_T} \circ x = Gx = G = A_{1_T}$. Moreover, if $i \neq 1_T$ and $j = 1_T$ then we obtain $A_i \circ x = \bigcup_{h \in G} a_{i,h} \circ x = \bigcup_{h \in G} \{a_{i,hx}\} = A_i$. By analogous arguments, we can deduce that $x \circ A_i = A_{ji}$. These simple remarks yield the basis of the following result, where we prove that $(H, \circ)$ is a G-hypergroup with some special properties.

Theorem 2. *In the previous notations, the hyperoperation $\circ$ defined in (1) is associative. Moreover, we have*

1. *for every integer $n \geq 3$ and for every $(z_1, z_2, \ldots, z_n) \in H^n$, there exists $r \in T$ such that $z_1 \circ z_2 \circ \ldots \circ z_n \subseteq A_r$;*
2. *for all $i \in T$ there exist $x, y \in H$ such that $x \circ y = A_i$;*
3. *$(H, \circ)$ is a hypergroup such that $\beta = \beta_2$;*
4. *$\omega_H = A_{1_T} = G$ and $\beta(x) = A_k$, for all $x \in A_k$ and $k \in T$;*
5. *$H/\beta \cong T$.*

Proof. Let $x \in A_i$, $y \in A_j$ and $z \in A_k$ with $i, j, k \in T$. If $i = j = k = 1_T$ then we have immediately $(x \circ y) \circ z = x \circ (y \circ z)$ since $A_{1_T} = G$ is a group. Otherwise, we have the following cases:

- Only two of the three elements i, j, k coincide with 1_T;
- Only one of the three elements i, j, k coincides with 1_T;
- $i, j, k \in T - \{1_T\}$.

In the first case, if we assume that $i = j = 1_T$ and $k \neq 1_T$, then we have $(x \circ y) \circ z = \{xy\} \circ z = A_k = x \circ A_k = x \circ (y \circ z)$.
If $i = k = 1_T$ and $j \neq 1_T$, we obtain $(x \circ y) \circ z = x \circ (y \circ z) = A_j$.
If $j = k = 1_T$ and $i \neq 1_T$, we have $(x \circ y) \circ z = x \circ (y \circ z) = A_i$.

In the second case, suppose $i = 1_T$ and $j, k \in T - \{1_T\}$, we have $(x \circ y) \circ z = A_j \circ z = A_{jk} = x \circ A_{jk} = x \circ (y \circ z)$.
If $j = 1_T$ and $i, k \in T - \{1_T\}$, we obtain $(x \circ y) \circ z = x \circ (y \circ z) = A_{ik}$ because $x \circ y \subseteq A_i$, $z \in A_k$ and $y \circ z = A_k$.
If $i, j \in T - \{1_T\}$ and $k = 1_T$, we deduce $(x \circ y) \circ z = x \circ (y \circ z) = A_{ij}$ as $x \circ y = A_{ij}$, $x \in A_i$ and $y \circ z \subseteq A_j$.

In the last case we have $(x \circ y) \circ z = x \circ (y \circ z) = A_{ijk}$. Thus, $\circ$ is associative. Now, we complete the proof of the remaining claims.

1. To prove this claim it suffices to proceed by induction on n, based on (2) and the associativity of hyperproduct $\circ$.
2. Let $i \in T$. If $i \neq 1_T$ then we have $x \circ y = A_i$, for all $x \in A_{1_T}$ and $y \in A_i$. If $i = 1_T$, since $|T| \geq 2$, there exists $j, k \in T - \{1_T\}$ such that $jk = 1_T$ and so $x \circ y = A_{jk} = A_{1_T}$, for all $x \in A_j$ and $y \in A_k$.
3. To prove that $(H, \circ)$ is a hypergroup we only need to prove reproducibility. Let $x \in A_i$. As $iT = T$, using (2) we obtain

$$x \circ H = x \circ \left(\bigcup_{j \in T} A_j \right) = \bigcup_{j \in T} x \circ A_j = \bigcup_{j \in T} A_{ij} = H.$$

Analogously, we can prove that $H \circ x = H$ for every $x \in H$. Now, being $(H, \circ)$ a hypergroup, we have the chain of inclusions

$$\beta_1 \subseteq \beta_2 \subseteq \beta_3 \subseteq \cdots \subseteq \beta_n \cdots$$

Thus, if $a\beta b$ then there exists $n \geq 3$ such that $a\beta_n b$. For points 1. and 2., there exist $r \in T$ and $x, y \in H$ such that $\{a, b\} \subseteq A_r = x \circ y$, so we obtain $x\beta_2 y$.

4. Clearly $A_1 = G$ is a subhypergroup of H. Moreover, G is conjugable as for all $x \in H - G$ and $x \in A_j$ there exists $x' \in A_{j-1}$ such that $x \circ x' = A_{1_T} = G$. By point 4. of Theorem 1, we have $\omega_H \subseteq G$. Moreover, $G \subseteq \omega_H$ because ω_H is a complete part of H and $G = x \circ x' \cap \omega_H \neq \emptyset$. Finally, by (2) we have $\beta(x) = \omega_H \circ x = G \circ x = A_k$, for all $x \in A_k$ and $k \in T$.
5. The application $f : T \mapsto H/\beta$ such that $f(k) = A_k$ is a group isomorphism.

□

3.2. If G Is a Torsion Group

In this subsection, we denote by ε the identity of the heart of a G-hypergroup $(H, \circ)$. Moreover, we denote by $\mathfrak{T}(H)$ the class of G-hypergroups whose heart is a torsion group. For each element x of a hypergroup $(H, \circ)$, we identify x^1 with the singleton $\{x\}$ and, for any integer $n \geq 2$, we set

$$x^n = \underbrace{x \circ x \circ \cdots \circ x}_{n \text{ times}}.$$

Moreover, define

$$\check{x} = \bigcup_{k=1}^{\infty} x^k.$$

The set $\check{x}$ is the cyclic semihypergroup generated by x. This hypercompositional analogue of cyclic semigroups has attracted the interest of many researchers, being a powerful tool for the construction and study of remarkable families of hypergroups. We point the interested reader to the detailed reviews in [23,24].

In what follows, we exploit cyclic sub-semihypergroups to derive some properties of hypergroups in $\mathfrak{T}(H)$. Specifically, we prove that the identity of the heart ω_H of a hypergroup $(H, \circ) \in \mathfrak{T}(H)$ is an invertible sub-hypergroup of $(H, \circ)$. We will use these properties in the subsequent section to describe the group actions of ω_H on families of hyperproducts $g \circ x$ and $x \circ g$ for $g \in \omega_H$ and $x \in H$.

Theorem 3. *Let $(H, \circ) \in \mathfrak{T}(H)$. Then, ε is an identity of $(H, \circ)$.*

Proof. Let $x \in H - G$. There exists $e \in \omega_H$ such that $x \in x \circ e$ by reproducibility of $(H, \circ)$. Moreover, $x \in x \circ e \subseteq (x \circ e) \circ e = x \circ (e \circ e) = x \circ e^2$ and, by an inductive argument, $x \in x \circ e^n$ for all $n \geq 1$. Finally, as ω_H is a torsion group, there exists $m \geq 1$ such that $e^m = \{\varepsilon\}$, thus $x \in x \circ \varepsilon$. By analogous arguments we also have $x \in \varepsilon \circ x$. □

Proposition 1. *Let $(H, \circ)$ an G-hypergroup and $x \in H$. The following conditions are equivalent:*
1. *$\varepsilon \circ y = \{y\}$ (resp., $y \circ \varepsilon = \{y\}$) for all $y \in \beta(x)$;*
2. *$|g \circ y| = 1$ (resp., $|y \circ g| = 1$) for all $g \in \omega_H$ and $y \in \beta(x)$.*

Proof. 1. $\Rightarrow$ 2. Let $g \in \omega_H$ and $y \in \beta(x)$. The thesis is obvious if $\beta(x) = \omega_H$, so let $x \in H - \omega_H$ and $a \in g \circ y$. We have $a \in g \circ y \subseteq \omega_H \circ y = \beta(y) = \beta(x)$ and so $\varepsilon \circ a = \{a\}$. Moreover, $g^{-1} \circ a \subseteq g^{-1} \circ (g \circ y) = (g^{-1} \circ g) \circ y = \varepsilon \circ y = \{y\}$. Hence $g^{-1} \circ a = \{y\}$. Consequently, $g \circ y = g \circ (g^{-1} \circ a) = (g \circ g^{-1}) \circ a = \varepsilon \circ a = \{a\}$. Therefore $|g \circ y| = 1$. In the same way we prove that $|y \circ g| = 1$ if $y \circ \varepsilon = \{y\}$ for all $y \in \beta(x)$.

The converse implication, 2. $\Rightarrow$ 1., is an immediate consequence of Theorem 3. □

Corollary 1. *Let $(H, \circ) \in \mathfrak{T}(H)$. Then ε is a left scalar identity (resp., right scalar identity) of $(H, \circ)$ if and only if $|g \circ x| = 1$ (resp., $|x \circ g| = 1$), for all $g \in \omega_H$ and $x \in H$.*

Theorem 4. *Let $(H, \circ) \in \mathfrak{T}(H)$. If S is a finite sub-semihypergroup of $(H, \circ)$ then we have:*
1. *$\varepsilon \in S$;*
2. *S is a sub-hypergroup of $(H, \circ)$.*

Proof. 1. Let $\varphi : H \to H/\beta$ be the canonical projection. As S is finite, there exists $x \in S$ such that $\check{x}$ has minimal size.

If $x \in x^2$ then $\varphi(x) = \varphi(x) \otimes \varphi(x)$. Hence $\varphi(x) = 1_{H/\beta}$ and $x \in \omega_H$. As ω_H is a torsion group, there exists a positive integer n such that $x^n = \{\varepsilon\}$ and so $\varepsilon \in S$.

If $x \notin x^2$ then there exists $y \in x^2$ such that $y \neq x$. Clearly, we have $\check{y} \subseteq \check{x}$ and consequently $\check{y} = \check{x}$ as $\check{x}$ has minimal size. Therefore, $x \in \check{y}$ and there exists a integer $n \geq 2$ such that $x \in y^n \subseteq (x^2)^n = x^{2n} = x \circ x^{2n-1}$. Therefore, there exists $a \in x^{2n-1}$ such that

$x \in x \circ a$ and $\varphi(x) = \varphi(x) \otimes \varphi(a)$. Thus, $\varphi(a) = 1_{H/\beta}$ and $a \in \omega_H$. Finally, there exists a integer positive m such that $a^m = \{\varepsilon\}$ and $\varepsilon \in (x^{2n-1})^m = x^{(2n-1)m} \subseteq \check{x} \subseteq S$.

2. We must show that $x \circ S = S \circ x = S$, for all $x \in S$. By point 1. and Theoreme 3, $\varepsilon \in S$ and ε is identity in $(H, \circ)$. Therefore, we have $S \subseteq \varepsilon \circ S \subseteq S \circ S \subseteq S$ and so $\varepsilon \circ S = S \circ S = S$. Now, if $x \in S$, the subset $x \circ \check{x}$ is a finite sub-semihypergroup of $(H, \circ)$ as S is finite, $x \circ \check{x} \subseteq S$ and $(x \circ \check{x}) \circ (x \circ \check{x}) = x \circ x \circ \check{x} \circ \check{x} \subseteq x^2 \circ \check{x} \subseteq x \circ \check{x}$. Thus, for point 1., we obtain $\varepsilon \in x \circ \check{x}$. Finally,

$$S = \varepsilon \circ S \subseteq x \circ \check{x} \circ S \subseteq x \circ S \circ S \subseteq x \circ S \subseteq S \circ S = S.$$

Therefore, $x \circ S = S$ for all $x \in S$. In the same way we prove that $S \circ x = S$. □

Theorem 5. *Let $(H, \circ) \in \mathfrak{T}(H)$. The singleton $S = \{\varepsilon\}$ is a invertible sub-hypergroup of $(H, \circ)$.*

Proof. We prove that $S = \{\varepsilon\}$ is invertible on the left, that is $x \in S \circ y \Rightarrow y \in S \circ x$, for all $x, y \in H$. In the same way, it is proved that S is invertible on the right. Let $x \in S \circ y = \varepsilon \circ y$. If $y \in \omega_H$, we have $x = y$ and $y \in \varepsilon \circ x = S \circ x$. Now, we suppose $y \in H - \omega_H$. Clearly, we have $\varepsilon \circ x \subseteq \varepsilon \circ y$. Moreover, we obtain $x \in \omega_H \circ y = \beta(y)$ and so $y \in \beta(x) = \omega_H \circ x$. Therefore, there exists $g \in \omega_H$ such that $y \in g \circ x$. Consequently, $y \in g \circ x \subseteq g \circ (\varepsilon \circ y) \subseteq g \circ (\varepsilon \circ (g \circ x)) = g^2 \circ x$ and $y \in g^2 \circ x$. By induction, we deduce that $y \in g^n \circ x$, for all integer $n \geq 1$. As ω_H is a torsion group, there exists a positive integer m such that $g^m = \{\varepsilon\}$ and so $y \in \varepsilon \circ x = S \circ x$. □

Remark 1. *The invertibility on the left (resp., on the right) of the sub-hypergroup $S = \{\varepsilon\}$ implies that the family of right cosets (resp., left cosets) of $S = \{\varepsilon\}$ is a partition of H. Since for each element y of a β-class $\beta(x)$ we have $\varepsilon \circ y \subseteq \omega_H \circ x = \beta(x)$ (resp., $y \circ \varepsilon \subseteq x \circ \omega_H = \beta(x)$), then every β-class is a disjoint union of right cosets of S (resp., left cosets of S).*

3.3. The Cosets of $\{\varepsilon\}$

As suggested by Remark 1, the families of right and left cosets of $S = \{\varepsilon\}$ are relevant to determine the structure of G-hypergroups in $\mathfrak{T}(H)$. In this subsection we deepen the knowledge of these cosets. We will only do proofs for right cosets because properties that are true for a hypergroup are also true for its transposed hypergroup.

Proposition 2. *Let $(H, \circ) \in \mathfrak{T}(H)$. For all $x \in H$ and $g \in \omega_H$ we have:*

1. $x \in g \circ x \Leftrightarrow g \circ x = \varepsilon \circ x$;
2. $g \circ x \cap \varepsilon \circ x \neq \emptyset \Leftrightarrow g \circ x = \varepsilon \circ x$;
3. $x \in x \circ g \Leftrightarrow x \circ g = x \circ \varepsilon$;
4. $x \circ g \cap x \circ \varepsilon \neq \emptyset \Leftrightarrow x \circ g = x \circ \varepsilon$;

Proof. 1. The implication $\Leftarrow$ is a consequence of Theorem 3. Now, suppose that $x \in g \circ x$. Clearly, we have $\varepsilon \circ x \subseteq \varepsilon \circ (g \circ x) = (\varepsilon \circ g) \circ x = g \circ x$. Moreover, $g \circ x \subseteq g \circ (g \circ x) = g^2 \circ x$ and, by induction, we obtain the chain of inclusions $\varepsilon \circ x \subseteq g \circ x \subseteq g^2 \circ x \subseteq \cdots \subseteq g^n \circ x \subseteq \cdots$. As ω_H is a torsion group, there exists a positive integer m such that $g^m = \{\varepsilon\}$ and so $\varepsilon \circ x \subseteq g \circ x \subseteq \varepsilon \circ x$. Therefore, $\varepsilon \circ x = g \circ x$.

Concerning point 2., it is enough to prove the implication $\Rightarrow$. Let $z \in \varepsilon \circ x \cap g \circ x$. As $S = \{\varepsilon\}$ is a invertible subhypergroup of H, we have $\varepsilon \circ x = \varepsilon \circ z$ and so $z \in g \circ x = g \circ \varepsilon \circ x = g \circ \varepsilon \circ z = g \circ z$. Therefore, by point 1., we obtain $\varepsilon \circ z = g \circ z$. Consequently, we deduce $\varepsilon \circ x = \varepsilon \circ z = g \circ z = g \circ \varepsilon \circ z = g \circ \varepsilon \circ x = g \circ x$.

Points 3. and 4. follow from 1. and 2. by considering the transposed hypergroup of $(H, \circ)$. □

Proposition 3. *Let $(H, \circ) \in \mathfrak{T}(H)$. For all $x, y \in H$ and $g, g' \in \omega_H$ we have*

1. $y \in g \circ x \Leftrightarrow \varepsilon \circ y = g \circ x$;
2. $g \circ x \cap g' \circ y \neq \emptyset \Leftrightarrow g \circ x = g' \circ y$;

3. if $y \in g \circ x$, then $\varepsilon \circ x \cap g \circ x = \emptyset \Leftrightarrow \varepsilon \circ y \cap g \circ y = \emptyset$;
4. $y \in x \circ g \Leftrightarrow y \circ \varepsilon = x \circ g$;
5. $x \circ g \cap y \circ g' \neq \emptyset \Leftrightarrow x \circ g = y \circ g'$;
6. if $y \in x \circ g$, then $x \circ \varepsilon \cap x \circ g = \emptyset \Leftrightarrow y \circ \varepsilon \cap y \circ g = \emptyset$.

Proof. 1. The implication $\Leftarrow$ is a consequence of the Theorem 3. Let $y \in g \circ x$. We have $g^{-1} \circ y \subseteq g^{-1} \circ g \circ x = \varepsilon \circ x$. Taking an element $z \in g^{-1} \circ y$, we obtain $\varepsilon \circ z \subseteq \varepsilon \circ x$. Therefore, for invertibility of subhypergroup $S = \{\varepsilon\}$ in $(H, \circ)$, $\varepsilon \circ z = \varepsilon \circ x$. Consequently, as $\varepsilon \circ z \subseteq g^{-1} \circ y \subseteq \varepsilon \circ x$, we deduce $g^{-1} \circ y = \varepsilon \circ x$ and so $\varepsilon \circ y = g \circ x$.

2. Let $z \in g \circ x \cap g' \circ y$. By point 1. of Proposition 3, we have $\varepsilon \circ z = g \circ x$ and $\varepsilon \circ z = g' \circ y$. Therefore, $g \circ x = g' \circ y$.

3. As $y \in g \circ x$, by point 1. of Proposition 3, we have $\varepsilon \circ y = g \circ x$ and so $\varepsilon \circ x = g^{-1} \circ (g \circ x) = g^{-1} \circ (\varepsilon \circ y) = g^{-1} \circ y$. Consequently, $\varepsilon \circ x \cap g \circ x = \emptyset \Leftrightarrow g^{-1} \circ y \cap \varepsilon \circ y = \emptyset \Leftrightarrow \varepsilon \circ y \cap g \circ y = \emptyset$.

Points 4., 5., and 6. follow from 1., 2., and 3. by considering the transposed hypergroup of $(H, \circ)$. □

4. Actions of ω_H

If $\phi : (g, e) \mapsto ge$ is a group action of G on the set E, the sets $O(e) = \{ge \mid g \in G\}$ and $Stab_G(e) = \{g \in G \mid ge = e\}$ are the orbit and the stabilizer of element $e \in E$, respectively. The orbits family $\{O(e)\}_{e \in E}$ is a partition of E and the stabilizer $Stab_G(e)$ is a subgroup of G. If e and e' belong to the same orbit the stabilizers are conjugates. Moreover, we have $|O(e)| = [G : Stab_G(e)]$ and when G is finite we obtain that $|O(e)|$ divides the size of G.

If $(H, \circ) \in \mathfrak{T}(H)$, we denote by $\mathfrak{L}$ and $\mathfrak{R}$ the following sets:

$$\mathfrak{L} = \{x \circ g \mid x \in H - \omega_H, g \in \omega_H\}, \quad \mathfrak{R} = \{g \circ x \mid x \in H - \omega_H, g \in \omega_H\}.$$

On $\mathfrak{L}$ and $\mathfrak{R}$ we consider the actions $\phi_l : \omega_H \times \mathfrak{L} \to \mathfrak{L}$ e $\phi_r : \omega_H \times \mathfrak{R} \to \mathfrak{R}$ such that

$$\phi_l(h, x \circ g) = x \circ (g \circ h) \quad \text{e} \quad \phi_r(h, g \circ x) = (h \circ g) \circ x,$$

for all $x \circ g \in \mathfrak{L}$, $g \circ x \in \mathfrak{R}$ and $h \in \omega_H$.

For simplicity, let $Stab_{\omega_H}(x \circ \varepsilon) = {}_xS$ and ${}_xO = O(x \circ \varepsilon) = \{x \circ g \mid g \in \omega_H\}$ be the stabilizer and the orbit of $x \in H - \omega_H$ with respect to the action ϕ_l, and let $Stab_{\omega_H}(x \circ \varepsilon) = {}_xS$ and ${}_xO = O(x \circ \varepsilon) = \{x \circ g \mid g \in \omega_H\}$ be those with respect to ϕ_r.

If $y \in \beta(x)$ then there exists $g \in \omega_H$ such that $y \in g \circ x$. By Proposition 3, we deduce $\varepsilon \circ y = g \circ x$. Conversely, again for the Proposition 3, if $g \circ x \in O_x$ and $y \in g \circ x$ we have $\varepsilon \circ y = g \circ x$ with $y \in \beta(x)$. Therefore, we obtain

$$O_x = \{g \circ x \mid g \in \omega_H\} = \{\varepsilon \circ y \mid y \in \beta(x)\}. \tag{3}$$

$${}_xO = \{x \circ g \mid g \in \omega_H\} = \{y \circ \varepsilon \mid y \in \beta(x)\}. \tag{4}$$

Next, we establish a connection between the sizes of O_x, ${}_xO$, ω_H, and $\beta(x)$. For brevity, we only expose results for the action ϕ_r. The corresponding results for the action ϕ_l follow trivially by recurring to transposed hypergroups.

Lemma 1. Let $(H, \circ) \in \mathfrak{T}(H)$ and $x \in H - \omega_H$.

1. $S_x = \{\varepsilon\}$ (resp., ${}_xS = \{\varepsilon\}$) if and only if $g \circ x \cap g' \circ x = \emptyset$ (resp., $x \circ g \cap x \circ g' = \emptyset$), for all $\{g, g'\} \subseteq \omega_H$ and $g \neq g'$;
2. if $S_x = \{\varepsilon\}$ (resp., ${}_xS = \{\varepsilon\}$) then $|O_x| = |\omega_H| \leq |\beta(x)|$ (resp., $|{}_xO| = |\omega_H| \leq |\beta(x)|$);
3. $S_x = \omega_H$ (resp., ${}_xS = \omega_H$) if and only if $\beta(x) = \varepsilon \circ x$ (resp., $\beta(x) = x \circ \varepsilon$).

Proof. 1. If $S_x = \{\varepsilon\}$ then $|O(x)| = [\omega_H : \{\varepsilon\}] = |\omega_H|$ and, by Proposition 3, $g \circ x \cap g' \circ x = \emptyset$, for all $\{g, g'\} \subseteq \omega_H$ and $g \neq g'$. Conversely, if $g \in S_x$ then $g \circ x = \varepsilon \circ x$ and $g = \varepsilon$ by hypothesis.

2. By point 1., we have $E_x = O_x$ and so $|E_x| = |O(x)| = [\omega_H : \{\varepsilon\}] = |\omega_H|$. Moreover, as the hyperproducts $g \circ x$ in $O(x)$ have size ≥ 1 and $g \circ x \subseteq \beta(x)$, for all $g \in \omega_H$, we deduce $|\omega_H| \leq |\beta(x)|$.

3. If $S_x = \omega_H$ then $g \circ x = \varepsilon \circ x$, for all $g \in \omega_H$. Therefore, $\beta(x) = \omega_H \circ x = \cup_{g \in \omega_H} g \circ x = \varepsilon \circ x$. Conversely, if $\beta(x) = \varepsilon \circ x$ then $\omega_H \circ x = \varepsilon \circ x$ and $g \circ x = g \circ (\varepsilon \circ x) = g \circ \omega_H \circ x = \omega_H \circ x = \varepsilon \circ x$, for all $g \in \omega_H$. Hence $S_x = \omega_H$. □

Proposition 4. *Let $(H, \circ) \in \mathfrak{T}(H)$ and let x be an element of $H - \omega_H$ such that $\varepsilon \circ y = \{y\}$ (resp., $y \circ \varepsilon = \{y\}$), for every $y \in \beta(x)$. Then $|\beta(x)| \leq |\omega_H|$ and equality holds if and only if $S_x = \{\varepsilon\}$ (resp., $_xS = \{\varepsilon\}$).*

Proof. By Proposition 1, we have $|g \circ y| = 1$, for every $g \in \omega_H$ and $y \in \beta(x)$. Therefore, $|\beta(x)| = |\omega_H \circ x| = |\cup_{g \in \omega_H} g \circ x| \leq |\omega_H|$. Now, if $S_x = \{\varepsilon\}$ then $|\beta(x)| = |\omega_H|$ by point 2. of Lemma 1. Conversely, if $|\beta(x)| = |\omega_H|$, then $g \circ x \cap g' \circ x = \emptyset$, for all $g, g' \in \omega_H$ and $g \neq g'$. Thus, by point 1. of Lemma 1, $S_x = \{\varepsilon\}$. □

A consequence of the previous proposition is the following result:

Theorem 6. *Let $(H, \circ) \in \mathfrak{T}(H)$ be such that ε is a left scalar identity and $S_x = \{\varepsilon\}$, for all $x \in H - \omega_H$ (resp., ε is a right scalar identity and $_xS = \{\varepsilon\}$, for all $x \in H - \omega_H$). Then $|\beta(x)| = |\omega_H|$, for all $x \in H$. Moreover, if $(H, \circ)$ is finite then $|H| = |H/\beta| \cdot |\omega_H|$.*

Now, if $(H, \circ) \in \mathfrak{T}(H)$ and $x, y \in H$, we denote by $L_x(y)$ and $_xL(y)$ the following sets: $L_x(y) = \{g \in \omega_H \mid g \circ x = \varepsilon \circ y\}$, $_xL(y) = \{g \in \omega_H \mid x \circ g = y \circ \varepsilon\}$. Clearly, we have $L_x(x) = S_x$ and $_xL(x) = {_xS}$.

Proposition 5. *If $(H, \circ) \in \mathfrak{T}(H)$ and $x, y \in H - \omega_H$ then the following conditions are equivalent:*
1. *$L_x(y) \neq \emptyset$ (resp., $_xL(y) \neq \emptyset$);*
2. *$\beta(x) = \beta(y)$;*
3. *$O_x = O_y$ (resp., $_xO = {_yO}$).*

Proof. 1. ⇔ 2. If $L_x(y) \neq \emptyset$ then there exists $g \in \omega_H$ such that $g \circ x = \varepsilon \circ y$, and so $\omega_H \circ x \cap \omega_H \circ y \neq \emptyset$. Thus $\beta(x) = \beta(y)$. On the other hand, if $\beta(x) = \beta(y)$ then $y \in \beta(x) = \omega_H \circ x$ and there exists $g \in \omega_H$ such that $y \in g \circ x$. By point 1. of Proposition 3, we have $g \circ x = \varepsilon \circ y$ and so $g \in L_x(y)$.

2. ⇔ 3. Let $\beta(x) = \beta(y)$. By (3), $\varepsilon \circ y \in O_x \cap O_y$ and $O_x = O_y$ since the orbits are a partition of $H - \omega_H$. Now, let $O_x = O_y$. There exist $g \circ x \in O_x$ and $h \circ y \in O_y$ such that $g \circ x = h \circ y$. Consequently, we have $\omega_H \circ x \cap \omega_H \circ y \neq \emptyset$ and $\beta(x) = \beta(y)$. □

Proposition 6. *Let $(H, \circ) \in \mathfrak{T}(H)$ and let $x, y \in H - \omega_H$ such that $\beta(x) = \beta(y)$. Then, we have*
1. *$|L_x(y)| = |S_x|$ (resp., $|_xL(y)| = |_xS|$);*
2. *the subgroups S_x and S_y (resp., $_xS$ and $_yS$) are conjugates;*
3. *if S_x or S_y (resp., $_xS$ or $_yS$) is a normal subgroup or ω_H is abelian, then $S_x = S_y$ (resp., $_xS = {_yS}$);*
4. *$|L_x(y)| = |S_x| = |S_y| = |L_y(x)|$;*
5. *$|_xL(y)| = |_xS| = |_yS| = |_yL(x)|$.*

Proof. 1. By Proposition 5, the sets $L_x(y)$ and $L_y(x)$ are not empty as $\beta(x) = \beta(y)$. Fixed an element $h \in L_y(x)$, we have $h \circ y = \varepsilon \circ x = g \circ x$ for all $g \in S_x$. Therefore, $(h^{-1} \circ g) \circ x = \varepsilon \circ y$ and $h^{-1} \circ g \in L_x(y)$. Clearly, the application $\varphi_{h^{-1}} : S_x \to L_x(y)$ such that $\varphi_{h^{-1}}(g) = h^{-1} \circ g$, for all $g \in S_x$, is injective and so $|S_x| \leq |L_x(y)|$. On the other hand, as $h \circ y = \varepsilon \circ x$, we obtain $\varepsilon \circ y = h^{-1} \circ x$. Therefore, $g \in L_x(y) \Rightarrow g \circ x = \varepsilon \circ y =$

$h^{-1} \circ x \Rightarrow h \circ g \circ x = \varepsilon \circ x \Rightarrow h \circ g \in S_x$. Finally, the application $\phi_h : L_x(y) \to S_x$ such that $\phi_h(g) = h \circ g$, for all $g \in L_x(y)$, is injective and so $|L_x(y)| \leq |S_x(\omega_H)|$.

2. By Proposition 5, we have $O_x = O_y$. Thus, the elements $\varepsilon \circ x, \varepsilon \circ y$ of $\mathfrak{R}$ belong to the same orbit. Consequently, the stabilizers S_x and S_y are conjugates.

Point 3. is an immediate consequence of 2., and points 4. and 5. follow from 1. and 2. because conjugated subgroups have the same size. □

An immediate consequence of point 3. in Proposition 6 is the following result:

Corollary 2. *Let $(H, \circ) \in \mathfrak{T}(H)$ and let x, y be elements of $H - \omega_H$ such that $\beta(x) = \beta(y)$. Then,*
1. $S_x = \{\varepsilon\} \Leftrightarrow S_y = \{\varepsilon\}$ *(resp.,* $_xS = \{\varepsilon\} \Leftrightarrow {_yS} = \{\varepsilon\}$*);*
2. $S_x = \omega_H \Leftrightarrow S_y = \omega_H$ *(resp.,* $_xS = \omega_H \Leftrightarrow {_yS} = \omega_H$*).*

5. Properties of the Hyperproducts $x \circ y \subseteq \omega_H$ with $x, y \in H - \omega_H$

In this section, we prove certain properties of products of elements $x, y \in H - \omega_H$ such that $x \circ y \subseteq \omega_H$. These properties will be utilized in the next section in the construction of G-hypergroups of small size. Note that $x \circ y \cap \omega_H \neq \emptyset \implies x \circ y \subseteq \omega_H$, for all $x, y \in H$ as ω_H is a complete part of H by point 4. of Theorem 1.

Proposition 7. *Let $(H, \circ) \in \mathfrak{T}(H)$ and let $x, y \in H - \omega_H$ such that $x \circ y \cap \omega_H \neq \emptyset$. If $S_x = \omega_H$ and $S_y \in \{\{\varepsilon\}, \omega_H\}$ (alternatively, if $_yS = \omega_H$ and $_xS \in \{\{\varepsilon\}, \omega_H\}$) then $x \circ y = y \circ x = \omega_H$.*

Proof. Let $S_x = \omega_H$ and $x \circ y \cap \omega_H \neq \emptyset$. By Lemma 1 we have $\beta(x) = \omega_H \circ x = \varepsilon \circ x$. Thus,
$$x \circ y = \varepsilon \circ (x \circ y) = (\varepsilon \circ x) \circ y = (\omega_H \circ x) \circ y = \omega_H \circ (x \circ y) = \omega_H.$$

Moreover, we have $y \circ x \subseteq \omega_H$ because ω_H is a reflexive subhypergroup of $(H, \circ)$. Now, by hypothesis, two cases are possible: $S_y = \omega_H$ or $S_y = \{\varepsilon\}$. If $S_y = \omega_H$ then $y \circ x = \omega_H$ follows by transposing the previous arguments, and the claim follows. On the other hand, if $S_y = \{\varepsilon\}$ then, by Lemma 1, we have $g \circ y \cap g' \circ y = \emptyset$ for all $\{g, g'\} \subseteq \omega_H$ and $g \neq g'$. Consequently, if by absurd we suppose that $\omega_H \neq y \circ x$ then we deduce the contradiction
$$\beta(y) = y \circ \omega_H = y \circ (x \circ y) = (y \circ x) \circ y = \bigcup_{g \in y \circ x} g \circ y \neq \bigcup_{t \in \omega_H} t \circ y = \omega_H \circ y = \beta(y).$$

Therefore, also in this case $y \circ x = \omega_H$ and $x \circ y = y \circ x = \omega_H$. When $_yS = \omega_H$ and $_xS \in \{\{\varepsilon\}, \omega_H\}$ the claim follows by transposition. □

Remark 2. *If the heart ω_H of a hypergroup $(H, \circ) \in \mathfrak{T}(H)$ is isomorphic to a group of size a prime number p then $S_x \in \{\{\varepsilon\}, \omega_H\}$, for every $x \in H - \omega_H$. In this case, if $x, y \in H - \omega_H$, $x \circ y \cap \omega_H \neq \emptyset$ and at least one of the subgroups S_x, S_y is different from $\{\varepsilon\}$, then $x \circ y = y \circ x = \omega_H$. This fact is not true if $S_x = S_y = \{\varepsilon\}$. For example, consider the hypergroup represented by the following table:*

∘	ε	b	c	d	e	f
ε	ε	b	c	d	e	f
b	b	ε	d	c	e	f
c	c	d	ε	b	f	e
d	d	c	b	ε	f	e
e	e	e	f	f	ε,b	c,d
f	f	f	e	e	c,d	ε,b

Here, $\omega_H = \{\varepsilon, b\} \cong \mathbb{Z}_2$, $S_c = S_d = \{\varepsilon\}$, and $c \circ d = d \circ c = \{b\} \neq \omega_H$. Recall that the 1-hypergroups are a special class of G-hypergroups and their sub-hypergroups are conjugable. The same property is not true if the heart of a G-hypergroup is not trivial. Indeed, if ε is the identity of the heart then $\{\varepsilon\}$ is a non-conjugable sub-hypergroup of H. On the other hand, there are G-hypergroups that have non-trivial non-conjugable sub-hypergroups. For instance, the hypergroup of the previous table has five non-trivial sub-hypergroups different from H and ω_H, that is, $G_1 = \{\varepsilon, c\}$, $G_2 = \{\varepsilon, d\}$, $G_3 = \{\varepsilon, b, c, d\}$, $K_1 = \{\varepsilon, b, e\}$, $K_2 = \{\varepsilon, b, f\}$. Note that G_1 and G_2 are isomorphic to $\mathbb{Z}_2$ and are not conjugable.

Proposition 8. *Let $(H, \circ) \in \mathfrak{T}(H)$ and let $x, y \in H - \omega_H$ such that $\varepsilon \in x \circ y$. Then, $S_x \cup {}_yS \subseteq x \circ y$.*

Proof. We have $x \circ y \subseteq \omega_H$ as $\varepsilon \in x \circ y$. If $g \in S_x$ then $g \circ x = \varepsilon \circ x$ and so $g \in g \circ \varepsilon \subseteq g \circ (x \circ y) = (g \circ x) \circ y = (\varepsilon \circ x) \circ y = \varepsilon \circ (x \circ y) = x \circ y$. Therefore, $S_x \subseteq x \circ y$. In the same way, we prove that ${}_yS \subseteq x \circ y$. □

An immediate consequence of Propositions 7 and 8 is the following:

Corollary 3. *Let $(H, \circ) \in \mathfrak{T}(H)$ and let $x, y \in H - \omega_H$ such that $\varepsilon \in x \circ y$. If ω_H is isomorphic to a group of size a prime number and $S_x \neq \{\varepsilon\}$ or ${}_yS \neq \{\varepsilon\}$, then $x \circ y = \omega_H$.*

Proposition 9. *Let $(H, \circ) \in \mathfrak{T}(H)$ and let $x, y \in H - \omega_H$ such that $x \circ y \subseteq \omega_H$. Then, $|a \circ b| = |x \circ y|$, for all $a \in \beta(x)$ and $b \in \beta(y)$.*

Proof. By hypothesis $x \circ y \subseteq \omega_H$. Moreover, as $a \in \beta(x) = \omega_H \circ x$ and $b \in \beta(y) = y \circ \omega_H$, there exist $h, k \in \omega_H$ such that $a \in h \circ x$ and $b \in y \circ k$. By Proposition 3, we have $\varepsilon \circ a = h \circ x$ and $b \circ \varepsilon = y \circ k$. As $a \circ b \subseteq \beta(x) \circ \beta(y) = x \circ \omega_H \circ y \circ \omega_H = x \circ y \circ \omega_H = \omega_H$, we have $a \circ b = \varepsilon \circ a \circ b \circ \varepsilon = h \circ x \circ y \circ k$. Finally, the application $f : x \circ y \to a \circ b$ such that $f(g) = h \circ g \circ k$, for all $g \in x \circ y$, is bijective and so $|x \circ y| = |a \circ b|$. □

Lemma 2. *Let $(H, \circ) \in \mathfrak{T}(H)$ and let P be a normal subgroup of ω_H. Moreover, let $a, b \in H - \omega_H$ and $h \in \omega_H$. Then, we have*

1. *if $a \circ b = h \circ P$, then for all $z \in \beta(b)$ there exists $z' \in \beta(a)$ such that $z' \circ z \subseteq P$;*
2. *if $|H/\beta| = 2$ and $P \neq \omega_H$ then $a \circ b \neq h \circ P$.*

Proof. 1. Let $a \circ b = h \circ P$. If $z \in \beta(b) = b \circ \omega_H$, there exists $k \in \omega_H$ such that $z \in b \circ k$. Now, taken $z' \in k^{-1} \circ h^{-1} \circ a$, we have $z' \in \omega_H \circ a = \beta(a)$. Moreover, as P is a normal subgroup and $a \circ b = h \circ P$, we deduce $z' \circ z \subseteq k^{-1} \circ h^{-1} \circ a \circ b \circ k \subseteq k^{-1} \circ h^{-1} \circ h \circ P \circ k = P$.

2. By absurdity, let $a \circ b = h \circ P$. As $|H/\beta| = 2$ and $a, b \in H - \omega_H$, we have $\beta(a) = \beta(b) = H - \omega_H$. Now, let $z \in H$. Clearly, if $z \in \omega_H$ then $z^{-1} \circ z = \{\varepsilon\} \subseteq P$. If $z \in H - \omega_H$, we have $z \in \beta(a)$ and, by point 1., there exists $z' \in \beta(b)$ such that $z' \circ z \subseteq P$. Hence, P is a conjugable subhypergroup of $(H, \circ)$ and we have $\omega_H \subseteq P \subseteq \omega_H$; impossible as $P \neq \omega_H$. □

Proposition 10. *Let $(H, \circ) \in \mathfrak{T}(H)$ such that $|H/\beta| = 2$ and $|\omega_H| \geq 2$. We have*

1. *$|a \circ b| \geq 2$, for all $a, b \in H - \omega_H$;*
2. *if there exists $x \in H - \omega_H$ such that $S_x = \omega_H$ or ${}_xS = \omega_H$, then $a \circ b = \omega_H$, for all $a, b \in H - \omega_H$;*
3. *if $|\omega_H| = 2$ then $a \circ b = \omega_H$, for all $a, b \in H - \omega_H$.*

Proof. 1. By hypothesis $S = \{\varepsilon\}$ is a proper normal subgroup of ω_H and $a \circ b \subseteq \omega_H$, for all $a, b \in H - \omega_H$. If there exist $a, b \in H - \omega_H$ such that $|a \circ b| = 1$, we can suppose that $a \circ b = \{h\}$, with $h \in \omega_H$. Therefore, we have $a \circ b = \{h\} = h \circ S$, that is impossible by point 2. of Lemma 2.

2. Let $x \in H - \omega_H$ and $S_x = \omega_H$. For reproducibility, there exists $y \in H - \omega_H$ such that $\varepsilon \in x \circ y$. By Proposition 8, we have $x \circ y = \omega_H$. Consequently, from Proposition 9, we deduce $a \circ b = \omega_H$, for all $a, b \in H - \omega_H$. We get the same result if ${}_xS = \omega_H$.

3. is an immediate consequence of 1. □

Corollary 4. Let $(H, \circ) \in \mathfrak{T}(H)$ and let x be an element of $H - \omega_H$ such that $|\beta(x)| < |\omega_H|$. If $(H, \circ)$ is finite and ω_H is a group of size a prime number p then $y \circ g = g \circ y = \beta(x)$, for all $y \in \beta(x)$ and $g \in \omega_H$. Moreover, if $|H/\beta| = 2$ then $x \circ y = y \circ x = \omega_H$, for all $x, y \in H - \omega_H$.

Proof. Since $|\omega_H|$ is a prime number p and $|\beta(x)| < |\omega_H|$, by Lemma 1, $S_x = {}_xS = \omega_H$ and $\varepsilon \circ x = x \circ \varepsilon = \beta(x)$. Now, if $y \in \beta(x)$, $g \in \omega_H$ and $a \in g \circ y$ (resp., $a \in y \circ g$), by Proposition 3, we have $\varepsilon \circ a = g \circ y = \beta(y) = \beta(x)$ (resp., $a \circ \varepsilon = y \circ g = \beta(y) = \beta(x)$). Furthermore, by Proposition 10, if $|H/\beta| = 2$ and $x, y \in H - \omega_H$ then $x \circ y = y \circ x = \omega_H$. □

Example 1. In the next table we show a hypergroup $(H, \circ) \in \mathfrak{T}(H)$ such that $|H/\beta| = 2$, $\omega_H \cong \mathbb{Z}_3$, $|H| = |\omega_H| \cdot |H/\beta|$ and all hyperproducts $a \circ b$ have size 2, for all $a, b \in H - \omega_H$.

∘	ε	b	c	d	e	f
ε	ε	b	c	d	e	f
b	b	c	ε	f	d	e
c	c	ε	b	e	f	d
d	d	f	e	ε,b	ε,c	b,c
e	e	d	f	ε,c	b,c	ε,b
f	f	e	d	b,c	ε,b	ε,c

According to Proposition 10, necessarily we have here $S_x \neq \omega_H$ and ${}_xS \neq \omega_H$, for all $x \in H - \omega_H$.

In the previous example, each element of the heart is contained in exactly six hyperproducts $x \circ y \subset \omega_H$. This fact finds full justification in the next proposition. A new notation is entered: For all $x, y \in H$ such that $x \circ y \subseteq \omega_H$ and $g \in \omega_H$, let $N_g^{x,y} = \{(a,b) \in \beta(x) \times \beta(y) \mid g \in a \circ b\}$. Clearly, $N_g^{x,y} \neq \emptyset$ as $\beta(x) \circ \beta(y) = x \circ \omega_H \circ y \circ \omega_H = x \circ y \circ \omega_H \circ \omega_H = \omega_H$.

Proposition 11. Let $(H, \circ) \in \mathfrak{T}(H)$, $x, y \in H - \omega_H$ and $x \circ y \subseteq \omega_H$. If $\varepsilon \circ a = \{a\}$, for all $a \in \beta(x)$ (resp., $b \circ \varepsilon = \{b\}$, for all $b \in \beta(y)$), then $|N_g^{x,y}|$ is the same for all $g \in \omega_H$.

Proof. Let $(a, b) \in N_g^{x,y}$. There exists $h \in \omega_H$ such that $\{g'\} = h \circ g$. Clearly, $h \circ a \subseteq \omega_H \circ a = \beta(a) = \beta(x)$, with $|h \circ a| = 1$ by Proposition 1. Moreover, if $h \circ a = \{a'\}$ then $\{g'\} = h \circ g \subseteq h \circ (a \circ b) = (h \circ a) \circ b = a' \circ b$ and so $(a', b) \in N_{g'}^{x,y}$. Finally, the application $\varphi_h : N_g^{x,y} \to N_{g'}^{x,y}$ such that $\varphi_h(a, b) = (a', b)$, with $h \circ a = \{a'\}$, is injective because $h \circ a_1 = h \circ a_2 \Leftrightarrow a_1 = a_2$, for all $a_1, a_2 \in \beta(x)$. Therefore, $|N_g^{x,y}| \leq |N_{g'}^{x,y}|$. Similarly, we have $|N_{g'}^{x,y}| \leq |N_g^{x,y}|$. □

Corollary 5. Let $(H, \circ)$ be a finite hypergroup in $\mathfrak{T}(H)$, and let $x, y \in H - \omega_H$ such that $x \circ y \subseteq \omega_H$. If $\varepsilon \circ a = \{a\}$, for all $a \in \beta(x)$ (resp., $b \circ \varepsilon = \{b\}$, for all $b \in \beta(y)$), then $|a \circ b| \cdot |\beta(x)| \cdot |\beta(y)| = |\omega_H| \cdot |N_g^{x,y}|$ for all $a \in \beta(x)$, $b \in \beta(y)$ and $g \in \omega_H$. In particular, if $|\omega_H|$ is a prime number then $a \circ b = \omega_H$ or $|\omega_H|$ divides $|\beta(x)|$ or $|\beta(y)|$.

Proof. Let $n^{x,y} = |N_g^{x,y}|$. By Proposition 9, $|a \circ b| = |x \circ y|$ for all $a \in \beta(x)$ and $b \in \beta(y)$. Thus, taking $a \in \beta(x)$ and $b \in \beta(y)$, by Proposition 11 and $\beta(x) \circ \beta(y) = \omega_H$, we obtain $|a \circ b| \cdot |\beta(x)| \cdot |\beta(y)| = |\omega_H| \cdot n^{x,y}$ counting in two different ways. Finally, as $a \circ b \subseteq \omega_H$, if $|\omega_H|$ is a prime number then $a \circ b = \omega_H$ or $|\omega_H|$ divides $|\beta(x)|$ or $|\beta(y)|$. □

In Proposition 11, the hypothesis $\varepsilon \circ a = \{a\}$ for all $a \in \beta(x)$ is essential. Indeed, consider the following hypergroup:

∘	ε	b	c	d	e	f	g	h
ε	ε	b	c	d,e	d,e	f,g	f,g	h
b	b	c	ε	f,g	f,g	h	h	d,e
c	c	ε	b	h	h	d,e	d,e	f,g
d	d,e	f,g	h	ε,b	ε,b	b,c	b,c	ε,c
e	d,e	f,g	h	ε,b	ε,b	b,c	b,c	ε,c
f	f,g	h	d,e	b,c	b,c	ε,c	ε,c	ε,b
g	f,g	h	d,e	b,c	b,c	ε,c	ε,c	ε,b
h	h	d,e	f,g	ε,c	ε,c	ε,b	ε,b	b,c

Here, we have $\omega_H \cong \mathbb{Z}_3$, $|H/\beta| = 2$, $|N_\varepsilon^{d,e}| = 16$ and $|N_b^{d,e}| = |N_c^{d,e}| = 17$. In this example, Proposition 11 cannot be applied because $\varepsilon \circ d \neq \{d\}$.

6. Hypergroups of Type $\mathcal{U}$ in $\mathfrak{T}(H)$

Among the best-known classes of hypergroups are undoubtedly those of type $\mathcal{U}$, type C, and the cogroups. A hypergroup of type $\mathcal{U}$ on the right is a hypergroup $(H, \circ)$ with a right scalar identity ε that fulfills the condition $a \in a \circ b \Rightarrow b = \varepsilon$, for all $a, b \in H$, see [25–27]. A hypergroup of type C on the right is a hypergroup $(H, \circ)$ of type $\mathcal{U}$ on the right that fulfills the condition $a \circ b \cap a \circ c \neq \emptyset \Rightarrow \varepsilon \circ b = \varepsilon \circ c$, for all $a, b, c \in H$, see [5,6]. A cogroup on the right is a hypergroup of type C on the right such that $|a \circ c| = |b \circ c|$ for all $a, b, c \in H$, see in [2–4]. The transposed of a hypergroup of type $\mathcal{U}$ on the right is a hypergroup of type type $\mathcal{U}$ on the left, and analogously for hypergroup of type C and cogroups. The purpose of this subsection is to characterize the hypergroups in $\mathfrak{T}(H)$ that are of type $\mathcal{U}$ on the right or cogroups on the right. We have the following result:

Theorem 7. Let $(H, \circ) \in \mathfrak{T}(H)$. Then, $(H, \circ)$ is of type $\mathcal{U}$ on the right if and only if $_xS = \{\varepsilon\}$ and $x \circ \varepsilon = \{x\}$, for all $x \in H - \omega_H$.

Proof. If $(H, \circ)$ is of type $\mathcal{U}$ on the right, $x \in H - \omega_H$, and $g \in {}_xS$ then $x \circ g = x \circ \varepsilon = \{x\}$ and so we have $g = \varepsilon$. Conversely, let $_xS = \{\varepsilon\}$ and $x \circ \varepsilon = \{x\}$, for all $x \in H - \omega_H$. If a, u are elements of H such that $a \in a \circ u$ then $u \in \omega_H$. Indeed, if $\varphi : H \to H/\beta$ is the canonical projection then $\varphi(a) = \varphi(a) \otimes \varphi(u)$ and $\varphi(u) = 1_{H/\beta}$. Clearly, if $a \in \omega_H$ then $u = \varepsilon$ because $a \in a \circ u$ and ω_H is isomorphic to a group. If $a \in H - \omega_H$ then, using Proposition 2, we have $a \circ \varepsilon = a \circ u$ and $u \in {}_aS = \{\varepsilon\}$. Thus, $u = \varepsilon$ and so $(H, \circ)$ is of type $\mathcal{U}$ on the right. □

We note that if $(H, \circ) \in \mathfrak{T}(H)$ is a 1-hypergroup of type $\mathcal{U}$ on the right then $\omega_H = \{\varepsilon\}$ and $H/\{\varepsilon\} \cong H$ as ε is a right scalar identity. In this case H is isomorphic to a group. Consequently, we have the following result.

Corollary 6. A hypergroup $(H, \circ) \in \mathfrak{T}(H)$ is isomorphic to a group if and only if $(H, \circ)$ is a 1-hypergroup of type $\mathcal{U}$ on the right.

In reference to Theorem 7 and the previous corollary, we note that the hypergroup shown in Example 1 is of type $\mathcal{U}$ both on the right and on the left. Indeed, in that hypergroup we have $|\omega_H| \geq 2$, $_xS = S_x = \{\varepsilon\}$ and $x \circ \varepsilon = \varepsilon \circ x = \{x\}$, for all $x \in H - \omega_H$. The next result provides a sufficient condition for a hypergroup of type $\mathcal{U}$ on the right to be also a cogroup.

Theorem 8. Let $(H, \circ) \in \mathfrak{T}(H)$ be of type $\mathcal{U}$ on the right. If $S_x = \omega_H$ for all $x \in H - \omega_H$ then $(H, \circ)$ is a cogroup.

Proof. The thesis is obvious if $(H, \circ)$ is a group. Therefore, we suppose that $|\omega_H| \geq 2$. Let $S_x = \omega_H$, for all $x \in H - \omega_H$. If $a \circ b \cap a \circ c \neq \emptyset$, we obtain $\varphi(b) = \varphi(c)$ and so $\beta(b) = \beta(c)$. Hence, $b \in \omega_H$ if and only if $c \in \omega_H$. Now, if $b \in H - \omega_H$, by point 3. of

Lemma 1, then $\varepsilon \circ b = \beta(b) = \beta(c) = \varepsilon \circ c$. If $b \in \omega_H$ and $a \in H - \omega_H$, by point 5. of Proposition 3, we obtain $a \circ b = a \circ c$ and so $b \circ c^{-1} \in {}_aS = \{\varepsilon\}$ as $(H, \circ)$ is of type U on the right. Therefore, $b = c$ and $\varepsilon \circ b = \varepsilon \circ c$. We get the same result if we suppose that $a, b \in \omega_H$. Thus, $(H, \circ)$ is of type C on the right. Now, we distinguish two cases to prove that $|b \circ a| = |c \circ a|$, for all $a, b, c \in H$. We note that, by Theorem 6, we have that $|\beta(x)| = |\omega_H|$, for all $x \in H$.

If $a \in \omega_H$ then we have $|b \circ a| = |c \circ a| = 1$ by Corollary 1. On the other hand, if $a \in H - \omega_H$ and $x \in H$ then, from point 3. of Lemma 1, we have $\varepsilon \circ a = \beta(a) = a \circ \omega_H$ and so

$$x \circ a = (x \circ \varepsilon) \circ a = x \circ (\varepsilon \circ a)$$
$$= x \circ (a \circ \omega_H)$$
$$= (x \circ a) \circ \omega_H = \bigcup_{y \in x \circ a} y \circ \omega_H = \bigcup_{y \in x \circ a} \beta(y).$$

Finally, as $\beta(y) = \beta(z)$, for all $y, z \in x \circ a$, we obtain that $x \circ a = \beta(y)$, for all $y \in x \circ a$. Therefore, $|x \circ a| = |\beta(y)| = |\omega_H|$. Thus, if $a \in H - \omega_H$ and $b, c \in H$ then $|b \circ a| = |\omega_H| = |c \circ a|$ and the proof is over. □

The hypothesis $S_x = \omega_H$ in Theorem 7 is sufficient but not necessary for a hypergroup of type U on the right to be a cogroup. Indeed, the following hypergroup is a cogroup on the right in $\mathfrak{T}(H)$ but $S_x \neq \omega_H$, for all $x \in H - \omega_H$.

∘	ε	b	c	d	e	f	g	h	i	l	m	n
ε	ε	b	c	d	e,g	f,h	e,g	f,h	i,m	l,n	i,m	l,n
b	b	c	d	ε	f,h	e,g	f,h	e,g	l,n	i,m	l,n	i,m
c	c	d	ε	b	e,g	f,h	e,g	f,h	i,m	l,n	i,m	l,n
d	d	ε	b	c	f,h	e,g	f,h	e,g	l,n	i,m	l,n	i,m
e	e	f	g	h	i,m	l,n	i,m	l,n	ε,c	b,d	ε,c	b,d
f	f	g	h	e	l,n	i,m	l,n	i,m	b,d	ε,c	b,d	ε,c
g	g	h	e	f	i,m	l,n	i,m	l,n	ε,c	b,d	ε,c	b,d
h	h	e	f	g	l,n	i,m	l,n	i,m	b,d	ε,c	b,d	ε,c
i	i	l	m	n	ε,c	b,d	ε,c	b,d	e,g	f,h	e,g	f,h
l	l	m	n	i	b,d	ε,c	b,d	ε,c	f,h	e,g	f,h	e,g
m	m	n	i	l	ε,c	b,d	ε,c	b,d	e,g	f,h	e,g	f,h
n	n	i	l	m	b,d	ε,c	b,d	ε,c	f,h	e,g	f,h	e,g

In this case the heart $\omega_H = \{\varepsilon, b, c, d\}$ is isomorphic to $\mathbb{Z}_4$ and $S_x = \{\varepsilon, c\}$ for all $x \in H - \omega_H$.

7. G-Hypergroups of Minimal Size

In [16] the authors classified the 1-hypergroups of size ≤ 6. Hereafter, we classify the G-hypergroups of size ≤ 5 and $|G| \geq 2$, apart of isomorphisms. Recall that ε denotes the identity of G. Furthermore, let $\mathfrak{T}(G, p, q)$ be the subclass of $\mathfrak{T}(H)$ such that $\omega_H = G$, $|H| = p$ and $|H/\beta| = q$. With these notations, using the results in Section 3, we classify the hypergroups of the subclasses $\mathfrak{T}(\mathbb{Z}_n, p, q)$ with $2 \leq n \leq 4$, $3 \leq p \leq 5$, $2 \leq q \leq 4$ and $\mathfrak{T}(\mathbb{Z}_2 \times \mathbb{Z}_2, 5, 2)$.

Apart of isomorphisms, the classes $\mathfrak{T}(\mathbb{Z}_2,3,2)$, $\mathfrak{T}(\mathbb{Z}_2,4,3)$, $\mathfrak{T}(\mathbb{Z}_3,4,2)$, $\mathfrak{T}(\mathbb{Z}_3,5,2)$, $\mathfrak{T}(\mathbb{Z}_3,5,3)$, $\mathfrak{T}(\mathbb{Z}_4,5,2)$, and $\mathfrak{T}(\mathbb{Z}_2 \times \mathbb{Z}_2,5,2)$ consist of only one hypergroup. We list their tables respecting the order in which the previous classes are written.

H_1:

∘	ε	b	c
ε	ε	b	c
b	b	ε	c
c	c	c	ε,b

H_2:

∘	ε	b	c	d
ε	ε	b	c	d
b	b	ε	c	d
c	c	c	d	ε,b
d	d	d	ε,b	c

H_3:

∘	ε	b	c	d
ε	ε	b	c	d
b	b	c	ε	d
c	c	ε	b	d
d	d	d	d	ε,b,c

H_4:

∘	ε	b	c	d	e
ε	ε	b	c	d	e
b	b	c	ε	d,e	d,e
c	c	ε	b	d,e	d,e
d	d	d,e	d,e	d,e	ε,b,c	ε,b,c
e	e	d,e	d,e	d,e	ε,b,c	ε,b,c

H_5:

∘	ε	b	c	d	e
ε	ε	b	c	d	e
b	b	c	ε	d	e
c	c	ε	b	d	e
d	d	d	d	e	ε,b,c
e	e	e	e	ε,b,c	d

H_6:

∘	ε	b	c	d	e
ε	ε	b	c	d	e
b	b	c	d	ε	e
c	c	d	ε	b	e
d	d	ε	b	c	e
e	e	e	e	e	ε,b,c,d

H_7:

∘	ε	b	c	d	e
ε	ε	b	c	d	e
b	b	ε	d	c	e
c	c	d	ε	b	e
d	d	c	b	ε	e
e	e	e	e	e	ε,b,c,d

We note that the table of hypergroup in $\mathfrak{T}(\mathbb{Z}_3,5,2)$ is a consequence of Corollary 4. The other tables are deduced by considering the quotient group H/β.

Class: $\mathfrak{T}(\mathbb{Z}_2,4,2)$. Using the Propositions 6 and 10, we have the following four hypergroups, apart of isomorphisms:

H_8:

∘	ε	b	c	d
ε	ε	b	c	d
b	b	ε	d	c
c	c	d	ε,b	ε,b
d	d	c	ε,b	ε,b

H_9:

∘	ε	b	c	d
ε	ε	b	c	d
b	b	ε	d	c
c	c,d	c,d	ε,b	ε,b
d	c,d	c,d	ε,b	ε,b

H_{10}:

∘	ε	b	c	d
ε	ε	b	c,d	c,d
b	b	ε	c,d	c,d
c	c	d	ε,b	ε,b
d	d	c	ε,b	ε,b

H_{11}:

∘	ε	b	c	d
ε	ε	b	c,d	c,d
b	b	ε	c,d	c,d
c	c,d	c,d	ε,b	ε,b
d	c,d	c,d	ε,b	ε,b

According to Theorems 7 and 8, and Corollary 6, H_8 is a hypergroups of type U on the right and on the left, H_9 is a cogroup on the left and H_{10} is a cogroup on the right. We note that if $(G,\cdot)$ is a group and S is a non-normal subgroup of G then the quotient G/S (resp. $S\backslash G$) is a hypergroup with hyperproduct $xh \otimes yh = \{zh \mid z \in xhyh\}$ (resp. $hx \otimes hy = \{hz \mid z \in hxhy\}$). These hypergroups are called D-hypergroups [3]. The hypergroups H_9 and H_{10} are isomorphic to $S\backslash D_4$ and D_4/S respectively, being D_4 is the dihedral group of size 8 and S is a non-normal subgroup of size 2. Moreover, H_{10} can be obtained from the construction shown in Section 3.1 with $G = T \cong \mathbb{Z}_2$.

Class: $\mathfrak{T}(\mathbb{Z}_2,5,2)$. The element ε is not a left scalar identity (resp., right scalar identity) otherwise, by Proposition 1, we have $|g \circ y| = 1$, for all $g \in \omega_H$ and $y \in H - \omega_H$. Consequently, as $|\omega_H| = 2$, if $y \in H - \omega_H$ then we have the contradiction $3 = |H - \omega_H| = |\beta(y)| = |\omega_H \circ y| = 2$. Furthermore, in this case, using the Propositions 6 and 10, we obtain the following four hypergroups, apart of isomorphisms:

$H_{12}:$

∘	ε	b	c	d	e
ε	ε	b	c	d,e	d,e
b	b	ε	d,e	c	c
c	c	d,e	ε,b	ε,b	ε,b
d	d,e	c	ε,b	ε,b	ε,b
e	d,e	c	ε,b	ε,b	ε,b

$H_{13}:$

∘	ε	b	c	d	e
ε	ε	b	c	d,e	d,e
b	b	ε	d,e	c	c
c	c,d,e	c,d,e	ε,b	ε,b	ε,b
d	c,d,e	c,d,e	ε,b	ε,b	ε,b
e	c,d,e	c,d,e	ε,b	ε,b	ε,b

$H_{14}:$

∘	ε	b	c	d	e
ε	ε	b	c,d,e	c,d,e	c,d,e
b	b	ε	c,d,e	c,d,e	c,d,e
c	c	d,e	ε,b	ε,b	ε,b
d	d,e	c	ε,b	ε,b	ε,b
e	d,e	c	ε,b	ε,b	ε,b

$H_{15}:$

∘	ε	b	c	d	e
ε	ε	b	c,d,e	c,d,e	c,d,e
b	b	ε	c,d,e	c,d,e	c,d,e
c	c,d,e	c,d,e	ε,b	ε,b	ε,b
d	c,d,e	c,d,e	ε,b	ε,b	ε,b
e	c,d,e	c,d,e	ε,b	ε,b	ε,b

Class: $\mathfrak{T}(\mathbb{Z}_2, 5, 3)$. If the β-classes are $\omega_H = \{\varepsilon, b\}$, $\beta(c) = \{c, d\}$ and $\beta(e) = \{e\}$, the quotient group H/β returns the partial table:

∘	ε	b	c	d	e
ε	ε	b			e
b	b	ε			e
c			e	e	ε,b
d			e	e	ε,b
e	e	e	ε,b	ε,b	c,d

By Propositions 2 and 6, we obtain the following four tables, apart of isomorphisms:

$H_{16}:$

∘	ε	b	c	d	e
ε	ε	b	c	d	e
b	b	ε	d	c	e
c	c	d	e	e	ε,b
d	d	c	e	e	ε,b
e	e	e	ε,b	ε,b	c,d

$H_{17}:$

∘	ε	b	c	d	e
ε	ε	b	c	d	e
b	b	ε	d	c	e
c	c,d	c,d	e	e	ε,b
d	c,d	c,d	e	e	ε,b
e	e	e	ε,b	ε,b	c,d

$H_{18}:$

∘	ε	b	c	d	e
ε	ε	b	c,d	c,d	e
b	b	ε	c,d	c,d	e
c	c	d	e	e	ε,b
d	d	c	e	e	ε,b
e	e	e	ε,b	ε,b	c,d

$H_{19}:$

∘	ε	b	c	d	e
ε	ε	b	c,d	c,d	e
b	b	ε	c,d	c,d	e
c	c,d	c,d	e	e	ε,b
d	c,d	c,d	e	e	ε,b
e	e	e	ε,b	ε,b	c,d

Class: $\mathfrak{T}(\mathbb{Z}_2, 5, 4)$. Apart of isomorphisms, we obtain two hypergroups according to that H/β is isomorphic to $\mathbb{Z}_4$ or $\mathbb{Z}_2 \times \mathbb{Z}_2$.

$H_{20}:$

∘	ε	b	c	d	e
ε	ε	b	c	d	e
b	b	ε	c	d	e
c	c	c	d	e	ε,b
d	d	d	e	ε,b	c
e	e	e	ε,b	c	d

$H_{21}:$

∘	ε	b	c	d	e
ε	ε	b	c	d	e
b	b	ε	c	d	e
c	c	c	ε,b	e	d
d	d	d	e	ε,b	c
e	e	e	d	c	ε,b

Therefore, the following result is obtained:

Theorem 9. *There are 21 non-isomorphic G-hypergroup of size ≤ 5 and $|\omega_H| \in \{2, 3, 4\}$, as summarized in Table 1.*

Table 1. The G-hypergroups with $|H| \leq 5$.

$|H|$	3	4	5
$|\omega_H| = 2$	H_1	$H_2, H_{8\ldots 11}$	$H_{12\ldots 21}$
$|\omega_H| = 3$	-	H_3	$H_{4,5}$
$|\omega_H| = 4$	-	-	$H_{6,7}$

The G-hypergroups with $|\omega_H| = 1$ are 1-hypergroups, which include groups. In [16], the authors classified the 1-hypergroups of size ≤ 6. In particular, those of size ≤ 5 are 27. Thus, we have the following result.

Corollary 7. *There are 48 non-isomorphic G-hypergroup of size ≤ 5.*

Remark 3. *In every G-hypergroup $(H, \circ)$ with $|H| \leq 5$ all subgroups $S \subset H$ satisfy the condition that $\omega_H \subseteq S$ or $S \subseteq \omega_H$. On the other hand, the hypergroup shown in Remark 2 has order 6 and contains a subgroup S such that neither $S \subseteq \omega_H$ nor $\omega_H \subseteq S$. Therefore, that hypergroup is minimal with respect to this property.*

8. Conclusions and Directions for Further Research

If $(H, \circ)$ is a hypergroup then the kernel ω_H of the canonical projection $\varphi : H \mapsto H/\beta$ is a sub-hypergroup called heart [10,12]. If $|\omega_H| = 1$ then $(H, \circ)$ is a 1-hypergroup [13,14,16]. However, very little is known about hypergroups that have a heart that does not consist of either a single element or the entire hypergroup. This paper provides a contribution to the knowledge of such hypergroups. To achieve this goal, we generalized the notion of 1-hypergroup to hypergroups whose heart is isomorphic to a group. We analyzed in detail this class of hypergroups, here called G-hypergroups, with a special emphasis on the sub-class $\mathfrak{T}(H)$ of G-hypergroups whose heart is a torsion group. In the future, these results can hopefully lead to a more general construction than the one presented in Section 3.1, allowing all G-hypergroups to be constructed.

Table 2. Number of non-isomorphic G-hypergroups with $|H| \leq 5$, depending on the size of their hearts.

$|H|$	1	2	3	4	5
$|\omega_H| = 1$	1	1	2	4	19
$|\omega_H| = 2$	-	-	1	5	10
$|\omega_H| = 3$	-	-	-	1	2
$|\omega_H| = 4$	-	-	-	-	2
Total	1	1	3	10	33

Among our main results, we characterized the G-hypergroups that are also of type U on the right or cogroups on the right. Furthermore, we enumerated all non-isomorphic G-hypergroups with $|H| \leq 5$. The results achieved in Section 7 describe all G-hypergroups with $|H| \in \{3, 4, 5\}$ and $|\omega_H| \in \{2, 3, 4\}$, and are condensed in Table 2. We note that the hypergroups H_9 and H_{10} are also cogroups. Cogroups are one of the best known classes of hypergroups. A most notable problem with them is characterizing cogroups that are also D-hypergroups, i.e., quotient hypergroups G/S of a group G with respect to a non-normal subgroup S. This problem was solved in greater generality by L. Haddad and Y. Sureau in [3,4] by considering the group of permutations σ of H such that $\sigma(x \circ y) = \sigma(x) \circ y$, for all $x, y \in H$. The cogroups H_9 and H_{10} are D-hypergroups isomorphic to $S \backslash D_4$ and D_4/S, respectively, being D_4 the dihedral group of size 8 and S a non-normal subgroup of size 2.

At the conclusion of this work we would like to indicate some possible topics for further investigation. First of all, it would be interesting to verify whether the cogroups in

$\mathfrak{T}(H)$ are also D-hypergroups. A challenging problem related to the research carried out, e.g., in [16], which is classifying G-hypergroups of size greater than 5.

Finally, we observe that all G-hypergroups $(H, \circ)$ produced by the construction shown in Section 3.1 are such that the identity of ω_H is also identity of $(H, \circ)$, also when ω_H is not a torsion group. At present, we are not able to prove or disprove that this is always the case. Hence, a problem that remains open after our findings can be formulated as the following conjecture: if $(H, \circ)$ is a G-hypergroup then the scalar identity of ω_H is also the identity of $(H, \circ)$.

Author Contributions: Conceptualization, investigation, writing—original draft: M.D.S. and D.F. (Domenico Freni), and G.L.F.; software, writing—review and editing: D.F. (Dario Fasino). All authors have read and agreed to the published version of the manuscript.

Funding: The research work of Mario De Salvo was funded by Università di Messina, Italy, grant FFABR Unime 2019. Giovanni Lo Faro was supported by INdAM-GNSAGA, Italy, and by Università di Messina, Italy, grant FFABR Unime 2020. The work of Dario Fasino was partially supported by INdAM-GNCS, Italy.

Institutional Review Board Statement: Not applicable.

Informed Consent Statement: Not applicable.

Data Availability Statement: Not applicable.

Conflicts of Interest: The authors declare no conflicts of interest.

References

1. Massouros, C. (Ed.) *Hypercompositional Algebra and Applications*; MDPI: Basel, Switzerland, 2021.
2. Comer, S. D. Polygroups derived from cogroups. *J. Algebra* **1984**, *89*, 394–405. [CrossRef]
3. Haddad, L.; Sureau, Y. Les cogroupes et les D-hypergroupes. *J. Algebra* **1988**, *108*, 446–476. [CrossRef]
4. Haddad, L.; Sureau, Y. Les cogroupes et la construction de Utumi. *Pacific J. Math.* **1990**, *145*, 17–58. [CrossRef]
5. Sureau, Y. Hypergroupes de type C. *Rend. Circ. Mat. Palermo* **1991**, *40*, 421–437. [CrossRef]
6. Gutan, M.; Sureau, Y. Hypergroupes de type C à petites partitions. *Riv. Mat. Pura Appl.* **1995**, *16*, 13–38.
7. Koskas, H. Groupoïdes, demi-hypergroupes et hypergroupes. *J. Math. Pures Appl.* **1970**, *49*, 155–192.
8. Freni, D. Une note sur le cœur d'un hypergroup et sur la clôture transitive β^* de β. *Riv. Mat. Pura Appl.* **1991**, *8*, 153–156.
9. Gutan, M. On the transitivity of the relation β in semihypergroups. *Rend. Circ. Mat. Palermo* **1996**, *45*, 189–200. [CrossRef]
10. Leoreanu, V. On the heart of join spaces and of regular hypergroups. *Riv. Mat. Pura Appl.* **1995**, *17*, 133–142.
11. Antampoufis, N.; Hošková-Mayerová, Š. A brief survey on the two different approaches of fundamental equivalence relations on hyperstructures. *Ratio Math.* **2017**, *33*, 47–60.
12. Corsini, P.; Freni, D. On the heart of hypergroups. *Math. Montisnigri* **1993**, *2*, 21–27.
13. Cristea, I. Complete hypergroups, 1-hypergroups and fuzzy sets. *An. Stiin. Univ. Ovidius Constanta Ser. Mat.* **2002**, *10*, 25–37.
14. Corsini, P.; Cristea, I. Fuzzy sets and non complete 1-hypergroups. *An. Stiint. Univ. Ovidius Constanta Ser. Mat.* **2005**, *13*, 27–53.
15. De Salvo, M.; Fasino, D.; Freni, D.; Lo Faro, G. On hypergroups with a β-class of finite height. *Symmetry* **2020**, *12*, 168. [CrossRef]
16. De Salvo, M.; Fasino, D.; Freni, D.; Lo Faro, G. 1-hypergroups of small sizes. *Mathematics* **2021**, *9*, 108. [CrossRef]
17. Corsini, P. *Prolegomena of Hypergroup Theory*; Aviani Editore: Tricesimo, Italy, 1993.
18. Davvaz, B. *Semihypergroup Theory*; Academic Press: London, UK, 2016.
19. Massouros, C.; Massouros, G. An overview of the foundations of the hypergroup theory. *Mathematics* **2021**, *9*, 1014. [CrossRef]
20. Vougiouklis, T. Fundamental relations in hyperstructures. *Bull. Greek Math. Soc.* **1999**, *42*, 113–118.
21. De Salvo, M.; Freni, D.; Lo Faro, G. Fully simple semihypergroups. *J. Algebra* **2014**, *399*, 358–377. [CrossRef]
22. De Salvo, M.; Fasino, D.; Freni, D.; Lo Faro, G. Fully simple semihypergroups, transitive digraphs, and Sequence A000712. *J. Algebra* **2014**, *415*, 65–87. [CrossRef]
23. Al Tahan, M.; Davvaz, B. On some properties of single power cyclic hypergroups and regular relations. *J. Algebra Appl.* **2017**, *16*, 1750214. [CrossRef]
24. Novák, M.; Křehlík, Š.; Cristea, I. Cyclicity in EL–hypergroups. *Symmetry* **2018**, *10*, 611. [CrossRef]
25. Freni, D. Structure des hypergroupes quotients et des hypergroupes de type U. *Ann. Sci. Univ. Clermont-Ferrand II Math.* **1984**, *22*, 51–77.
26. Fasino, D.; Freni, D. Existence of proper semihypergroups of type U on the right. *Discrete Math.* **2007**, *307*, 2826–2836. [CrossRef]
27. Freni, D. Minimal order semi-hypergroupes of type U on the right, II. *J. Algebra* **2011**, *340*, 77–89. [CrossRef]

Article

v-Regular Ternary Menger Algebras and Left Translations of Ternary Menger Algebras

Anak Nongmanee [1,†] **and Sorasak Leeratanavalee** [2,*,†]

1. Department of Mathematics, Faculty of Science, Chiang Mai University, Chiang Mai 50200, Thailand; anak_nongmanee@cmu.ac.th
2. Research Group in Mathematics and Applied Mathematics, Department of Mathematics, Faculty of Science, Chiang Mai University, Chiang Mai 50200, Thailand
* Correspondence: sorasak.l@cmu.ac.th
† These authors contributed equally to this work.

Abstract: Let n be a fixed natural number. Ternary Menger algebras of rank n, which was established by the authors, can be regarded as a suitable generalization of ternary semigroups. In this article, we introduce the notion of v-regular ternary Menger algebras of rank n, which can be considered as a generalization of regular ternary semigroups. Moreover, we investigate some of its interesting properties. Based on the concept of n-place functions (n-ary operations), these lead us to construct ternary Menger algebras of rank n of all full n-place functions. Finally, we study a special class of full n-place functions, the so-called left translations. In particular, we investigate a relationship between the concept of full n-place functions and left translations.

Keywords: ternary Menger algebras; v-regular ternary Menger algebras; left translations

Citation: Nongmanee, A.; Leeratanavalee, S. v-Regular Ternary Menger Algebras and Left Translations of Ternary Menger Algebras. *Mathematics* **2021**, *9*, 2691. https://doi.org/10.3390/math9212691

Academic Editors: Dario Fasino and Domenico Freni

Received: 9 September 2021
Accepted: 21 October 2021
Published: 22 October 2021

Publisher's Note: MDPI stays neutral with regard to jurisdictional claims in published maps and institutional affiliations.

Copyright: © 2021 by the authors. Licensee MDPI, Basel, Switzerland. This article is an open access article distributed under the terms and conditions of the Creative Commons Attribution (CC BY) license (https://creativecommons.org/licenses/by/4.0/).

1. Introduction

Let X be a nonempty set. A unary function which is defined on X, that is a mapping from X into X, is called a *transformation*. Based on the concept of unary functions, the study of *multiplace functions* (which are also said to be functions of many elements (or many variables)) arose in various fields of mathematics. In 1946, Menger, K [1] studied the algebraic property of the composition of multiplace functions, the so-called a *superassociative law*. The algebras of multiplace functions, which are called *Menger algebras*, are studied in different ways and other branches of theoretical mathematics (c.f., e.g., [2,3]).

Let G be a nonempty set. An algebraic structure (G, o) is called a *Menger algebra of rank* n if the $(n+1)$-ary operation o, which is defined on G, satisfies the *superassociative law*, i.e.,

$$o(o(x, y_1, \ldots, y_n), z_1, \ldots, z_n) = o(x, o(y_1, z_1, \ldots, z_n), \ldots, o(y_n, z_1, \ldots, z_n)),$$

for every $x, y_i, z_i \in G, i = 1, \ldots, n$. Moreover, the algebraic structure is an arbitrary semigroup if $n = 1$.

The theory of Menger algebras of rank n and its applications are developed by Dudek, W. A. and Trokhimenko, V. S. who presented the concept of subtraction Menger algebras in 2012 (see [4]). Moreover, they studied more results which are related to subtraction Menger algebras (see [5]). Up to 2014, they introduced some types of congruences on Menger algebras of rank n, which can be considered as the generalizations of principal right and left congruences on ordinary semigroups (see [6]). Furthermore, a generalization of regular semigroups, which is called v-regular Menger algebras of rank n, and its interesting properties were established and studied by Trokhimenko V. S. in 1997 (see [7]).

Nowadays, the concept of Menger algebras is extended to study in various research topics by algebraists and semigroup theorists. Denecke, K. [8] used it to establish the notion of Menger algebras and clones of terms. Recently, Denecke, K. and Hounnon, H. [9] presented the notion of partial Menger algebras of terms in 2021. Moreover, Menger algebra of terms induced by order-decreasing transformations and some of its properties were investigated by Wattanatripop, K. and Changphas, T. in 2021 (see [10]). In 2021, Kumduang, T. and Leeratanavalee, S. [11] introduced the concept of left translations on Menger algebras of rank n and investigated its isomorphism theorems. By using the concept of Menger algebras and semihypergroups (or hypersemigroups), they established the so-called *Menger hyperalgebras of rank n*, which can be regarded as a canonical generalization of arbitrary semihypergroups (see [12]).

The set X^n is denoted to the n-th Cartesian product of the set X. Any mapping $f : X^n \longrightarrow X$ is called a *full n-place function* or an *n-ary operation* if it is defined for all elements of X. The set of all full n-place functions (n-ary operations) is denoted by $\mathcal{T}(X^n, X)$. On the based set $\mathcal{T}(X^n, X)$, we can consider the Menger's superposition (i.e., mappings which map some such functions into other ones and forms a new function in the following way) $\mathcal{O} : \mathcal{T}(X^n, X)^{n+1} \longrightarrow \mathcal{T}(X^n, X)$ which is defined by

$$\mathcal{O}(f, g_1, \ldots, g_n)(x_1, \ldots, x_n) = f(g_1(x_1, \ldots, x_n), \ldots, g_n(x_1, \ldots, x_n))$$

where $f, g_i \in \mathcal{T}(X^n, X), i = 1, \ldots, n$. Please note that, if $n = 1$ the Menger's superposition can be reduced to the usual composition of functions. Moreover, the based set $\mathcal{T}(X^n, X)$ is closed with respect to the Menger's superposition $\mathcal{O}$ and such a algebraic structure is called an *algebra of n-place functions*. Furthermore, the Menger's superposition $\mathcal{O}$ satisfies the superassociative law which was confirmed in [11]. Similar to the concept of partial transformation on semigroups, Menger algebras $\mathcal{F}(X^n, X)$ of all partial n-place functions were constructed. Such an algebraic structure is the set $\mathcal{F}(X^n, X)$ of all partial n-place functions together with the Menger's superposition.

The fundamental algebraic properties of Menger algebras of full n-place functions were proved by Dicker, R. M. in 1963 (see [13]). Moreover, he proved that a Menger algebra of rank n is isomorphic to some Menger algebras of full n-place functions, and the particular case of this fact was received by Whitlock, H. in 1964 [14]. Up to 1988, Länger, H. [15] presented a characterization of full function algebras. For more information related to the concept of Menger algebras and full n-place functions, see [16–22].

The notion of ternary semigroups was known to Banach, S. (see [23]) who was credited with an example of ternary semigroup which does not reduced to a (binary) semigroup. Similar to the theory of (binary) semigroups, the algebraic properties and applications of ternary semigroups were investigated by many mathematicians. In 1932, Lehmer, D. H. [24] presented the definition of ternary semigroups and investigated its algebraic properties. Subsequently, Von Neumann, J. [25] introduced and studied the notion of regularities in 1936. In 2008, Dutta, T. K., Kar, S. and Maity, B. K. [26] studied the properties of (completely, intra) regular ternary semigroups. Up to 2010, Santiago, M. L. and Sri Bala, S. [27] investigated some interesting properties of regular ternary semigroups. Furthermore, there were interesting results related to ternary semigroups and regular ternary semigroups (see [28–31]).

A ternary semigroup $(T, \star)$ is a pair of a nonempty set T together with the ternary operation $\star : T^3 \longrightarrow T$ which satisfies the *ternary associative law* as follows: for each $a, b, c, d, e \in T$,

$$\star(\star(a, b, c), d, e) = \star(a, \star(b, c, d), e) = \star(a, b, \star(c, d, e)).$$

On the other hand, a ternary semigroup can be considered as a special case of the so-called *n-ary semigroups*, which are regarded as a suitable generalization of ternary semigroups where $n = 3$ (c.f. [32]). According to the important result, ternary semigroups as a special case of n-ary semigroups were studied by many authors (see [33–35]).

Moreover, (c.f. [36]) ternary semigroups and ternary algebras are interesting for their applications in some problems of modern mathematical physics, such as the *Nambu mechanics* which was introduced by Nambu, Y. [37] in 1973. For other physical applications (see [38–40]). Furthermore, the theory of functional equations and the stability of functional equations in ternary (n-ary) algebraic structures were studied by many authors (e.g., [41–44]).

According to the algebraic structure of Menger algebras of rank n, the authors discovered that such an algebraic structure is a generalization of (binary) semigroups, while it is not a generalization of arbitrary ternary semigroups. Based on this important result, the notion of *ternary Menger algebras of rank n* where n is a fixed natural number was first introduced by the authors in 2021 (see [45]). In particular, the isomorphism theorems and the reduction of ternary Menger algebras of rank n into Menger algebras of rank n were investigated. Ternary Menger algebras of rank n can be considered as a canonical generalization of arbitrary ternary semigroups, and it can be reduced to ternary semigroups if $n = 1$.

In this article, we start by recalling some important results on ternary semigroups and ternary Menger algebras of rank n in Section 2. In Section 3, we introduce the notion of v-regular ternary Menger algebras of rank n, which can be regarded as a generalization of regular ternary semigroups. In addition, we investigate some of its algebraic properties. In Section 4, we establish the concept of left translations on ternary Menger algebras of rank n. Furthermore, we complete this section by showing a relationship between the set of all full n-place functions, the set of all left translations and left zero ternary Menger algebras of rank n. The conclusions and future works are provided in the last section.

2. Preliminaries

In this section, we recall some specific notations and results of ternary semigroups and ternary Menger algebras of rank n.

Definition 1. *Let $(T, \star)$ be a ternary semigroup. An element $a \in T$ is said to be*

(i) *an idempotent element if $\star(a, a, a) = a$;*
(ii) *a regular element if there exist $x \in T$ such that $\star(a, x, a) = a$.*

Please note that a ternary semigroup $(T, \star)$ is called *regular* if every element of T is regular.

Definition 2 ([46]). *Let $(T, \star)$ be a ternary semigroup. An element $a \in T$ is inverse to $b \in T$ if $\star(a, b, a) = a$ and $\star(b, a, b) = b$.*

Definition 3 ([45]). *A $(2n + 1)$-ary groupoid $(T, \bullet)$, i.e., the nonempty based set T together with a $(2n + 1)$-ary operation $\bullet$ defined on T, is called a ternary Menger algebra of rank n if the $(2n + 1)$-ary operation $\bullet$ satisfies the ternary superassociative law as follows: for every $a, b_i, c_i, d_i, e_i \in T, i = 1, \ldots, n,$*

$$\bullet(\bullet(a, b_1, \ldots, b_n, c_1, \ldots, c_n), d_1, \ldots, d_n, e_1, \ldots, e_n)$$
$$= \bullet(a, \bullet(b_1, c_1, \ldots, c_n, d_1, \ldots, d_n), \ldots, \bullet(b_n, c_1, \ldots, c_n, d_1, \ldots, d_n), e_1, \ldots, e_n)$$
$$= \bullet(a, b_1, \ldots, b_n, \bullet(c_1, d_1, \ldots, d_n, e_1, \ldots, e_n), \ldots, \bullet(c_n, d_1, \ldots, d_n, e_1, \ldots, e_n)).$$

According to Definition 3, if $n = 1$, then we immediately obtain that the ternary Menger algebra $(T, \bullet)$ is reduced to a ternary semigroup. Here and throughout in this article, a sequence of elements $b_1, \ldots, b_n$ of T is denoted by $\bar{b}$. Moreover, we write $a[\bar{b}\,\bar{c}]$ instead of $\bullet(a, b_1, \ldots, b_n, c_1, \ldots, c_n)$. Moreover, we write $a[b^n\,c^n]$ instead of $\bullet(a, \underbrace{b, \ldots, b}_{n \text{ terms}}, \underbrace{c, \ldots, c}_{n \text{ terms}})$.

For convenience, the ternary superassociative law can be written as follows:

$$a[\bar{b}\,\bar{c}][\bar{d}\,\bar{e}] = a[b_1[\bar{c}\,\bar{d}]\ldots b_n[\bar{c}\,\bar{d}]\bar{e}] = a[\bar{b}c_1[\bar{d}\,\bar{e}]\ldots c_n[\bar{d}\,\bar{e}]].$$

Similar to the concept of ternary subsemigroups, a ternary Menger subalgebra $(S, \bullet)$ of rank n of the ternary nary Menger algebras $(T, \bullet)$ of rank n is defined analogously. That is, let S be a nonempty subset of T, the set S under the $(2n+1)$-ary operation $\bullet$ is called ternary Menger subalgebra of rank n if for every $x, y_i, z_i \in S, i = 1, \ldots, n$, then $x[\bar{y}\,\bar{z}] \in S$.

Definition 4. Let 0 be an element of a ternary Menger algebra $(T, \bullet)$ of rank n. Then, 0 is said to be a left zero element if $0[\bar{y}\,\bar{z}] = 0$ holds for every $y_i, z_i \in T, i = 1, \ldots, n$.

We call a ternary Menger algebra of rank n with all elements are left zero elements as a *left zero ternary Menger algebra of rank n*.

Example 1. *Several examples of ternary Menger algebras of rank n are provided.*

(i) Consider on the set of all positive real numbers ([45]) $\mathbb{R}_+$ together with a $(2n+1)$-ary operation $\bullet$ defined by

$$x[\bar{y}\,\bar{z}] = x \cdot \sqrt[n]{y_1 \cdots y_n \cdot z_1 \cdots z_n} \quad \text{for all } x, y_i, z_i \in \mathbb{R}_+, i = 1, \ldots, n$$

where $\cdot$ is the usual (binary) multiplication. Then, $(\mathbb{R}_+, \bullet)$ forms a ternary Menger algebra of rank n.

(ii) Let $\mathbb{R}$ be the set of all real numbers ([45]). Define a $(2n+1)$-ary operation $\bullet$ on $\mathbb{R}$ by

$$x[\bar{y}\,\bar{z}] = x + \frac{y_1 + \cdots + y_n + z_1 + \cdots + z_n}{n} \quad \text{for all } x, y_i, z_i \in T, i = 1, \ldots, n$$

where $+$ is the usual (binary) addition. Then, $(\mathbb{R}, \bullet)$ is a ternary Menger algebra of rank n. Furthermore, $(\mathbb{R}_+, \bullet)$ is a ternary Menger subalgebra of rank n of $(\mathbb{R}, \bullet)$.

(iii) Let T be a nonempty set and a $(2n+1)$-ary operation $\bullet$ be defined as follows:

$$x[\bar{y}\,\bar{z}] = x \quad \text{for all } x, y_i, z_i \in T, i = 1, \ldots, n.$$

Every element of T is left zero. Consequently, $(T, \bullet)$ forms a left zero ternary Menger algebra of rank n.

(iv) Consider on the set of all real numbers $\mathbb{R}$ under one ternary operation $\diamond$ defined as follows:

$$\diamond(x, y, z) = x - y + z \quad \text{for all } x, y, z \in \mathbb{R},$$

where $-$ and $+$ are usual (binary) subtraction and (binary) addition. $(\mathbb{R}, \diamond)$ forms a ternary semigroup. Now, we define a $(2n+1)$-ary operation $\bullet$ on the same based set $\mathbb{R}$ by

$$x[\bar{y}\,\bar{z}] = \diamond(x, y_1, z_1) \quad \text{for all } a, y_i, z_i \in \mathbb{R}, i = 1, \ldots, n.$$

By the ternary associativity of the ternary operation $\diamond$, the $(2n+1)$-ary operation $\bullet$ is ternary superassociative and hence $(\mathbb{R}, \bullet)$ forms a ternary Menger algebra of rank n.

(v) Let $(T, \star)$ be an arbitrary ternary semigroup. For each $p, q \in \{1, \ldots, n\}$, we define $(2n+1)$-ary operations $\bullet_{pq}$ on T as follows:

$$\bullet_{pq}(x, y_1, \ldots, y_n, z_1, \ldots, z_n) = \star(x, y_p, z_q) \quad \text{for all } x, y_i, z_i \in T, i = 1, \ldots, n.$$

By the ternary associativity of the ternary operation $\star$ on T, the $(2n+1)$-ary operation $\bullet_{pq}$ is ternary superassociative. Therefore, $(T, \bullet_{pq})$ forms a ternary Menger algebra of rank n for every $p, q \in \{1, \ldots, n\}$.

Based on the algebraic structures of Menger algebras and ternary Menger algebras, we obtain an important remark as follows:

Remark 1. Let (G, o) be a Menger algebra of rank n under an $(n+1)$-ary operation o defined by $(x, y_1, \ldots, y_n) \mapsto o(x, y_1, \ldots, y_n)$ ([45]). Then the based set G together with a $(2n+1)$-ary operation $\bullet$, which is defined by $\bullet(x, y_1, \ldots, y_n, z_1, \ldots, z_n) \mapsto o(o(x, y_1, \ldots, y_n), z_1, \ldots, z_n)$, forms a ternary Menger algebra of rank n, while ternary Menger algebras of rank n do not necessarily reduce to Menger algebras of rank n.

Example 2. Let $\mathbb{Z}_-$ be the set of all negative integers. While $\mathbb{Z}_-$ together with a ternary multiplication $\bullet$, which is defined by $\bullet(x, y, z) = \cdot(\cdot(x, y), z)$ where $\cdot$ is the usual (binary) multiplication, forms a ternary Menger algebra of rank 1, the set $\mathbb{Z}_-$ under the usual multiplication $\cdot$ is not a Menger algebra of rank 1.

According to Definition 3, for each ternary Menger algebra $(T, \bullet)$ of rank n we can construct a new ternary operation induced by the $(2n+1)$-ary operation $\bullet$ defined on T. Then, the based set T under such a ternary operation forms a ternary semigroup.

Definition 5. *Let $(T, \bullet)$ be a ternary Menger algebra of rank n ([45]). A ternary operation $* : T^3 \longrightarrow T$ is defined on T by*

$$*(x, y, z) = x[y^n\ z^n] \quad \text{for all } x, y, z \in T. \tag{1}$$

Then $(T, *)$ forms a ternary semigroup.

According to Definition 5, we call the ternary semigroup $(T, *)$ together with the ternary operation $*$, which is induced by the $(2n+1)$-ary operation $\bullet$ of the ternary Menger algebra $(T, \bullet)$ of rank n, a *diagonal ternary semigroup*. For more information, see [45].

3. v-Regular Ternary Menger Algebras

In this section, we present the notion of v-regular ternary Menger algebras of rank n, which can be considered as a generalization of regular ternary semigroups. Moreover, we give some of its examples. We complete the section by investigating some of its interesting algebraic properties.

Definition 6. *Let $(T, \bullet)$ be a ternary Menger algebra of rank n. An element $(t_1, \ldots, t_n) \in T^n$ is called*

(i) *idempotent if it satisfies the following equation*

$$t_i[\bar{t}\ \bar{t}] = t_i \quad \text{for all } i = 1, \ldots, n;$$

(ii) *v-regular if there exists $x \in T$ such that*

$$t_i[x^n\ \bar{t}] = t_i \quad \text{for all } i = 1, \ldots, n.$$

A ternary Menger algebra $(T, \bullet)$ of rank n is called v-regular if every element $(t_1, \ldots, t_n) \in T^n$ is v-regular.

Definition 7. *Let x be an element of a ternary Menger algebra $(T, \bullet)$ of rank n and $(t_1, \ldots, t_n) \in T^n$. Then x is called an inverse of $(t_1, \ldots, t_n) \in T^n$ if it satisfies the following equations:*

$$x[\bar{t}\ x^n] = x \quad \text{and} \quad t_i[x^n\ \bar{t}] = t_i \quad \text{for all } i = 1, \ldots, n.$$

Please note that if $n = 1$, then we immediately obtain that Definitions 1 and 6 are the same thing. Similarly, Definitions 2 and 7 are the same in case $n = 1$.

Example 3. *(i) Consider on the set of all real numbers $\mathbb{R}$ together with a $(2n + 1)$-ary operation $\bullet$ defined by the following:*

$$x[\bar{y}\ \bar{z}] = x \quad \text{for all } x, y_i, z_i \in \mathbb{R}, i = 1, \ldots, n.$$

$(\mathbb{R}, \bullet)$ forms a v-regular ternary Menger algebra of rank n. Furthermore, each element $x \in \mathbb{R}$ is an inverse element of each element $(r_1, \ldots, r_n)$ of $\mathbb{R}^n$. All elements $(r_1, \ldots, r_n) \in \mathbb{R}^n$ are idempotent.

(ii) Let $\mathbb{N}_+$ be the set of all nonzero natural numbers. Define a $(2n+1)$-ary operation $\bullet$ on the set $\mathbb{N}_+$ by

$$x[\bar{y}\ \bar{z}] = \min\{x, y_1, \ldots, y_n, z_1, \ldots, z_n\} \quad \text{for all } x, y_i, z_i \in \mathbb{N}_+, i = 1, \ldots, n.$$

It implies that $(\mathbb{N}_+, \bullet)$ forms a ternary Menger algebra of rank n. Moreover, for every element $(t, t, \ldots, t) \in \mathbb{N}_+^n$ there exists element $x \in \mathbb{N}_+$ such that $t[x^n\ t^n] = t$ for all $(t, t, \ldots, t) \in \mathbb{N}_+^n$. Hence, $(\mathbb{N}_+, \bullet)$ forms a ternary Menger algebra of rank n with all elements $(t, t, \ldots, t) \in \mathbb{N}_+^n$ are v-regular elements and also idempotent elements.

Proposition 1. *Let $(T, \bullet)$ be a v-regular ternary Menger algebra of rank n. Then, a diagonal ternary semigroup $(T, *)$ of $(T, \bullet)$ forms a regular ternary semigroup.*

Proof. Firstly, we assume that $(T, *)$ is a diagonal ternary semigroup of a v-regular ternary Menger algebra $(T, \bullet)$ of rank n, where the ternary operation $*$ on T is defined as in (1), i.e.,

$$*(x, y, z) = x[y^n\ z^n] \quad \text{for all } x, y, z \in T.$$

Lastly, let $t \in T$. So, we immediately get $(t, t, \ldots, t) \in T^n$. Since $(T, \bullet)$ is v-regular, we obtain that the element $(t, t, \ldots, t)$ of T^n is a v-regular element, which yield that there exists $x \in T$ such that

$$t[x^n\ t^n] = t \quad \text{for all } (t, t, \ldots, t) \in T^n.$$

By the definition of the ternary operation $*$ on T and the above equality, we have

$$*(t, x, t) = t[x^n\ t^n] = t.$$

It implies that for every $t \in T$ there exists $x \in T$ such that $*(t, x, t) = t$. Consequently, $(T, *)$ forms a regular ternary semigroup. □

Proposition 2. *Let $(T, \bullet)$ be a ternary Menger algebra of rank n. Then, every v-regular element of $(T, \bullet)$ has an inverse element.*

Proof. Let $(t_1, \ldots, t_n)$ be a v-regular element of a ternary Menger algebra $(T, \bullet)$ of rank n. Hence, there is $x \in T$ such that

$$o(t_i, x, \ldots, x, t_1, \ldots t_n) = t_i \quad \text{for all } i = 1, \ldots, n. \tag{2}$$

Firstly, we choose the element $y = o(x, t_1, \ldots, t_n, x, \ldots, x) \in T$. Next, we will show that y is an inverse element of the element $(t_1, \ldots, t_n) \in T^n$ by showing that $o(y, t_1, \ldots, t_n, y, \ldots, y) = y$ and $o(t_i, y, \ldots, y, t_1, \ldots, t_n) = t_i$ for all $i = 1, \ldots, n$. By the ternary superassociative law of the $(2n+1)$-ary operation $\bullet$ on T and the Equation (2), we have

$$\begin{aligned}
o(y, t_1, \ldots, t_n, y, \ldots, y) &= o(o(x, t_1, \ldots, t_n, x, \ldots, x), t_1, \ldots, t_n, o(x, t_1, \ldots, t_n, x, \ldots, x) \\
&\quad, \ldots, o(x, t_1, \ldots, t_n, x, \ldots, x)) \\
&= o(x, o(t_1, x, \ldots, x, t_1, \ldots, t_n), \ldots, o(t_n, x, \ldots, x, t_1, \ldots, t_n), \\
&\quad o(x, t_1, \ldots, t_n, x, \ldots, x), \ldots, o(x, t_1, \ldots, t_n, x, \ldots, x)) \\
&= o(x, t_1, \ldots, t_n, o(x, t_1, \ldots, t_n, x, \ldots, x), \ldots, o(x, t_1, \ldots, t_n, x, \ldots, x)) \\
&= o(x, o(t_1, x, \ldots, x, t_1, \ldots, t_n), \ldots, o(t_n, x, \ldots, x, t_1, \ldots, t_n), x, \ldots, x) \\
&= o(x, t_1, \ldots, t_n, x, \ldots, x) \\
&= y.
\end{aligned}$$

Again, by the ternary superassociative law of the $(2n+1)$-ary operation $\bullet$ on T and the Equation (2), we obtain that

$$\begin{aligned}
o(t_i, y, \ldots, y, t_1, \ldots, t_n) &= o(t_i, o(x, t_1, \ldots, t_n, x, \ldots, x), \ldots, o(x, t_1, \ldots, t_n, x, \ldots, x) \\
&\quad, t_1, \ldots, t_n) \\
&= o(o(t_i, x, \ldots, x, t_1, \ldots, t_n), x, \ldots, x, t_1, \ldots, t_n) \\
&= o(t_i, x, \ldots, x, t_1, \ldots, t_n) \\
&= t_i,
\end{aligned}$$

hold for all $i = 1, \ldots, n$. Therefore, the element $y = o(x, t_1, \ldots, t_n, x, \ldots, x) \in T$ is an inverse element of the element $(t_1, \ldots, t_n) \in T^n$. □

Theorem 1. *For every ternary Menger algebra $(T, \bullet)$ of rank n, the following statement (a) implies (b).*

(a) *Every element of $(T, \bullet)$ is an inverse element for each element of T^n.*
(b) *For all $x, y, y_i \in T, i = 1, \ldots, n$*

$$x[y^n \ y^n] = y[x^n \ x^n] \implies x = y \quad \text{and} \tag{3}$$

$$y_i[x^n \ \bar{y}] = y_i \quad i = 1, \ldots, n. \tag{4}$$

Proof. Let (a) be satisfied. By definition 7, we immediately obtain that (4) holds. Now, we assume that $x[y^n \ y^n] = y[x^n \ x^n]$ for all $x, y \in T$. Next, we shall show that $x = y$. By our assumption, we have

$$x[y^n \ y^n][(y[x^n \ x^n])^n \ (y[x^n \ x^n])^n] = y[x^n \ x^n][(x[y^n \ y^n])^n \ (x[y^n \ y^n])^n].$$

Since (a) holds, we obtain that elements x and y of T are inverse elements of elements $(y[y^n \ x^n], \ldots, y[y^n \ x^n])$ and $(y[x^n \ x^n], \ldots, y[x^n \ x^n])$ of T^n, respectively. So, we have

$$\begin{aligned}
x[y^n \ y^n][(y[x^n \ x^n])^n \ (y[x^n \ x^n])^n] &= x[y^n \ y^n][(y[x^n \ x^n])^n \ y^n][x^n \ x^n] \\
&= x[y^n \ (y[(y[x^n \ x^n])^n \ y^n])^n][x^n \ x^n] \\
&= x[y^n \ y^n][x^n \ x^n] \\
&= x[(y[y^n \ x^n])^n \ x^n] \\
&= x.
\end{aligned}$$

Again, since (a) holds, elements x and y of T are inverse elements of elements $(x[y^n \ y^n], \ldots, x[y^n \ y^n])$ and $(x[x^n \ y^n], \ldots, x[x^n \ y^n])$ of T^n, respectively. Now we consider

$$\begin{aligned}
y[x^n \ x^n][(x[y^n \ y^n])^n \ (x[y^n \ y^n])^n] &= y[x^n \ x^n][(x[y^n \ y^n])^n \ x^n][y^n \ y^n] \\
&= y[x^n \ (x[(x[y^n \ y^n])^n \ x^n])^n][y^n \ y^n] \\
&= y[x^n \ x^n][y^n \ y^n] \\
&= y[(x[x^n \ y^n])^n \ y^n] \\
&= y.
\end{aligned}$$

Consequently, $x = y$ and hence the statement (3) holds. □

For illustrating Theorem 1, the following examples are provided.

Example 4. (i) Example 1 (iii) and Example 3 (i) satisfy Theorem 1.
(ii) Every left zero ternary Menger algebra of rank n satisfies Theorem 1.
(iii) Let $(T, \star)$ be a left zero ternary semigroup, i.e., T has the property $\star(x, y, z) = x$ for all $x, y, z \in T$. For each $j \in \{1, \ldots, n\}$, we define $(2n + 1)$-ary operations $\bullet_j$ on T by

$$\bullet_j(x, y_1, \ldots, y_n, z_1, \ldots, z_n) = \star(x, y_j, z_j) \quad \text{for all } x, y_i, z_i \in T, i = 1, \ldots, n.$$

By the ternary associativity of the ternary operation $\star$ on T, the $(2n + 1)$-ary operation $\bullet_j$ is ternary superassociative. Therefore, $(T, \bullet_j)$ forms a ternary Menger algebra of rank n for every $j \in \{1, \ldots, n\}$. Moreover, $(T, \bullet_j)$ satisfies Theorem 1 for every $j \in \{1, \ldots, n\}$, it follows from the definition of left zero ternary semigroups.

Theorem 2. *Let $(T, \bullet)$ be a v-regular ternary Menger algebra of rank n. Then, every element of $(T, \bullet)$ has exactly one inverse element if and only if*

$$t_i[x^n \ \bar{t}] = t_i \text{ and } t_i[y^n \ \bar{t}] = t_i, i = 1, \ldots, n \implies x[\bar{t} \ x^n] = y[\bar{t} \ y^n],$$

for all $x, y, t_i \in T, i = 1, \ldots, n$.

Proof. ($\Rightarrow$) Firstly, we assume that each element $(t_1, \ldots, t_n)$ of a v-regular Menger algebra $(T, \bullet)$ of rank n has exactly one inverse element. Now, let $x, y, t_i \in T, i = 1, \ldots, n$ be such that
$$t_i[x^n \, \bar{t}] = t_i \text{ and } t_i[y^n \, \bar{t}] = t_i \quad \text{for all } i = 1, \ldots, n.$$

According to the proof of Proposition 2, we obtain that $x[\bar{t} \, x^n]$ and $y[\bar{t} \, y^n]$ are inverse elements of the element $(t_1, \ldots, t_n)$ of T^n. By our assumption, we have $x[\bar{t} \, x^n] = y[\bar{t} \, y^n]$.

($\Leftarrow$) Suppose that the given condition holds. Now we assume that x and y are inverse elements of an element $(t_1, \ldots, t_n)$ of T^n. We will show that $x = y$. By Definition 7, we immediately obtain that the following equations are true:
$$t_i[x^n \, \bar{t}] = t_i \text{ and } t_i[y^n \, \bar{t}] = t_i \quad \text{for all } i = 1, \ldots, n.$$

It implies that $x[\bar{t} \, x^n] = y[\bar{t} \, y^n]$, which follows from the given condition. Again, by Definition 7, we immediately obtain that
$$x[\bar{t} \, x^n] = x \text{ and } y[\bar{t} \, y^n] = y.$$

Consequently, $x = y$ and hence $(T, \bullet)$ has exactly one inverse element. This completes the proof. $\square$

4. Left Translation on Ternary Menger Algebras

In this section, we introduce the notion of the ternary Menger algebras of all full n-place functions and its $(2n+1)$-ary operation which satisfies the ternary superassociative law. Moreover, we present the concept of left translations on ternary Menger algebras of rank n and investigate some of its interesting properties. Furthermore, the relationship between the previous concepts is investigated.

Let X be a nonempty set. Now, we define the following $(2n+1)$-ary operation $\mathcal{O}^*$ on the set $\mathcal{T}(X^n, X)$ of all *full n-place functions* or *n-ary operations* by

$$\mathcal{O}^*(f, \bar{g}, \bar{h})(\bar{x}) = f(g_1(h_1(\bar{x}), \ldots, h_n(\bar{x})), \ldots, g_n(h_1(\bar{x}), \ldots, h_n(\bar{x}))) \tag{5}$$

for all $x_i \in X, i = 1, \ldots, n$. We call the $(2n+1)$-ary operation $\mathcal{O}^*$ on $\mathcal{T}(X^n, X)$, a *ternary Menger's superposition*.

Based on the definition of the ternary Menger's superposition, we can remark that the ternary Menger's superposition can be reduced to the ternary composition of functions, if $n = 1$, i.e., the Equation (5) is reduced to the following equation:

$$\mathcal{O}^*(f, g_1, h_1)(x_1) = f(g_1(h_1(x_1))). \tag{6}$$

According to the definition of the ternary Menger's superposition, we obtain the following important result.

Theorem 3. *Let X be a nonempty set. The ternary Menger's superposition $\mathcal{O}^*$ of full n-place functions, which are defined on X, is ternary superassociative, i.e., for each $\lambda, \alpha_i, \beta_i, \gamma_i, \delta_i \in \mathcal{T}(X^n, X), i = 1, \ldots, n$,*

$$\mathcal{O}^*(\mathcal{O}^*(\lambda, \bar{\alpha}, \bar{\beta}), \bar{\gamma}, \bar{\delta}) = \mathcal{O}^*(\lambda, \mathcal{O}^*(\alpha_1, \bar{\beta}, \bar{\gamma}), \ldots, \mathcal{O}^*(\alpha_n, \bar{\beta}, \bar{\gamma}), \bar{\delta})$$
$$= \mathcal{O}^*(\lambda, \bar{\alpha}, \mathcal{O}^*(\beta_1, \bar{\gamma}, \bar{\delta}), \ldots, \mathcal{O}^*(\beta_n, \bar{\gamma}, \bar{\delta})).$$

Proof. $\mathcal{O}^*$ is a $(2n+1)$-ary operation on the set $\mathcal{T}(X^n, X)$. Indeed, for each $\lambda, \alpha_i, \beta_i, \gamma_i, \delta_i \in \mathcal{T}(X^n, X), i = 1, \ldots, n$, we have

$$\mathcal{O}^*(\mathcal{O}^*(\lambda, \bar{\alpha}, \bar{\beta}), \bar{\gamma}, \bar{\delta})(\bar{x}) = \mathcal{O}^*(\lambda, \bar{\alpha}, \bar{\beta})(\gamma_1(\delta_1(\bar{x}), \ldots, \delta_n(\bar{x})), \ldots, \gamma_n(\delta_1(\bar{x}), \ldots, \delta_n(\bar{x})))$$
$$= \lambda(\alpha_1(\beta_1(\gamma_1(\delta_1(\bar{x}), \ldots, \delta_n(\bar{x})), \ldots, \gamma_n(\delta_1(\bar{x}), \ldots, \delta_n(\bar{x}))), \ldots, \beta_n(\gamma_1(\delta_1(\bar{x}), \ldots, \delta_n(\bar{x}))$$
$$, \ldots, \gamma_n(\delta_1(\bar{x}), \ldots, \delta_n(\bar{x})))), \ldots, \alpha_n(\beta_1(\gamma_1(\delta_1(\bar{x}), \ldots, \delta_n(\bar{x})), \ldots, \gamma_n(\delta_1(\bar{x}), \ldots, \delta_n(\bar{x})))$$
$$, \ldots, \beta_n(\gamma_1(\delta_1(\bar{x}), \ldots, \delta_n(\bar{x})), \ldots, \gamma_n(\delta_1(\bar{x}), \ldots, \delta_n(\bar{x})))))$$
$$= \lambda(\mathcal{O}^*(\alpha_1, \bar{\beta}, \bar{\gamma})(\delta_1(\bar{x}), \ldots, \delta_n(\bar{x})), \ldots, \mathcal{O}^*(\alpha_n, \bar{\beta}, \bar{\gamma})(\delta_1(\bar{x}), \ldots, \delta_n(\bar{x})))$$
$$= \mathcal{O}^*(\lambda, \mathcal{O}^*(\alpha_1, \bar{\beta}, \bar{\gamma}), \ldots, \mathcal{O}^*(\alpha_n, \bar{\beta}, \bar{\gamma}), \bar{\delta})(\bar{x}) \quad \text{and}$$
$$\mathcal{O}^*(\mathcal{O}^*(\lambda, \bar{\alpha}, \bar{\beta}), \bar{\gamma}, \bar{\delta})(\bar{x}) = \mathcal{O}^*(\lambda, \bar{\alpha}, \bar{\beta})(\gamma_1(\delta_1(\bar{x}), \ldots, \delta_n(\bar{x})), \ldots, \gamma_n(\delta_1(\bar{x}), \ldots, \delta_n(\bar{x})))$$
$$= \lambda(\alpha_1(\beta_1(\gamma_1(\delta_1(\bar{x}), \ldots, \delta_n(\bar{x})), \ldots, \gamma_n(\delta_1(\bar{x}), \ldots, \delta_n(\bar{x}))), \ldots, \beta_n(\gamma_1(\delta_1(\bar{x}), \ldots, \delta_n(\bar{x}))$$
$$, \ldots, \gamma_n(\delta_1(\bar{x}), \ldots, \delta_n(\bar{x})))), \ldots, \alpha_n(\beta_1(\gamma_1(\delta_1(\bar{x}), \ldots, \delta_n(\bar{x})), \ldots, \gamma_n(\delta_1(\bar{x}), \ldots, \delta_n(\bar{x})))$$
$$, \ldots, \beta_n(\gamma_1(\delta_1(\bar{x}), \ldots, \delta_n(\bar{x})), \ldots, \gamma_n(\delta_1(\bar{x}), \ldots, \delta_n(\bar{x})))))$$
$$= \lambda(\alpha_1(\mathcal{O}^*(\beta_1, \bar{\gamma}, \bar{\delta})(\bar{x}), \ldots, \mathcal{O}^*(\beta_n, \bar{\gamma}, \bar{\delta})(\bar{x})), \ldots, \alpha_n(\mathcal{O}^*(\beta_1, \bar{\gamma}, \bar{\delta})(\bar{x})$$
$$, \ldots, \mathcal{O}^*(\beta_n, \bar{\gamma}, \bar{\delta})(\bar{x})))$$
$$= \mathcal{O}^*(\lambda, \bar{\alpha}, \mathcal{O}^*(\beta_1, \bar{\gamma}, \bar{\delta}), \ldots, \mathcal{O}^*(\beta_n, \bar{\gamma}, \bar{\delta}))(\bar{x}).$$

Consequently, the ternary Menger's superposition $\mathcal{O}^*$ of full n-place functions is ternary superassociative. □

By Theorem 3, we immediately obtain the following important corollary.

Corollary 1. *Let X be a nonempty set. Then, $\mathcal{T}(X^n, X)$ forms a ternary Menger algebra of rank n under the ternary Menger's superposition $\mathcal{O}^*$.*

According to Corollary 1, a *ternary Menger algebra of all full n-place functions (n-ary operations)* is referred to the pair of the set of all n-place functions (n-ary operations), which are defined on a nonempty set X, and the ternary Menger's superposition $\mathcal{O}^*$ of full n-place functions (n-ary operations) satisfying the ternary superassociative law. For convenience, a ternary Menger algebra of full n-place functions (n-ary operations) is referred to each ternary Menger subalgebra of $(\mathcal{T}(X^n, X), \mathcal{O}^*)$.

Now, we introduce an important class of the set of all full n-place functions $\mathcal{T}(X^n, X)$ defined on a nonempty set X.

Definition 8. *Let $(T, \bullet)$ be a ternary Menger algebra of rank n. A mapping $\lambda : T^n \longrightarrow T$ is called a left translation of T^n if it satisfies the following equation:*

$$\lambda(x_1[\bar{y}\,\bar{z}], \ldots, x_n[\bar{y}\,\bar{z}]) = \lambda(\bar{x})[\bar{y}\,\bar{z}] \quad \text{for all } x_i, y_i, z_i \in T, i = 1, \ldots, n.$$

Please note that Definition 8 can be considered as a natural generalization concept of left translations in ternary semigroups by setting $n = 1$, i.e.,

$$\lambda(\bullet(x, y, z)) = \bullet(\lambda(x), y, z)$$

holds for every element x, y, z of the ternary semigroup $(T, \bullet)$.

Example 5. *Let $(T, \bullet)$ a Ternary Menger algebra of rank n. The projection function π_j is defined on the n-th Cartesian product T^n by,*

$$\pi_j(x_1, \ldots, x_j, \ldots, x_n) = x_j \quad \text{for all } 1 \leq j \leq n.$$

Indeed, for each $x_1, \ldots, x_j, \ldots, x_n \in T$ we have

$$\pi_j(x_1[\bar{y}\,\bar{z}], \ldots, x_j[\bar{y}\,\bar{z}], \ldots, x_n[\bar{y}\,\bar{z}]) = x_j[\bar{y}\,\bar{z}] = \pi_j(x_1, \ldots, x_j, \ldots, x_n)[\bar{y}\,\bar{z}].$$

It implies that the projection π_j is a left translation.

Lemma 1. Let $(T, \bullet)$ be a ternary Menger algebra of rank n and

$$\Lambda(T) = \{\lambda : T^n \longrightarrow T \mid \lambda \text{ is a left translation of } T^n\}.$$

Then, $\Lambda(T)$ is a ternary Menger algebra of rank n under the ternary Menger's superposition $\mathcal{O}^*$.

Proof. It is clearly that $\Lambda(T) \neq \emptyset$. Let an n-place function $\lambda : T^n \longrightarrow T$ be defined by

$$\lambda(x_1, \ldots, x_n) = x_n \quad \text{for all } x_i \in T, i = 1, \ldots, n.$$

So, we obtain that $\lambda(x_1[\bar{y}\,\bar{z}], \ldots, x_n[\bar{y}\,\bar{z}]) = x_n[\bar{y}\,\bar{z}] = \lambda(\bar{x})[\bar{y}\,\bar{z}]$ and hence $\Lambda(T) \neq \emptyset$. Now, let $\lambda, \alpha_i, \beta_i \in \Lambda(T), i = 1, \ldots, n$. Indeed, for each $x_i, y_i, z_i \in T, i = 1, \ldots, n$ we get

$$\mathcal{O}^*(\lambda, \bar{\alpha}, \bar{\beta})(x_1[\bar{y}\,\bar{z}], \ldots, x_n[\bar{y}\,\bar{z}])$$
$$= \lambda(\alpha_1(\beta_1(x_1[\bar{y}\,\bar{z}], \ldots, x_n[\bar{y}\,\bar{z}]), \ldots, \beta_n(x_1[\bar{y}\,\bar{z}], \ldots, x_n[\bar{y}\,\bar{z}]))$$
$$, \ldots, \alpha_n(\beta_1(x_1[\bar{y}\,\bar{z}], \ldots, x_n[\bar{y}\,\bar{z}]), \ldots, \beta_n(x_1[\bar{y}\,\bar{z}], \ldots, x_n[\bar{y}\,\bar{z}])))$$
$$= \lambda(\alpha_1(\beta_1(\bar{x})[\bar{y}\,\bar{z}], \ldots, \beta_n(\bar{x})[\bar{y}\,\bar{z}]), \ldots, \alpha_n(\beta_1(\bar{x})[\bar{y}\,\bar{z}], \ldots, \beta_n(\bar{x})[\bar{y}\,\bar{z}]))$$
$$= \lambda(\alpha_1(\beta_1(\bar{x}), \ldots, \beta_n(\bar{x}))[\bar{y}\,\bar{z}], \ldots, \alpha_n(\beta_1(\bar{x}), \ldots, \beta_n(\bar{x}))[\bar{y}\,\bar{z}])$$
$$= \lambda(\alpha_1(\beta_1(\bar{x}), \ldots, \beta_n(\bar{x})), \ldots, \alpha_n(\beta_1(\bar{x}), \ldots, \beta_n(\bar{x})))[\bar{y}\,\bar{z}]$$
$$= \mathcal{O}^*(\lambda, \bar{\alpha}, \bar{\beta})(\bar{x})[\bar{y}\,\bar{z}].$$

It implies that $\mathcal{O}^*(\lambda, \bar{\alpha}, \bar{\beta}) \in \Lambda(T)$, and hence, the ternary Menger's superposition $\mathcal{O}^*$ is a $(2n+1)$-ary operation on $\Lambda(T)$. By Theorem 3, we conclude that $(\Lambda(T), \mathcal{O}^*)$ forms a ternary Menger algebra of rank n. □

By the definition of the set of all left translations $\Lambda(T)$ defined on a ternary Menger algebra $(T, \bullet)$ of rank n, we obtain that $\Lambda(T) \subseteq \mathcal{T}(T^n, T)$. Moreover, we immediately obtain the following corollary.

Corollary 2. Let $(T, \bullet)$ be a ternary Menger algebra of rank n. Then, the algebraic structure $(\Lambda(T), \mathcal{O}^*)$ is a ternary Menger subalgebra of rank n of the ternary Menger algebra of all full n-place functions $(\mathcal{T}(T^n, T), \mathcal{O}^*)$.

Now, we give a relationship between the set of all left translations $\Lambda(T)$ defined on a ternary Menger algebra $(T, \bullet)$ of rank n and the set of all full n-place functions $\mathcal{T}(T^n, T)$ by using the concept of left zero ternary Menger algebras of rank n.

Proposition 3. A left translation λ maps a left zero element 0 of a ternary Menger algebra of rank n, if it exists, to a left zero element, i.e.,

$$\lambda(0^n)[\bar{y}\,\bar{z}] = \lambda(0^n)$$

holds for all $y_i, z_i \in T, i = 1, \ldots, n$.

Proof. Assume that 0 is a left zero element of a ternary Menger algebra $(T, \bullet)$ of rank n. By our assumption, we already have $0[\bar{y}\,\bar{z}] = 0$ for all $y_i, z_i \in T, i = 1, \ldots, n$. Now, let λ be a left translation. Indeed, for each $y_i, z_i \in T, i = 1, \ldots, n$ we get

$$\lambda(0^n)[\bar{x}\,\bar{z}] = \lambda(0[\bar{y}\,\bar{z}], \ldots, 0[\bar{y}\,\bar{z}]) = \lambda(0^n).$$

This completes the proof. □

Theorem 4. Let $(T, \bullet)$ be a ternary Menger algebra of rank n. $\Lambda(T) = \mathcal{T}(T^n, T)$ if and only if T is a left zero ternary Menger algebra of rank n.

Proof. ($\Rightarrow$) Suppose that $\Lambda(T) = \mathcal{T}(T^n, T)$. Now, we assume that there are $x, y_i, z_i \in T$ such that $x[\bar{y}\,\bar{z}] \neq x$. We can choose an n-place function $\lambda \in \mathcal{T}(T^n, T)$ such that

$$\lambda(x^n) = x \text{ and } \lambda(x[\bar{y}\,\bar{z}], \ldots, x[\bar{y}\,\bar{z}]) = x.$$

By our assumption, λ is also a left translation of T^n. It implies that

$$x = \lambda(x[\bar{y}\,\bar{z}], \ldots, x[\bar{y}\,\bar{z}]) = \lambda(x^n)[\bar{y}\,\bar{z}] = x[\bar{y}\,\bar{z}].$$

It is a contradiction with the assumption that $x[\bar{y}\,\bar{z}] \neq x$. So, we have $x[\bar{y}\,\bar{z}] = x$ for all $x, y_i, z_i \in T$. Therefore, $(T, \bullet)$ forms a left zero ternary Menger algebra of rank n.

($\Leftarrow$) We assume that T is a left zero ternary Menger algebra of rank n. Note that $\Lambda(T) \subseteq \mathcal{T}(T^n, T)$. We only show that $\mathcal{T}(T^n, T) \subseteq \Lambda(T)$. By our assumption, we already have $x[\bar{y}\,\bar{z}] = x$ for all $x, y_i, z_i \in T, i = 1, \ldots, n$. Now, let $\lambda \in \mathcal{T}(T^n, T)$ and $x_i, y_i, z_i \in T, i = 1, \ldots, n$. Then

$$\lambda(x_1[\bar{y}\,\bar{z}], \ldots, x_n[\bar{y}\,\bar{z}]) = \lambda(x_1, \ldots, x_n) = \lambda(x_1, \ldots, x_n)[\bar{y}\,\bar{z}].$$

It implies that $\lambda(x_1[\bar{y}\,\bar{z}], \ldots, x_n[\bar{y}\,\bar{z}]) = \lambda(\bar{x})[\bar{y}\,\bar{z}]$ and hence λ is a left translation of T^n for every $\lambda \in \mathcal{T}(T^n, T)$. Thus $\lambda \in \Lambda(T)$ and this shows that $\mathcal{T}(T^n, T) \subseteq \Lambda(T)$. □

5. Conclusions and Future Works

In this article, there are two important purposes. Firstly, we introduced the notion of classical algebraic structure the so-called v-regular ternary Menger algebras of rank n, which can be considered as a generalization of regular ternary semigroups, and investigated its interesting properties. To receive the results, the significant knowledge and basic results of ternary semigroups and ternary Menger algebras of rank n are presented in Section 2. Lastly, we used the classical idea on the theory of semigroups the so-called left translations to construct a left translation on ternary Menger algebras of rank n. Moreover, some algebraic properties are investigated.

Based on the results of the article, there are interesting research questions to study in the future works. Can we extend some results on regular ternary semigroups, completely regular ternary semigroups and intraregular ternary semigroups, which were already presented in [26], to study in v-regular ternary Menger algebras of rank n? Furthermore, the following problems are interesting to study:

(i) Defining a right translation and an inner-left (right) translation for ternary Menger algebras of rank n. Investigating a characterization of ternary Menger algebras of rank n via this concept.

(ii) Studying some theories of (ideal) extensions of ternary Menger algebras of rank n via translation as we defined in this article.

Author Contributions: Conceptualization, A.N. and S.L.; Investigation, A.N. and S.L.; Supervision, S.L.; Writing–original draft, A.N.; Writing–review and editing, S.L. All authors contributed equally to all aspects of this work. All authors have read and agreed to the published version of the manuscript.

Funding: This research received no external funding.

Data Availability Statement: Not applicable.

Acknowledgments: The authors are highly grateful to the referees for their valuable comments and suggestions for improving the article. This research was supported by Chiang Mai University, Chiang Mai, 50200 Thailand.

Conflicts of Interest: The authors declare no conflict of interest.

References

1. Menger, K. General algebra of analysis. *Rep. Math. Colloq. Notre Dame Univ.* **1946**, *7*, 46–60.
2. Menger, K. The algebra of functions: Past, present, future. *Rend. Math.* **1961**, *20*, 409–430.
3. Menger, K. Superassociative systems and logical functors. *Math. Ann.* **1964**, *157*, 278–295. [CrossRef]
4. Dudek, W.A.; Trokhimenko, V.S. Subtraction Menger algebras. *Semigroup Forum* **2012**, *85*, 111–128. [CrossRef]
5. Dudek, W.A.; Trokhimenko, V.S. On some subtraction Menger algebras of multiplace functions. *Semigroup Forum* **2016**, *93*, 375–386. [CrossRef]
6. Dudek, W.A.; Trokhimenko, V.S. Congruences in Menger algebras. *Commun. Algebra* **2014**, *42*, 3407–3426. [CrossRef]

7. Trokhimenko, V.S. v-Regular Menger algebras. *Algebra Univers.* **1997**, *38*, 150—164. [CrossRef]
8. Denecke, K. Menger algebras and clones of terms. *East-West J. Math.* **2003**, *5*, 179–193.
9. Denecke, K.; Hounnon, H. Partial Menger algebras of terms. *Asian-Eur. J. Math.* **2021**, *14*, 2150092. [CrossRef]
10. Wattanatripop, K.; Changphas, T. The Menger algebra of terms induced by order-decreasing transformations. *Commun. Algebra* **2021**, *49*, 3114–3123. [CrossRef]
11. Kumduang, T.; Leeratanavalee, S. Left translations and isomorphism theorems for Menger algebras of rank n. *Kyungpook Math. J.* **2021**, *61*, 223–237.
12. Kumduang, T.; Leeratanavalee, S. Menger hyperalgebras and their representations. *Commun. Algebra* **2020**, *49*, 1513–1533. [CrossRef]
13. Dicker, R.M. The substitutive law. *Proc. London Math. Soc.* **1963**, *3*, 493–510. [CrossRef]
14. Whitlock, H. A composition algebra for multiplace functions. *Math. Ann.* **1964**, *157*, 167–178. [CrossRef]
15. Länger, H. A characterization of full function algebras. *J. Algebra* **1988**, *119*, 261–264. [CrossRef]
16. Dudek, W.A.; Trokhimenko, V.S. *Algebras of Multiplace Functions*; Walter de Gruyter GmbH & Co. KG.: Berlin, Germany; Boston, MA, USA, 2012; pp. 21–84.
17. Dudek, W.A.; Trokhimenko, V.S. De Morgan $(2,n)$-semigrops of n-place functions. *Commun. Algebra* **2016**, *44*, 4430–4437. [CrossRef]
18. Dudek, W.A.; Trokhimenko, V.S. Menger algebras of k-commutative n-place functions. *Georgian Math. J.* **2019**, *28*, 355–361. [CrossRef]
19. Dudek, W.A.; Trokhimenko, V.S. On σ-commutativity in Menger algebras of n-place functions. *Commun. Algebra* **2017**, *45*, 4557–4568. [CrossRef]
20. Schein, B.M.; Trokhimenko, V.S. Algebras of multiplace functions. *Semigroup Forum* **1979**, *17*, 1–64. [CrossRef]
21. Kumduang, T.; Leeratanavalee, S. Semigroups of terms, tree languages, Menger algebra of n-ary functions and their embedding theorems. *Symmetry* **2021**, *13*, 558. [CrossRef]
22. Wattanatripop, K.; Kumduang, T.; Changphas, T.; Leeratanavalee, S. Power Menger algebras of terms induced by order-decreasing transformations and superpositions. *Int. J. Math. Comput. Sci.* **2021**, *16*, 1697–1707.
23. Los, J. On the extending of models I. *Fund. Math.* **1955**, *42*, 38–54. [CrossRef]
24. Lehmer, D.H. A ternary analogue of abelian groups. *Am. J. Math.* **1932**, *54*, 329–338. [CrossRef]
25. Von Neumann, J. On regular rings. *Proc. Natl. Acad. Sci. USA* **1936**, *22*, 707–713. [CrossRef] [PubMed]
26. Dutta, T.K.; Kar, S.; Maity, B.K. On ideals in regular ternary semigroups. *Discuss. Math. Gen. Algebra Appl.* **2008**, *28*, 147–159. [CrossRef]
27. Santiago, M.L.; Sri Bala, S. Ternary semigroups. *Semigroup Forum* **2010**, *81*, 380–388. [CrossRef]
28. Pornsurat, P.; Pibaljommee, B. Regularities in ordered ternary semigroups. *Quasigr. Relat. Syst.* **2019**, *27*, 107–118.
29. Lekkoksung, N.; Sanpan, H.; Lekkoksung, S. Characterizations of some regularities in ordered ternary semigroups in terms of fuzzy subsets. *Commun. Math. Appl.* **2021**, *12*, 325–333.
30. Nongmanee, A.; Leeratanavalee, S. Quaternary rectangular bands and representations of ternary semigroups. *Thai J. Math.* **2021**, accepted.
31. Peposhi, A. A Note on ideals and regularity in ternary semigroups. *Int. J. Innov. Res. Eng. Manag.* **2020**, *7*, 78–81. [CrossRef]
32. Monk, S.; Sioson, F.M. m-Semigroups, semigroups and functions representations. *Fundam. Math.* **1966**, *59*, 191–201. [CrossRef]
33. Sioson, F.M. On regular algebraic systems. *Proc. Jpn Acad.* **1963**, *39*, 283–286.
34. Dudek, W.A.; Grozdinska, I. On ideals in regular n-ary semigroups. *Matematichki Bilten (Skopje)* **1980**, *4*, 25–44.
35. Trokhimenko, V.S. On n-groupoids in which all transformations are endomorphisms. *Quasigr. Relat. Syst.* **2011**, *19*, 349–352.
36. Russias, T.M.; Brzdek, J. *Functional Equations in Mathematics Analysis*; Springer: New York, NY, USA, 2012; pp. 372–416.
37. Nambu, Y. Generalized Hamiltonian dynamics. *J. Phys. Rev.* **1973**, *7*, 2405. [CrossRef]
38. Ataguema, H.; Makhlouf, A. Notes on cohomologies of ternary algebras of associative type. *J. Gen. Lie Theory Appl.* **2009**, *3*, 157–174. [CrossRef]
39. Amyari, M.; Moslehian, M.S. Approximate Homomorphisms of Ternary Semigroups. *Lett. Math. Phys.* **2006**, *77*, 1–9. [CrossRef]
40. Gordji, M.E.; Jabbari, A.; Ebadian, A.; Ostadbashi, S. Automatic continuity of 3-homomorphisms on ternary Banach algebras. *Int. J. Geom. Methods Mod. Phys.* **2013**, *10*, 13200012.
41. Chronowski, A. The Pexider equation on n-semigroups and n-groups. *Publ. Math. Debrecen* **1990**, *37*, 121–130.
42. Park, C.G.; Rassias, T.M. Homomorphisms in C^*-ternary algebras and JB^*-triples. *J. Math. Anal. Appl.* **2008**, *337*, 13–20. [CrossRef]
43. Savadkouhi, M.B.; Gordji, M.E.; Rassias, J.M.; Ghobadipour, N. Approximate ternary Jordan derivations on Banach ternary algebras. *J. Math. Phys.* **2009**, *50*, 042303. [CrossRef]
44. Moslehian, M.S.; Rassias, T.M. Generalized Hyers–Ulam stability of mappings on normed Lie triple systems. *Math. Inequal. Appl.* **2008**, *11*, 371–380. [CrossRef]
45. Nongmanee, A.; Leeratanavalee, S. Ternary Menger algebras: A generalization of ternary semigroups. *Mathematics* **2021**, *9*, 553. [CrossRef]
46. Sheeja, G.; Sri Bala, S. Orthodox ternary semigroups. *Quasigr. Relat. Syst.* **2011**, *19*, 339–348.

Article

Taxonomy of Polar Subspaces of Multi-Qubit Symplectic Polar Spaces of Small Rank [†]

Metod Saniga [1,*], Henri de Boutray [2], Frédéric Holweck [3] and Alain Giorgetti [2]

[1] Astronomical Institute of the Slovak Academy of Sciences, SK-05960 Tatranská Lomnica, Slovakia
[2] Institut FEMTO-ST, DISC–UFR-ST, Université Bourgogne Franche-Comté, F-25030 Besançon, France; henri.deboutray@femto-st.fr (H.d.B.); alain.giorgetti@femto-st.fr (A.G.)
[3] Laboratoire Interdisciplinaire Carnot de Bourgogne, ICB/UTBM, UMR 6303 CNRS, Université Bourgogne Franche-Comté, F-90010 Belfort, France; frederic.holweck@utbm.fr
* Correspondence: msaniga@astro.sk; Tel.: +421-52-78791-28
[†] This paper is an extended version of the oral presentation given at the 8th European Congress of Mathematics, Portorož, Slovenia, 20–26 June 2021.

Citation: Saniga, M.; de Boutray, H.; Holweck, F.; Giorgetti, A. Taxonomy of Polar Subspaces of Multi-Qubit Symplectic Polar Spaces of Small Rank. *Mathematics* 2021, 9, 2272. https://doi.org/10.3390/math9182272

Academic Editors: Dario Fasino, Domenico Freni and Yang-Hui He

Received: 26 July 2021
Accepted: 13 September 2021
Published: 16 September 2021

Publisher's Note: MDPI stays neutral with regard to jurisdictional claims in published maps and institutional affiliations.

Copyright: © 2021 by the authors. Licensee MDPI, Basel, Switzerland. This article is an open access article distributed under the terms and conditions of the Creative Commons Attribution (CC BY) license (https://creativecommons.org/licenses/by/4.0/).

Abstract: We study certain physically-relevant subgeometries of binary symplectic polar spaces $W(2N-1,2)$ of small rank N, when the points of these spaces canonically encode N-qubit observables. Key characteristics of a subspace of such a space $W(2N-1,2)$ are: the number of its negative lines, the distribution of types of observables, the character of the geometric hyperplane the subspace shares with the distinguished (non-singular) quadric of $W(2N-1,2)$ and the structure of its Veldkamp space. In particular, we classify and count polar subspaces of $W(2N-1,2)$ whose rank is $N-1$. $W(3,2)$ features three negative lines of the same type and its $W(1,2)$'s are of five different types. $W(5,2)$ is endowed with 90 negative lines of two types and its $W(3,2)$'s split into 13 types. A total of 279 out of 480 $W(3,2)$'s with three negative lines are composite, i.e., they all originate from the two-qubit $W(3,2)$. Given a three-qubit $W(3,2)$ and any of its geometric hyperplanes, there are three other $W(3,2)$'s possessing the same hyperplane. The same holds if a geometric hyperplane is replaced by a 'planar' tricentric triad. A hyperbolic quadric of $W(5,2)$ is found to host particular sets of seven $W(3,2)$'s, each of them being uniquely tied to a Conwell heptad with respect to the quadric. There is also a particular type of $W(3,2)$'s, a representative of which features a point each line through which is negative. Finally, $W(7,2)$ is found to possess 1908 negative lines of five types and its $W(5,2)$'s fall into as many as 29 types. A total of 1524 out of 1560 $W(5,2)$'s with 90 negative lines originate from the three-qubit $W(5,2)$. Remarkably, the difference in the number of negative lines for any two distinct types of four-qubit $W(5,2)$'s is a multiple of four.

Keywords: N-qubit observables; binary symplectic polar spaces; distinguished sets of doilies; geometric hyperplanes; Veldkamp lines

1. Introduction

Some fifteen years ago, it was discovered (see, e.g., [1–4]) that there exists a deep connection between the structure of the N-qubit Pauli group and that of the binary symplectic polar space of rank N, $W(2N-1,2)$, where commutation relations between elements of the group are encoded in collinearity relations between points of $W(2N-1,2)$. This connection has subsequently been used to obtain a deeper insight into, for example, finite geometric nature of observable-based proofs of quantum contextuality (for a recent review, see [5]), properties of certain black-hole entropy formulas [6] and the so-called black-hole/qubit correspondence [7], leading to a finite-geometric underpinning of four distinct Hitchin's invariants and the Cartan invariant of form theories of gravity [8] and even to an intriguing finite-geometric toy model of space-time [9]. This group-geometric connection was further strengthened by making use of the concept of geometric hyperplane and that of Veldkamp space of $W(2N-1,2)$ [10]. As per quantum contextuality, famous two-qubit Mermin–Peres magic squares were found to be isomorphic to a special class of geometric hyperplanes of $W(3,2)$ called grids [11], whereas three-qubit Mermin pentagrams were found to have their natural settings in the magic Veldkamp line of $W(5,2)$ [12], being also isomorphic—under the Grassmannian correspondence of type $\mathrm{Gr}(2,4)$—to ovoids of $W(3,2)$ [13]. Concerning the black-hole/qubit correspondence, here a key role is played by the geometric hyperplane isomorphic to an elliptic quadric of $W(5,2)$. Interestingly, form theories of gravity seem to indicate that a certain part of the magic Veldkamp line in the four-qubit symplectic polar space, $W(7,2)$, and the associated extended geometric hyperplanes are of physical relevance as well.

From the preceding paragraph it is obvious that revealing finer traits of the structure of binary symplectic polar spaces of small rank can be vital for further physical applications of these spaces. Having this in view, we will focus on sets of $W(2N-3,2)$'s located in $W(2N-1,2)$, for $N=2,3,4$, providing their comprehensive observable-based taxonomy. Key parameters of our classification of such subspaces of $W(2N-1,2)$ will be: the number of negative lines they contain (which is also an important parameter when it comes to quantum contextuality), the distribution of different types of observables they feature, the character of the geometric hyperplane a subspace of a given type shares with the distinguished (non-singular) quadric of $W(2N-1,2)$ and, in the case of refined 'decomposition' of three-qubit $W(3,2)$'s, also the very structure of their Veldkamp lines.

The paper is organized as follows. Section 2 provides the reader with the necessary finite-geometric background and notation. Section 3 deals with $W(3,2)$ and the hierarchy of its triads. Section 4 addresses the three-qubit $W(5,2)$ and its $W(3,2)$'s; here we classify $W(3,2)$'s in two distinct ways and illustrate the fact that there are four $W(3,2)$'s sharing a geometric hyperplane, or a specific tricentric triad. Section 5 focuses on prominent septuplets of $W(3,2)$'s that are closely related to Conwell heptads with respect to a hyperbolic quadric of $W(5,2)$. Section 6 classifies $W(5,2)$'s living in the four-qubit $W(7,2)$ and furnishes a couple of examples of their composite types. Finally, Section 7 is devoted to concluding remarks.

2. Finite Geometry Background

Given a d-dimensional projective space $\mathrm{PG}(d,2)$ over $\mathrm{GF}(2)$, a *polar space* $\mathcal{P}$ in this projective space consists of the projective subspaces that are *totally isotropic/singular* with respect to a given non-singular bilinear form; $\mathrm{PG}(d,2)$ is called the *ambient projective space* of $\mathcal{P}$. A projective subspace of maximal dimension in $\mathcal{P}$ is called a *generator*; all generators have the same (projective) dimension $r-1$. One calls r the *rank* of the polar space.

Polar spaces of relevance for us are of three types (see, for example, [14,15]): symplectic, hyperbolic and elliptic. The *symplectic* polar space $W(2N-1,2)$, $N \geq 1$, consists of all the points of $\mathrm{PG}(2N-1,2)$, $\{(x_1,x_2,\ldots,x_{2N}) : x_j \in \{0,1\}, j \in \{1,2,\ldots,2N\}\} \setminus \{(0,0,\ldots,0)\}$, together with the totally isotropic subspaces with respect to the standard symplectic form

$$\sigma(x,y) = x_1 y_{N+1} - x_{N+1} y_1 + x_2 y_{N+2} - x_{N+2} y_2 + \cdots + x_N y_{2N} - x_{2N} y_N. \tag{1}$$

This space features

$$|W|_p = 4^N - 1 \tag{2}$$

points and

$$|W|_g = (2+1)(2^2+1)\cdots(2^N+1) \tag{3}$$

generators. The *hyperbolic* orthogonal polar space $\mathcal{Q}^+(2N-1,2)$, $N \geq 1$, is formed by all the subspaces of $\mathrm{PG}(2N-1,2)$ that lie on a given non-singular hyperbolic quadric, with the standard equation

$$x_1 x_{N+1} + x_2 x_{N+2} \ldots + x_N x_{2N} = 0. \tag{4}$$

Each $\mathcal{Q}^+(2N-1,2)$ contains

$$|\mathcal{Q}^+|_p = (2^{N-1}+1)(2^N-1) \tag{5}$$

points and there are

$$|W|_{Q^+} = |Q^+|_p + 1 = (2^{N-1} + 1)(2^N - 1) + 1 \tag{6}$$

copies of them in $W(2N - 1, 2)$. Finally, the *elliptic* orthogonal polar space $Q^-(2N - 1, 2)$, $N \geq 2$, comprises all points and subspaces of $PG(2N - 1, 2)$ satisfying the standard equation

$$f(x_1, x_{N+1}) + x_2 x_{N+2} + \cdots + x_N x_{2N} = 0, \tag{7}$$

where f is an irreducible quadratic polynomial over $GF(2)$. Each $Q^-(2N - 1, 2)$ contains

$$|Q^-|_p = (2^{N-1} - 1)(2^N + 1) \tag{8}$$

points and there are

$$|W|_{Q^-} = |Q^-|_p + 1 = (2^{N-1} - 1)(2^N + 1) + 1 \tag{9}$$

copies of them in $W(2N - 1, 2)$. For both symplectic and hyperbolic polar spaces $r = N$, whereas for the elliptic one $r = N - 1$. The smallest non-trivial symplectic polar space is the $N = 2$ one, $W(3, 2)$, often referred to as the *doily*. It features 15 points (see Equation (2)) and the same number of lines (that are also its generators, see Equation (3)), with three points per line and three lines through a point; it is a self-dual 15_3-configuration and the only one out of 245,342 such configurations that is triangle-free, being, in fact, isomorphic to the generalized quadrangle of order two (GQ(2,2)). This symplectic polar space features ten $Q^+(3,2)$'s (by Equation (6)) and six $Q^-(3,2)$'s (by Equation (9)). A $Q^+(3,2)$ contains nine points and six lines forming a 3×3 grid, so it is also called a grid. A $Q^-(3,2)$ features five pairwise non-collinear points, hence it is an ovoid. A triple of mutually non-collinear points of $W(3,2)$ is called a *triad* and a point collinear with all the three points of a triad is called a *center* of the triad; $W(3,2)$ contains 60 unicentric and 20 tricentric triads.

The N-qubit observables we will be dealing with belong to the set

$$\mathcal{S}_N = \{G_1 \otimes G_2 \otimes \cdots \otimes G_N : G_j \in \{I, X, Y, Z\}, j \in \{1, 2, \ldots, N\}\} \setminus \{\mathcal{I}_N\} \tag{10}$$

where $\mathcal{I}_N \equiv I_{(1)} \otimes I_{(2)} \otimes \ldots \otimes I_{(N)}$,

$$X = \begin{pmatrix} 0 & 1 \\ 1 & 0 \end{pmatrix}, \quad Y = \begin{pmatrix} 0 & -i \\ i & 0 \end{pmatrix}, \quad \text{and} \quad Z = \begin{pmatrix} 1 & 0 \\ 0 & -1 \end{pmatrix} \tag{11}$$

are the Pauli matrices, I is the identity matrix and '$\otimes$' stands for the tensor product of matrices. $\mathcal{S}_N$, whose elements are simply those of the N-qubit Pauli group if the global phase is disregarded, features two kinds of observables, namely *symmetric* (i.e., observables featuring an even number of Y's) and *skew-symmetric*; the number of symmetric observables is $(2^{N-1} + 1)(2^N - 1)$. We shall further employ a finer classification where an observable having $N - 1, N - 2, N - 3, \ldots I$'s will be, respectively, of type $A, B, C, \ldots$; also, whenever it is clear from the context, $G_1 \otimes G_2 \otimes \cdots \otimes G_N$ will be short-handed to $G_1 G_2 \cdots G_N$.

For a particular value of N, the $4^N - 1$ elements of $\mathcal{S}_N$ can be bijectively identified with the same number of points of $W(2N - 1, 2)$ in such a way that the images of two commuting elements lie on the same line of this polar space, and *generators* of $W(2N - 1, 2)$ correspond to maximal sets of mutually commuting elements. If we take the symplectic form defined by Equation (1), then this bijection acquires the form

$$G_j \leftrightarrow (x_j, x_{j+N}), \quad j \in \{1, 2, \ldots, N\}, \tag{12}$$

assuming that

$$I \leftrightarrow (0, 0), \quad X \leftrightarrow (0, 1), \quad Y \leftrightarrow (1, 1), \quad \text{and} \quad Z \leftrightarrow (1, 0). \tag{13}$$

Employing the above-introduced bijection (for more details see, e.g., [12]), it can be shown that given an observable O, the set of symmetric observables commuting with O together with the set of skew-symmetric observables not commuting with O will lie on a certain non-degenerate quadric of $W(2N-1,2)$, this quadric being hyperbolic (resp. elliptic) if O is symmetric (resp. skew-symmetric). We can express this important property by making, whenever appropriate, this associated observable explicit in a subscript, $\mathcal{Q}^{\pm}_{(O)}(2N-1,2)$, noting that there exists a particular hyperbolic quadric associated with $\mathcal{I}$:

$$\mathcal{Q}^{+}_{(\mathcal{I})}(2N-1,2) := \{(x_1, x_2, \ldots, x_{2N}) \in W(2N-1,2) \mid x_1 x_{N+1} + x_2 x_{N+2} + \ldots + x_N x_{2N} = 0\}. \tag{14}$$

Given a point-line incidence geometry $\Gamma(P,L)$, a *geometric hyperplane* of $\Gamma(P,L)$ is a subset of its point set such that a line of the geometry is either *fully* contained in the subset or has with it just a *single* point in common. The *Veldkamp space* $\mathcal{V}(\Gamma)$ of $\Gamma(P,L)$ is the space in which [16]: (i) a point is a geometric hyperplane of Γ and (ii) a line is the collection, denoted $\overline{H'H''}$, of all geometric hyperplanes H of Γ such that $H' \cap H'' = H' \cap H = H'' \cap H$ or $H = H', H''$, where H' and H'' are distinct points of $\mathcal{V}(\Gamma)$. For a $\Gamma(P,L)$ with three points on a line, all Veldkamp lines are of the form $\{H', H'', \overline{H' \Delta H''}\}$ where $\overline{H' \Delta H''}$ is the complement of symmetric difference of H' and H'', i.e., they form a vector space over $GF(2)$. As demonstrated in [10], $\mathcal{V}(W(2N-1,2)) \cong PG(2N,2)$. Its points are both hyperbolic and elliptic quadrics of $W(2N-1,2)$, as well as its perp-sets. Given a point x of $W(2N-1,2)$, the *perp-set* $\widehat{\mathcal{Q}}_{(x)}(2N-1,2)$ of x consists of all the points collinear with it,

$$\widehat{\mathcal{Q}}_{(x)}(2N-1,2) := \{y \in W(2N-1,2) \mid \sigma(x,y) = 0\}; \tag{15}$$

the point x being referred to as the *nucleus* of $\widehat{\mathcal{Q}}_{(x)}(2N-1,2)$.

We shall briefly recall basic properties of the Veldkamp space of the doily, $\mathcal{V}(W(3,2)) \simeq PG(4,2)$, whose in-depth description can be found in [11]. The 31 points of $\mathcal{V}(W(3,2))$ comprise fifteen perp-sets, ten grids and six ovoids—as also illustrated in Figure 1. The 155 lines of $\mathcal{V}(W(3,2))$ split into five distinct types as summarized in Table 1 and depicted in Figure 2. (Table 1, as well as Figures 1 and 2, were taken from [17].)

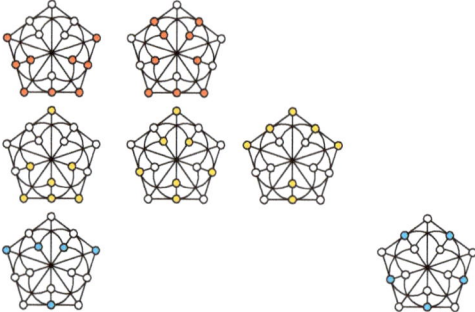

Figure 1. The three kinds of geometric hyperplanes of $W(3,2)$. The 15 points of the doily are represented by small circles and its 15 lines are illustrated by the straight segments as well as by the segments of circles; note that not every intersection of two segments counts for a point of the doily. The upper panel shows grids (red bullets), the middle panel perp-sets (yellow bullets) and the bottom panel ovoids (blue bullets). Each picture—except that located in the bottom right-hand corner—stands for five different hyperplanes, the four others being obtained from it by its successive rotations through 72 degrees around the center of the pentagon.

Table 1. An overview of the properties of the five different types of lines of $\mathcal{V}(W(3,2))$ in terms of the *core* (i.e., the set of points common to all the three hyperplanes forming the line) and the types of geometric hyperplanes featured by a generic line of a given type. The last column gives the total number of lines per each type.

Type	Core	Perps	Ovoids	Grids	#
I	Two Secant Lines	1	0	2	45
II	Single Line	3	0	0	15
III	Tricentric Triad	3	0	0	20
IV	Unicentric Triad	1	1	1	60
V	Single Point	1	2	0	15

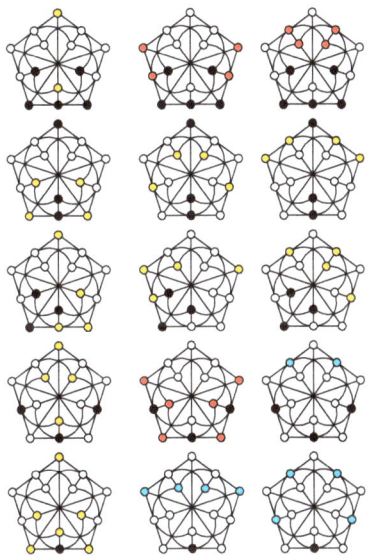

Figure 2. An illustrative portrayal of representatives (rows) of the five (numbered consecutively from top to bottom) different types of lines of $\mathcal{V}(W(3,2))$, each being uniquely determined by the properties of its core (black bullets).

In what follows, we will mainly be focused on $W(2N-3,2)$'s that are located in $W(2N-1,2)$. These are, in general, of two different kinds [10]. A $W(2N-3,2)$ of the first kind, to be called *linear*, is isomorphic to the intersection of two perp-sets with non-collinear nuclei and their number in $W(2N-1,2)$ is

$$|W|_{W_l} = \frac{1}{3}4^{N-1}(4^N - 1). \tag{16}$$

A $W(2N-3,2)$ of the second kind, to be called *quadratic*, is isomorphic to the intersection of a hyperbolic quadric and an elliptic quadric and $W(2N-1,2)$ features

$$|W|_{W_q} = 4^{N-1}(4^N - 1) \tag{17}$$

of them. By way of example, in $W(3,2)$ a linear (resp. quadratic) $W(1,2)$ corresponds to a tricentric (resp. unicentric) triad.

In the sequel, when referring to $W(2N-1,2)$ and its subspaces, we will always have in mind the $W(2N-1,2)$ and its subspaces whose points are labelled by N-qubit observables from the set $\mathcal{S}_N$ as expressed by Equations (12) and (13). Moreover, a linear subspace of such $W(2N-1,2)$ will be called *positive* or *negative* according as the (ordinary) product of the observables located in it is $+\mathcal{I}_N$ or $-\mathcal{I}_N$, respectively. Let us illustrate this point, taking again the $N = 2$ case. Up to isomorphism, there is just one type of the two-qubit doily. Its six observables of type A are IX, XI, IY, YI, IZ and ZI and its nine ones of type B are XX, $XY, XZ, YX, YY, YZ, ZX, ZY$ and ZZ, the latter lying on a particular hyperbolic quadric, $\mathcal{Q}^+_{(YY)}(3,2)$. Among the fifteen lines only the three lines $\{XX, YY, ZZ\}, \{XY, YZ, ZX\}$ and $\{XZ, YX, ZY\}$ are negative, forming also one system of generators of $\mathcal{Q}^+_{(YY)}(3,2)$.

3. W(3,2) and Its Two-Qubit W(1,2)'s

This is a rather trivial case. As already mentioned in Section 2, the doily contains three negative lines, which are all of the same $(B - B - B)$ type. Among its $W(1,2)$'s, we find two types of linear ones and three types of quadratic ones whose properties are summarized in Table 2.

Table 2. Classification of $W(1,2)$'s living in $W(3,2)$. Column one (T) shows the type, columns two and three (O_A and O_B) indicate the number of observables of corresponding types and columns four (W_l) and five (W_q) yield, respectively, the number of 'linear' and 'quadratic' $W(1,2)$'s of a given type.

T	O_A	O_B	W_l	W_q
1	0	3	—	6
2	1	2	—	36
3	1	2	18	—
4	2	1	—	18
5	3	0	2	—

It is worth noticing that the six quadratic $W(1,2)$'s (i.e., unicentric triads) of Type 1 lie on the distinguished quadric $\mathcal{Q}^+_{(YY)}(3,2)$, being in fact its six ovoids.

4. W(5,2) and Its Three-Qubit Doilies

The space $W(5,2)$ contains 63 points, 315 lines and 135 generators, the latter being all Fano planes. Among the 63 canonical three-qubit observables associated to the points, nine are of type A, twenty-seven are type B and twenty-seven are of type C. Through an observable of type C, there pass six negative lines, all being of type $C - C - B$; thus the total number of negative lines of this type is $\frac{27 \times 6}{2} = 81$. Through an observable of type B, there pass four negative lines. Of them, three are of the above-mentioned type and the fourth one is of type $B - B - B$; the total number of negative lines of the latter type is $\frac{27 \times 1}{3} = 9$. As no negative line features an observable of type A, one finds that the $W(5,2)$ accommodates as many as $(81 + 9 =)$ 90 negative lines.

When we pass to $W(3,2)$'s, we find a (much) richer structure, because alongside the types of observables we can employ one more parameter, namely the number of negative lines a given $W(3,2)$ contains. In fact, we find that the 336 linear doilies (see Equation (16)) fall into six different types and the 1008 quadratic ones (see Equation (17)) into seven types; we note in passing that Type 9 splits further into two subtypes depending on whether the two observables of type A do (Type 9A, 162 members) or do not (Type 9B, 54 members) commute. This classification is summarized in Table 3 and is also pictorially illustrated in Figure 3. It is worth noticing here that there are two different types of doilies (Type 3 and Type 6) exhibiting an even number of negative lines.

Figure 3. Representatives—numbered consecutively from left to right, top to bottom—of the 13 different types of three-qubit doilies; Type 1 is top left, Type 13 bottom middle; we also distinguish between Type 9*A* (3rd row, right) and Type 9*B* (4th row, left). The three different types of observables are distinguished by different colors and the negative lines are drawn heavy.

Table 3. Classification of doilies living in $W(5,2)$. Column one (T) shows the type, column two (C^-) the number of negative lines in a doily of the given type, columns three to five (O_A to O_C) indicate the number of observables of corresponding types and columns six (D_l) and seven (D_q) yield, respectively, the number of 'linear' and 'quadratic' doilies of a given type.

T	C^-	O_A	O_B	O_C	D_l	D_q
1	7	0	7	8	—	81
2	7	0	9	6	27	—
3	6	1	5	9	—	108
4	5	2	5	8	162	—
5	5	2	7	6	—	162
6	4	3	5	7	—	324
7	3	0	9	6	9	—
8	3	0	15	0	—	36
9	3	2	7	6	—	216
10	3	2	9	4	81	—
11	3	4	5	6	54	—
12	3	4	7	4	—	81
13	3	6	9	0	3	—

The 27 observables of type B lie on an elliptic quadric of $W(5,2)$, which can be defined as follows:

$$\mathcal{Q}^-_{(YYY)}(5,2) := x_1^2 + x_1x_4 + x_4^2 + x_2^2 + x_2x_5 + x_5^2 + x_3^2 + x_3x_6 + x_6^2 = 0. \tag{18}$$

Here, we took a coordinate basis of $W(5,2)$ in which the symplectic form $\sigma(x,y)$ is given by Equation (1),

$$\sigma(x,y) = (x_1y_4 - x_4y_1) + (x_2y_5 - x_5y_2) + (x_3y_6 - x_6y_3),$$

so that the correspondence between the 63 three-qubit observables (see Equation (10))

$$\mathcal{S}_3 = \{G_1 \otimes G_2 \otimes G_3 : G_j \in \{I,X,Y,Z\}, j \in \{1,2,3\}\} \setminus \mathcal{I}_3$$

and the 63 points of $W(5,2)$ is of the form (see Equation (12))

$$G_j \leftrightarrow (x_j, x_{j+3}), j \in \{1,2,3\},$$

taking also into account Equation (13).

This special quadric $\mathcal{Q}^-_{(YYY)}(5,2)$, as any non-degenerate quadric, is a *geometric hyperplane* of $W(5,2)$. As a doily is also a *subgeometry* of $W(5,2)$, it either lies fully in $\mathcal{Q}^-_{(YYY)}(5,2)$ (Type 8), or shares with $\mathcal{Q}^-_{(YYY)}(5,2)$ a set of points that form a geometric hyperplane; an ovoid (Types 3, 4, 6 and 11), a perp-set (Types 1, 5, 9 and 12) and a grid (Types 2, 7, 10 and 13). One also observes that no quadratic doily shares a grid with $\mathcal{Q}^-_{(YYY)}(5,2)$.

In addition to the distinguished elliptic quadric, there are also three distinguished hyperbolic quadrics in $W(5,2)$, namely: the quadric whose 35 observables feature either two X's or no X,

$$\mathcal{Q}^+_{(ZZZ)}(5,2) := x_4^2 + x_5^2 + x_6^2 + x_1x_4 + x_2x_5 + x_3x_6 = 0, \tag{19}$$

the one whose 35 observables feature either two Y's or no Y (see Equation (14)),

$$\mathcal{Q}^+_{(III)}(5,2) := x_1x_4 + x_2x_5 + x_3x_6 = 0, \tag{20}$$

and the one whose 35 observables feature either two Z's or no Z,

$$Q^+_{(XXX)}(5,2) := x_1^2 + x_2^2 + x_3^2 + x_1x_4 + x_2x_5 + x_3x_6 = 0. \tag{21}$$

Accordingly, there are three distinguished doilies of Type 8, namely the ones the quadric $Q^-_{(YYY)}(5,2)$ shares with these three hyperbolic quadrics.

Take the two-qubit doily. Add formally to each observable, at the same position, the same mark from the set $\{X, Y, Z\}$. Pick up a geometric hyperplane in this three-qubit labeled doily and replace by I the added mark in each observable that belongs to this geometric hyperplane. One obviously gets a three-qubit doily. Now, there are 31 geometric hyperplanes in the doily, three possibilities (X, Y, Z) to pick up a mark and three possibilities (left, middle, right) where to insert the mark; so there will be $31 \times 3 \times 3 = 279$ doilies created this way. In particular, out of the $15 \times 9 = 135$ doilies 'induced' by perp-sets, 81 are of Type 10 and 54 of Type 11; out of the $10 \times 9 = 90$ doilies 'generated' by grids, eighty-one are of Type 12 and nine of Type 8; finally, the $6 \times 9 = 54$ doilies stemming from ovoids are all of the same type 9B. So, if we look at Table 3, all doilies of Types 1 to 7, 27 doilies of Type 8 and all doilies of Type 9A can be regarded as 'genuine' three-qubit guys, nine doilies of Type 8 that originate from grids (henceforth referred to as Type 8') and all doilies of Types 9B to 13 can be viewed as 'built from the two-qubit guy'; with Type 13 doilies being even more two-qubit-like.

This stratification of three-qubit doilies can also be spotted in a different way. Take a representative doily of a particular type, for example, that of Type 3 depicted in Figure 4, top. From its three-qubit labels, keep first only the left mark (bottom left figure), then the middle mark (bottom middle figure) and, finally, the right mark (bottom right figure). In each of these three 'residual' doilies it is easy to see that if you take the points featuring a given non-trivial mark (i.e., X, Y or Z) together with the points featuring I, these always form a geometric hyperplane and the whole set form a Veldkamp line of the doily where the points featuring I represent its core. Employing Table 1 we readily see that this Veldkamp line is of type V (the core is a single point) for the left residual doily, type III (the core is a tricentric triad) for the middle doily and of type IV (the core is a unicentric triad) for the right one. To account this way for the 13 types of three-qubit doilies, we also need the concept of a trivial Veldkamp line of the doily, i.e., a line consisting of a geometric hyperplane counted twice and the full doily, which exactly accounts for those doilies 'generated' by the two-qubit doily. This classification is summarized in Table 4. Here, columns two to six give the number of ordinary Veldkamp lines of a given type, columns seven to nine show the same for trivial Veldkamp lines and the last column corresponds to the degenerate case when all the points of a residual doily bear the label I. Note that all doilies stemming from the two-qubit doily (i.e., Types 8' to 13) feature ordinary Veldkamp lines of the same type.

Using a computer, we have also found out a very interesting property that given a doily and any geometric hyperplane in it, there are three other doilies having the same geometric hyperplane. Figure 5 serves as a visualisation of this fact when the common geometric hyperplane is an ovoid. The four doilies sharing a geometric hyperplane, however, do not stand on the same footing. This is quite easy to spot from our example depicted in Figure 5. A point of the doily is collinear with three distinct points of an ovoid, the three points forming a unicentric triad. Let us pick up such a triad, say $\{ZYI, XYI, YYI\}$ and look for its centers in each of the four doilies. These are IYI (top doily), IIX (left doily), IIY (right doily) and IYZ (bottom doily). We see that the last three observables are mutually anticommuting, whereas the first observable commutes with each of them. This property is found to hold for each of $\binom{5}{3} = 10$ triads contained in an ovoid. Hence, the top doily of Figure 5 has indeed a different footing than the remaining three. A similar $3 + 1$ split up is also observed in any quadruple of doilies having a grid in common because a point of the doily is also collinear with three points of a grid that form a unicentric triad. However, when the shared hyperplane is a perp-set, one gets a different, namely a $2 + 2$ split, because in this case the corresponding triple of points forms a tricentric triad.

Table 4. A refined classification of doilies living in $W(5,2)$. We use the following abbreviations for the cores of Veldkamp lines: 2cl—two concurrent lines, le—line, ttr—tricentric triad, utr—unicentric triad, pt—point, ov—ovoid, ps—perp-set, gr—grid and fl stands for the full doily.

T	2cl	le	ttr	utr	pt	ov	ps	gr	fl
1	1	—	—	—	2	—	—	—	—
2	—	3	—	—	—	—	—	—	—
3	—	—	1	1	1	—	—	—	—
4	—	1	2	—	—	—	—	—	—
5	1	—	—	2	—	—	—	—	—
6	1	—	1	1	—	—	—	—	—
7	—	3	—	—	—	—	—	—	—
8	3	—	—	—	—	—	—	—	—
9A	1	—	—	2	—	—	—	—	—
8′	—	—	2	—	—	—	—	1	—
9B	—	—	2	—	—	1	—	—	—
10	—	—	2	—	—	—	1	—	—
11	—	—	2	—	—	—	1	—	—
12	—	—	2	—	—	—	—	1	—
13	—	—	2	—	—	—	—	—	1

Figure 4. A formal decomposition of a three-qubit doily (top) into three 'single-qubit residuals' (bottom). In each doily of the bottom row, the three geometric hyperplanes forming a Veldkamp line are distinguished by different color, with their common points being drawn black; also, the nuclei of perp-sets are represented by double circles.

A tricentric triad of a linear resp. quadratic doily of $W(5,2)$ defines a line resp. plane in the ambient $PG(5,2)$. The latter type of a triad is found to be shared by four quadratic doilies. Given the three observables of such a triad, there are seven observables commuting with each of them, the corresponding seven points lying in a Fano plane (namely in the polar plane to the plane defined by the triad) in the ambient $PG(5,2)$. One of the seven observables has a distinguished footing as it commutes with each of the remaining six ones, with these six observables forming three commuting pairs. Out of the six observables, one can form just four tricentric triads of which each is complementary to the triad we started with and thus defines with the latter a unique quadratic doily. These properties are also illustrated in Figure 6.

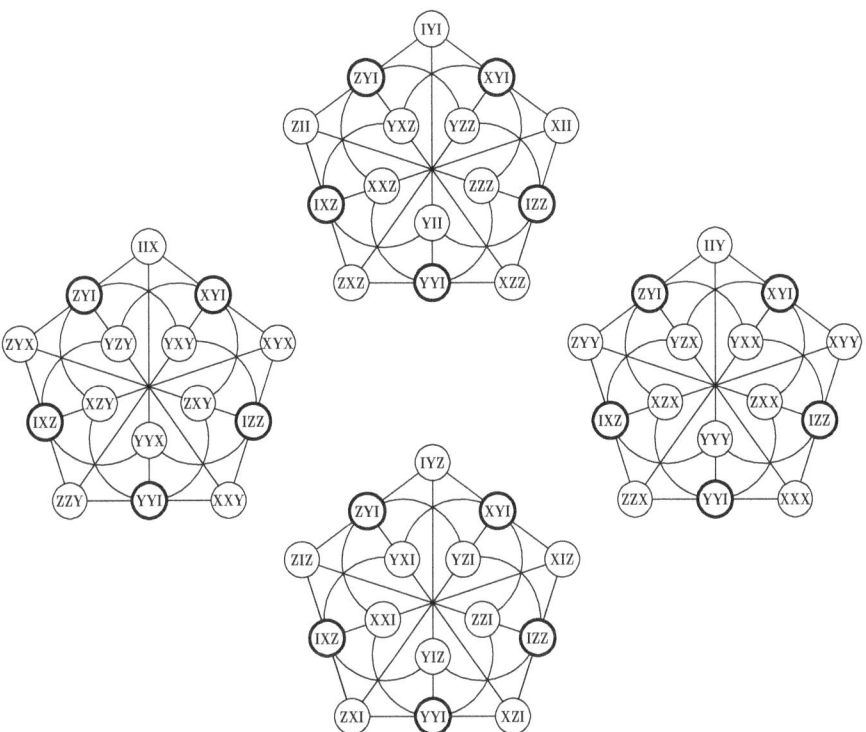

Figure 5. An illustration of the case when four different doilies share an ovoid (boldfaced). The top doily is of Type 11, the bottom one of Type 8, and both the left and right doilies are of Type 3.

Among the 13 different types of three-qubit doilies, there is one type, namely Type 3, which has two remarkable properties. The first property is that there is one point (to be called a deep point) such that all three lines passing through it are negative. Let us take a representative doily of such a type shown in Figure 3, 1st row right. The deep point is ZIZ. Then one sees that there are just two points (to be called zero-points) such that neither of them lies on a negative line; one is IIY and the other is XIZ. These two points and the deep point form in the doily a tricentric triad, hence a copy of 'linear' $W(1,2)$. The second property is related to the fact that through each observable of type B there pass four negative lines. Three of them are such that each features one observable of type B and two observables of type C, whereas the remaining one consists of all observables of type B. Written vertically, the four negative lines passing through our deep point ZIZ are:

$$\begin{array}{|ccc|c|} ZIZ & ZIZ & ZIZ & ZIZ \\ XXX & XYX & XZX & XIX \\ YXY & YYY & YZY & YIY \end{array}$$

We see that the three lines that are located in the doily are of the same type, viz. $B - C - C$. If we also include the fourth negative line, viz. the $B - B - B$ one, we obtain what we can call a 'doily with a tail.' Taking into account the above-mentioned four-doilies-per-hyperplane property, we see that there are altogether 12 doilies, four per each observable, having the same tail and all being of Type 3.

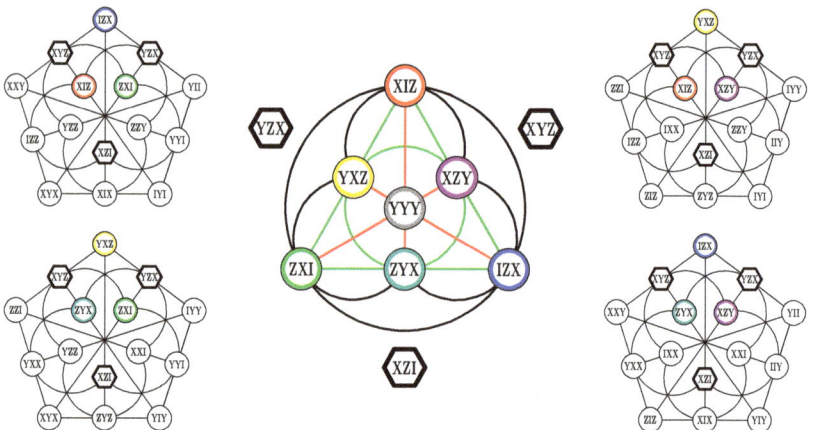

Figure 6. Four three-qubit doilies on a 'planar' tricentric triad (represented by hexagons). The seven observables commuting with the three hexagonal ones are, for better illustration, colored differently. The three red lines of the Fano plane that meet at the distinguished observable (gray) are totally isotropic, whilst the remaining four (depicted green) are not. The four complementary triads (of observables) are illustrated by a full black circle and three half-circles.

5. 'Conwell' Heptads of Doilies in $W(5,2)$

Recall Sylvester's famous construction of $W(3,2)$, see [18]. Given a six-element set $M_6 \equiv \{1,2,3,4,5,6\}$, a duad is an unordered pair $(ij) \in M_6, i \neq j$ and a syntheme is a set of three pairwise disjoint duads, i.e., a set $\{(ij),(kl),(mn)\}$ where $i,j,k,l,m,n \in M_6$ are all distinct. The point-line incidence structure whose points are duads and whose lines are synthemes, with incidence being inclusion, is isomorphic to $W(3,2)$, as also illustrated in Figure 7.

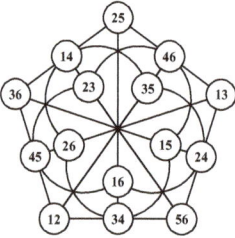

Figure 7. A duad-syntheme model of $W(3,2)$.

Next, take a seven-element set, $M_7 \equiv \{1,2,3,4,5,6,7\}$. One can form from it $\binom{7}{3} = 35$ unordered triples $(ijk), i \neq j \neq k \neq i$. From each set of fifteen triples having the same element in common, we can create a doily using the duad-syntheme construction on that six-element subset of M_7 where the common element is omitted. So, we achieve seven different doilies, one per each element, as depicted in Figure 8. Any two of them have an ovoid in common; because each ovoid is characterized by two elements, say a and b, and it is of the form $\{(abc),(abd),(abe),(abf),(abg)\}$, where $a,b,c,d,e,f,g \in M_7$ are all different, hence it belongs to both the a-doily and the b-doily. Also, any triple is shared by three doilies.

Figure 8. An abstract heptad of doilies on a seven-element set.

A remarkable fact is that this abstract heptad of doilies has a neat realization in our three-qubit $W(5,2)$. To see this, we have to introduce the notion of a *Conwell heptad* of $PG(5,2)$. Given a $\mathcal{Q}^+(5,2)$ of $PG(5,2)$, a Conwell heptad [19] (in the modern language [20] also known as a *maximal exterior set*) with respect to $\mathcal{Q}^+(5,2)$ is a set of seven off-quadric points such that each line joining two distinct points of the heptad is skew to the $\mathcal{Q}^+(5,2)$. There are exactly 8 heptads with respect to $\mathcal{Q}^+(5,2)$. Any two of them have exactly one point in common and any point off $\mathcal{Q}^+(5,2)$ is exactly in two heptads; also any six points of a heptad are linearly independent in $PG(5,2)$. Next [21], let P be a point on $\mathcal{Q}^+(5,2)$. The tangent hyperplane of $\mathcal{Q}^+(5,2)$ at P intersects a heptad C in exactly three points P_1, P_2 and P_3 such that the points P, P_1, P_2 and P_3 are coplanar and P_1, P_2 and P_3 are not collinear; that is, the points P_1, P_2 and P_3 represent a conic in the plane and the point P is its knot (the common intersection of its tangents). Hence, there exists a bijection from the set of the 35 points of $\mathcal{Q}^+(5,2)$ onto the set of the 35 triples of points of C.

Now, let us take a $\mathcal{Q}^+(5,2)$ that belongs to $W(5,2)$, for example, $\mathcal{Q}^+_{(III)}(5,2)$ (see Equation (20)) that accommodates all symmetric observables from $\mathcal{S}_3$. The eight Conwell heptads with respect to this distinguished hyperbolic quadric, expressed in terms of three-qubit observables, are:

1	2	3	4	5	6	7	8
ZYX	YZI	YIZ	YZI	YIZ	YXI	XYI	YII
YIX	YXZ	YZX	YXI	YIX	YZZ	ZYZ	ZYI
YZZ	YXX	YXX	IYZ	XYI	YZX	ZYX	XYZ
XYX	IYI	IYX	IYX	IZY	IYI	ZIY	XYX
IYZ	IXY	ZYZ	ZIY	IXY	IZY	XZY	XIY
YXZ	XZY	IIY	YYY	YYY	XXY	XXY	ZZY
IIY	ZZY	XZY	XIY	ZYI	ZXY	YII	ZXY

We see that each Conwell heptad entails seven pairwise anticommuting observables and so, in fact, corresponds to a set of generators of a seven-dimensional Clifford algebra [22]. Let us pick up one of them, say the heptad number $\boxed{1}$, and associate its observables with the elements of M_7 as follows:

$$1 \leftrightarrow ZYX, 2 \leftrightarrow YIX, 3 \leftrightarrow YZZ, 4 \leftrightarrow XYX, 5 \leftrightarrow IYZ, 6 \leftrightarrow YXZ, 7 \leftrightarrow IIY.$$

From the above-described relation between tangent hyperplanes to a hyperbolic quadric and a Conwell heptad it follows that any unordered triple (ijk), $i, j, k \in M_7$, will be associated with a particular point on $\mathcal{Q}^+_{(III)}(5, 2)$ and its associated observable is the (ordinary) product of the observables associated with elements/points i, j and k; for example, $146 \leftrightarrow ZYX.XYX.YXZ = IXZ$. Hence, all seven doilies of the heptad lie fully in $\mathcal{Q}^+_{(III)}(5, 2)$ and, since no two of them share a line, they partition the set of 105 lines of $\mathcal{Q}^+_{(III)}(5, 2)$. Figure 9 serves as a visualization of this particular 'Conwell' heptad of doilies. As $W(5, 2)$ contains 36 hyperbolic quadrics (see Equation (6)), it features altogether $36 \times 8 = 288$ such heptads of doilies. It is also worth mentioning that employing the well-known Klein correspondence between the points of $\mathcal{Q}^+(5, 2)$ and the lines of PG(3, 2) (see, e.g., Table 15.10 of [23] for more details) and taking into account that the doily is a self-dual object, any Conwell heptad of doilies corresponds to a heptad of mutually azygetic doilies in PG(3,2) (see, e.g., [24]).

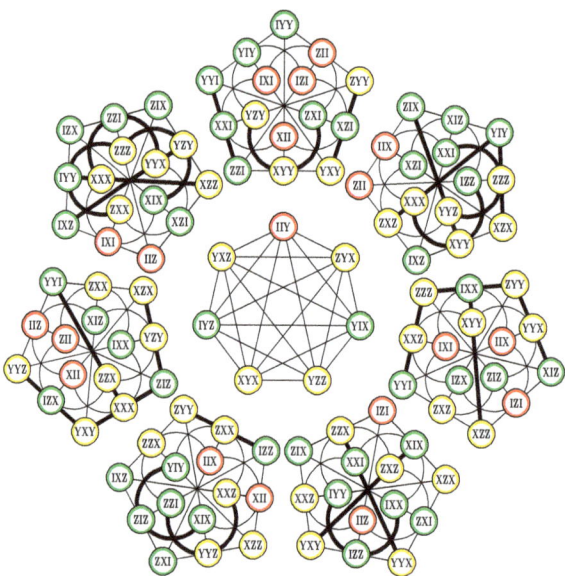

Figure 9. A 'Conwell' heptad of doilies in the three-qubit $W(5,2)$. Following our convention, different types of observables are distinguished by different colors and negative lines are shown in bold.

6. $W(7,2)$ and Its Four-Qubit $W(5,2)$'s

The space $W(7,2)$ possesses 255 points, 5355 lines, 11,475 planes and 2295 generators, the latter being all PG(3,2)'s. Among the 255 canonical four-qubit observables associated to the points, 12 are of type A, 54 of type B, 108 of type C and 81 of type D. Through an observable of type D there pass: four negative lines of type $D - D - D$, totaling to $\frac{81 \times 4}{3} = 108$; twelve negative lines of type $D - D - B$, totaling to $\frac{81 \times 12}{2} = 486$; and twelve negative lines of type $D - C - C$, totaling to $81 \times 12 = 972$. Through an observable of type C there pass, apart from the above-mentioned lines of type $D - C - C$, six negative lines of type $C - C - B$, totaling to $\frac{108 \times 6}{2} = 324$. Through an observable of type B there passes, apart from the already discussed two types of lines, a single negative line of type $B - B - B$, the total number of such lines being $\frac{54 \times 1}{3} = 18$. Since no negative line can contain an observable of type A, the four-qubit $W(7,2)$ thus exhibits five distinct types of negative lines whose total number is $(108 + 486 + 972 + 324 + 18 =) 1908$.

When it comes to $W(5,2)$'s, we find 11 types among their 5440 linear members and as many as 18 types among their 16,320 quadratic cousins, as summarized in Table 5. It represents no difficulty to check that 54 observables of type B and 81 of type D lie on a particular hyperbolic quadric in $W(7,2)$, to be referred to as the distinguished hyperbolic quadric $\mathcal{Q}^+_{(YYYY)}(7,2)$, which is also a geometric hyperplane in the latter space. A $W(5,2)$ either lies fully in this quadric (Types 2 and 21) or shares with it a set of points that forms a geometric hyperplane. Hence, the sum of O_B and O_D in each row of Table 5 must be one of the following numbers: 27 (when the hyperplane of $W(5,2)$ is an elliptic quadric), 31 (a perp-set) and/or 35 (a hyperbolic quadric); for the reader's convenience, the type of such geometric hyperplane is explicitly listed in column 9 of Table 5. One sees that no linear $W(5,2)$ shares with $\mathcal{Q}^+_{(YYYY)}(7,2)$ a perp-set and no quadratic $W(5,2)$ cuts this distinguished quadric in an elliptic quadric. Comparing Table 5 with Table 3, one readily discerns that whereas $W(3,2)$'s in $W(5,2)$ are endowed with both an even and odd number of negative lines, for $W(5,2)$'s in $W(7,2)$ this number is always even; in addition, the difference in C^- for any two distinct types of four-qubit $W(5,2)$'s is a multiple of four.

Let us have a closer look at $W(5,2)$'s featuring 90 (i.e., the smallest possible number of) negative lines. We can easily show that almost all of them originate from the three-qubit $W(5,2)$. First, by adding I to each three-qubit observable at the same position we achieve the four trivial four-qubit $W(5,2)$'s of Type 29. Next, adding to each observable at the same position a mark from the set $\{X, Y, Z\}$, picking up a geometric hyperplane in this four-qubit labeled $W(5,2)$ and replacing by I the added mark of each observable in the geometric hyperplane one gets a four-qubit $W(5,2)$ with 90 negative lines. Now, there are 28 (# of elliptic quadrics) + 36 (# of hyperbolic quadrics) + 63 (# of perp-sets) = 127 geometric hyperplanes in the $W(5,2)$, three possibilities (X, Y, Z) to pick up a mark and four possibilities (left, middle-left, middle-right, right) where to insert the mark. So, there will be $127 \times 3 \times 4 = 1524$ four-qubit $W(5,2)$'s created this way, which only falls short by 36 the total number of $W(5,2)$'s endowed with 90 negative lines (the four guys of Type 29 being, of course, disregarded). A concise summary is given in the last column of Table 5, where the type of geometric hyperplane is further specified by the character/type of the associated (three-qubit) observable. One observes that Type 23 is the only irreducible type of $W(5,2)$'s having 90 negative lines.

We shall illustrate this process by a couple of examples. Let us start with the perp-set of the three-qubit $W(5,2)$ whose nucleus is an observable of type A, say XII. Out of 31 observables commuting with this observable there are seven of type A ($XII, IXI, IIX, IYI, IIY, IZI$ and IIZ), fifteen of type B ($IXX, IXY, IXZ, XXI, XIX, IYX, IYY, IYZ, XYI, XIY, IZX, IZY, IZZ, XZI$, and XIZ) and nine of type C ($XXX, XXY, XXZ, XYX, XYY, XYZ, XZX, XZY$, and XZZ). Hence, out of 32 observables off the perp, there will be $9 - 7 = 2$ of type A, $27 - 15 = 12$ of type B and $27 - 9 = 18$ of type C:

$\widehat{\mathcal{Q}}_{(XII)}$	O_A	O_B	O_C
on	7	15	9
off	2	12	18

Next, each observable of the perp-set acquires a trivial mark I and hence goes into the four-qubit observable of the same type. However, an observable lying off the perp-set gets a non-trivial label X, Y or Z and so yields the four-qubit observable of the subsequent type; that is, $O_A^{(3)} \to O_B^{(4)}$, $O_B^{(3)} \to O_C^{(4)}$ and $O_C^{(3)} \to O_D^{(4)}$. Hence, in our case, we get:

$(\widehat{\mathcal{Q}}_{(XII)})$	O_A	O_B	O_C	O_D
(on – type intact)	7	15	9	0
(off – type shifted)	0	2	12	18
Total	7	17	21	18

Table 5. Classification of $W(5,2)$'s living in $W(7,2)$. Column one (T) shows the type, column two (C^-) the number of negative lines in a $W(5,2)$ of the given type, columns three to six (O_A to O_D) indicate the number of observables featuring three I's, two I's, one I or no I, respectively, columns seven (W_l) and eight (W_q) yield, respectively, the number of 'linear' and 'quadratic' $W(5,2)$'s of a given type, the last but one column depicts the type of intersection of a representative $W(5,2)$ with the distinguished hyperbolic quadric and the last column indicates the type of geometric hyperplane featuring the trivial mark (I) for composite $W(5,2)$'s.

T	C^-	O_A	O_B	O_C	O_D	W_l	W_q	Int	GH
1	130	3	9	33	18	108	—	ell	— — —
2	126	0	24	0	39	—	108	full	— — —
3	126	1	13	27	22	—	1944	hyp	— — —
4	126	2	10	30	21	—	1620	perp	— — —
5	122	1	15	27	20	972	—	hyp	— — —
6	122	2	10	30	21	—	648	perp	— — —
7	118	0	16	32	15	—	324	perp	— — —
8	118	3	9	33	18	648	—	ell	— — —
9	118	3	11	25	24	—	1296	hyp	— — —
10	114	1	15	27	20	324	—	hyp	— — —
11	114	1	17	27	18	—	216	hyp	— — —
12	114	3	13	25	22	1944	—	hyp	— — —
13	114	4	12	28	19	—	1944	perp	— — —
14	110	3	15	25	20	—	1944	hyp	— — —
15	110	5	11	23	24	648	—	hyp	— — —
16	106	5	13	23	22	—	1944	hyp	— — —
17	102	1	21	27	14	—	648	hyp	— — —
18	102	2	18	30	13	—	324	perp	— — —
19	102	3	15	25	20	—	648	hyp	— — —
20	102	4	12	28	19	—	1944	perp	— — —
21	90	0	36	0	27	—	12	full	ell: $O = YYY$
22	90	2	22	30	9	—	108	perp	hyp: all 9 O's featuring two Y's
23	90	3	9	33	18	36	—	ell	— — —
24	90	3	21	25	14	324	—	hyp	perp: all 27 O's of type C
25	90	4	16	28	15	—	324	perp	ell: all 27 O's featuring one Y
26	90	5	15	31	12	324	—	ell	perp: all 27 O's of type B
27	90	6	18	26	13	—	324	perp	hyp: 26 O's having no $Y + III$
28	90	7	17	21	18	108	—	hyp	perp: all 9 O's of type A
29	90	9	27	27	0	4	—	ell	full $W(5,2)$

Comparing with Table 5 we see that this is a four-qubit $W(5,2)$ of Type 28.

As the second example we shall take the case when the geometric hyperplane of $W(5,2)$ is an elliptic quadric generated by an antisymmetric observable of type B, say YXI. This quadric, $\mathcal{Q}^-_{(YXI)}(5,2)$, consists of all symmetric observables that commute with YXI and all antisymmetric observables that anticommute with YXI. In particular, it contains 4 observables of type A (IXI, IIX, IIZ and IYI), 11 observables of type B (XZI, ZZI, YIY, IXX, IXZ, YZI, IYX, IYZ, XIY, ZIY and IZY) and 12 observables of type C (XZX, ZZX, XZZ, ZZZ, YXY, XYY, ZYY, YZX, YZZ, XXY, ZXY and YYY). So, out of 36 observables off the quadric, there will be 5, 16 and 15 of type A, B and C, respectively. In a succinct form,

$\mathcal{Q}^-_{(YXI)}(5,2)$	O_A	O_B	O_C
on	4	11	12
off	5	16	15

From this it follows that the corresponding four-qubit $W(5,2)$ is of Type 25:

$(\mathcal{Q}^-_{(YXI)}(5,2))$	O_A	O_B	O_C	O_D
(on – type intact)	4	11	12	0
(off – type shifted)	0	5	16	15
Total	4	16	28	15

7. Conclusions

We have introduced a remarkable observable-based taxonomy of subspaces of $W(2N-1,2)$, $2 \leq N \leq 4$, whose rank is just one less than that of the ambient space. Alongside the distribution of various types of observables, an important parameter of the classification was the number of negative lines contained in a subspace. As already mentioned in the introduction, this latter parameter is essential in checking whether a given finite geometric configuration is contextual or not. For example, our preliminary analysis shows that all three-qubit and four-qubit doilies are, as their two-qubit sibling, contextual. In a separate paper we plan to address this question in more detail, also employing the degree of contextuality for a variety of other symplectic subspaces. However, when approaching subspaces of higher rank this way, it would be natural to include as parameters the number of negative linear subspaces of every viable dimension from 1 to $N-2$, i.e., consider negative lines, negative planes, . . . , negative generators; so, already in the case of $N=4$ we can add one more parameter, the number of negative planes a four-qubit $W(5,2)$ is endowed with, to achieve an interesting refinement of our Table 5. As the three-qubit $W(5,2)$ features 54 negative planes [25], each composite four-qubit $W(5,2)$ must have the same number of negative planes; in connection with this fact, it would be interesting to check whether each irreducible four-qubit $W(5,2)$ having 90 lines (Type 23) also enjoys this property.

Another interesting extension/variation of our taxonomy would be to take into account the number of negative lines passing through a point of the subspace. Let us call this number the order of a point and for each subspace $W(2s-1,2)$ define the following string of parameters $[p_0, p_1, p_2, \ldots, p_{4^{s-1}-1}]$, where p_k, $0 \leq k \leq 4^{s-1}-1$ stands for the number of points of order k the subspace contains. Applying this to three-qubit doilies ($s=2$), we find the following five patterns (as readily discerned from Figure 3): $[0,9,6,0]$ (Types 1 and 2), $[2,9,3,1]$ (Type 3), $[5,5,5,0]$ (Types 4 and 5), $[6,6,3,0]$ (Type 6) and $[6,9,0,0]$ (Types 7 to 13).

A slightly different possibility of employing our strategy is to analyse other distinguished subgeometries of $W(2N-1,2)$ such as, for example, the split Cayley hexagon of order two [26]. This generalized polygon can be embedded into $W(5,2)$, and in two different ways [27], classical and skew. We have already discerned two distinct kinds of the former and as many as thirteen different types of the latter. Yet a full understanding of the case requires a more rigorous computer-assisted approach and will, therefore, be treated in a separate paper.

Author Contributions: Conceptualization, M.S., H.d.B., F.H. and A.G.; methodology, F.H. and A.G.; software, H.d.B. and A.G.; validation, M.S. and H.d.B.; formal analysis, M.S., H.d.B. and F.H.; data curation, H.d.B. and A.G.; writing—original draft preparation, M.S.; writing—review and editing, M.S.; funding acquisition, M.S., F.H. and A.G. All authors have read and agreed to the published version of the manuscript.

Funding: This work was supported by the Slovak VEGA Grant Agency, Project # 2/0004/20, the French "Investissements d'Avenir" programme, project ISITE-BFC (contract ANR-15-IDEX-03) and by the EIPHI Graduate School (contract ANR-17-EURE-0002).

Institutional Review Board Statement: Not applicable.

Informed Consent Statement: Not applicable.

Data Availability Statement: The computations have been performed on the supercomputer facilities of the Mésocentre de calcul de Franche-Comté; the code is freely available at https://quantcert.github.io/Magma-contextuality/.

Acknowledgments: We are indebted to Zsolt Szabó and Petr Pracna for their help with the figures and thank Péter Vrana for providing us with a list of doilies of $W(5,2)$. We also appreciate constructive remarks/suggestions from four anonymous referees.

Conflicts of Interest: The authors declare no conflict of interest. The funding agencies had no role in the design, execution, interpretation or writing of the study.

References

1. Saniga, M.; Planat, M. Multiple Qubits as Symplectic Polar Spaces of Order Two. *Adv. Stud. Theor. Phys.* **2007**, *1*, 1–4.
2. Planat, M.; Saniga, M. On the Pauli Graph of N-Qudits. *Quantum Inf. Comput.* **2008**, *8*, 127–146.
3. Havlicek, H.; Odehnal, B.; Saniga, M. Factor-Group-Generated Polar Spaces and (Multi-)Qudits. *SIGMA* **2009**, *5*, 096. [CrossRef]
4. Thas, K. The Geometry of Generalized Pauli Operators of N-qudit Hilbert Space, and an Application to MUBs. *Europhys. Lett.* **2009**, *86*, 60005. [CrossRef]
5. Holweck, F. Geometric constructions over $\mathcal{C}$ and $\mathcal{F}_2$ for quantum information. In *Quantum Physics and Geometry*; Ballico, E., Bernardi, A., Carusotto, I., Mazzucchi, S., Moretti, V., Eds.; Lecture Notes of the Unione Matematica Italiana; Springer Nature: Cham, Switzerland, 2019; Volume 25, pp. 87–124.
6. Lévay, P.; Saniga, M.; Vrana, P.; Pracna, P. Black Hole Entropy and Finite Geometry. *Phys. Rev. D* **2009**, *79*, 084036. [CrossRef]
7. Borsten, L.; Duff, M.; Lévay, P. The Black-Hole/Qubit Correspondence: An Up-to-Date Review. *Class. Quantum Grav.* **2012**, *29*, 224008. [CrossRef]
8. Lévay, P.; Holweck, F.; Saniga, M. The Magic Three-Qubit Veldkamp Line: A Finite Geometric Underpinning for Form Theories of Gravity and Black Hole Entropy. *Phys. Rev. D* **2017**, *96*, 026018. [CrossRef]
9. Lévay, P.; Holweck, F. Finite Geometric Toy Model of Spacetime as An Error Correcting Code. *Phys. Rev. D* **2019**, *99*, 086015. [CrossRef]
10. Vrana, P.; Lévay, P. The Veldkamp Space of Multiple Qubits. *J. Phys. A Math. Theor.* **2010**, *43*, 125303. [CrossRef]
11. Saniga, M.; Planat, M.; Pracna, P.; Havlicek, H. The Veldkamp Space of Two-Qubits. *SIGMA* **2007**, *3*, 075. [CrossRef]
12. Lévay, P.; Szabó, Z. Mermin Pentagrams Arising from Veldkamp Lines for Three Qubits. *J. Phys. A Math. Theor.* **2017**, *50*, 095201. [CrossRef]
13. Saniga, M.; Lévay, P. Mermin's Pentagram as an Ovoid of PG(3,2). *Europhys. Lett.* **2012**, *97*, 50006. [CrossRef]
14. Hirschfeld, J.W.P.; Thas, J.A. *General Galois Geometries*; Oxford University Press: Oxford, UK, 1991.
15. Cameron, P.J. Projective and Polar Spaces. In *QMW Maths Notes 13*; Queen Mary and Westfield College, School of Mathematical Sciences: London, UK, 1991.
16. Buekenhout, F.; Cohen, A.M. *Diagram Geometry: Related to Classical Groups and Buildings*; Springer: Berlin/Heidelberg, Germany, 2013.
17. Saniga, M.; Lévay, P.; Planat, M.; Pracna, P. Geometric Hyperplanes of the Near Hexagon $L_3 \times GQ(2,2)$. *Lett. Math. Phys.* **2010**, *91*, 93–104. [CrossRef]
18. Sylvester, J.J. Elementary researches in the analysis of combinatorial aggregation. *Phil. Mag.* **1844**, *24*, 285–296. [CrossRef]
19. Conwell, G.M. The 3-space PG(3,2) and its group. *Ann. Math.* **1910**, *11*, 60–76. [CrossRef]
20. Thas, J.A. Maximal exterior sets of hyperbolic quadrics: The complete classification. *J. Combin. Theory Ser. A* **1991**, *56*, 303–308. [CrossRef]
21. De Clerck, F.; Thas, J.A. Exterior sets with respect to the hyperbolic quadric in PG($2n-1,q$). In *Finite Geometries*; Lecture Notes in Pure and Applied Mathematics; Dekker: New York, NY, USA, 1985; Volume 103, pp. 83–91.
22. Lévay, P.; Planat, M.; Saniga, M. Grassmannian connection between three-and four-qubit observables, Mermin's contextuality and black holes. *J. High Energy Phys.* **2013**, *9*, 037. [CrossRef]
23. Hirschfeld, J.W.P. *Finite Projective Spaces of Three Dimensions*; Oxford University Press: Oxford, UK, 1985.
24. Edge, W.L. The Geometry of the Linear Fractional Group LF(4,2). *Proc. Lond. Math. Soc.* **1954**, *4*, 317–342. [CrossRef]
25. Saniga, M.; Holweck, F.; Jaffali, H. Taxonomy of Three-Qubit Mermin Pentagram. *Symmetry* **2020**, *12*, 534. [CrossRef]
26. Polster, B.; Schroth, A.E.; van Maldeghem, H. Generalized Flatland. *Math. Intell.* **2001**, *23*, 33–47. [CrossRef]
27. Coolsaet, K. The Smallest Split Cayley Hexagon Has Two Symplectic Embeddings. *Finite Fields Their Appl.* **2010**, *16*, 380–384. [CrossRef]

Article

The Reducibility Concept in General Hyperrings

Irina Cristea [1,*,†] and Milica Kankaraš [2,†]

1. Centre for Information Technologies and Applied Mathematics, University of Nova Gorica, 5000 Nova Gorica, Slovenia
2. Department of Mathematics, University of Montenegro, 81000 Podgorica, Montenegro; milica.k@ucg.ac.me
* Correspondence: irina.cristea@ung.si or irinacri@yahoo.co.uk; Tel.: +386-0533-15-395
† These authors contributed equally to this work.

Abstract: By using three equivalence relations, we characterize the behaviour of the elements in a hypercompositional structure. With respect to a hyperoperation, some elements play specific roles: their hypercomposition with all the elements of the carrier set gives the same result; they belong to the same hypercomposition of elements; or they have both properties, being essentially indistinguishable. These equivalences were first defined for hypergroups, and here we extend and study them for general hyperrings—that is, structures endowed with two hyperoperations. We first present their general properties, we define the concept of reducibility, and then we focus on particular classes of hyperrings: the hyperrings of formal series, the hyperrings with P-hyperoperations, complete hyperrings, and (H,R)-hyperrings. Our main aim is to find conditions under which these hyperrings are reduced or not.

Keywords: general hyperring; reducibility; fundamental relation; equivalence

1. Introduction

Algebraic hypercompositional structures, i.e., structures where the result of the synthesis of two elements is a subset of the carrier set, are natural generalizations of the classical algebraic structures, and thus many properties of groups, rings, fields, modules, vector spaces, etc., are extended to hypergroups, hyperrings, hyperfields, hypermodules, vector hyperspaces, etc., more or less in a canonical way. The powerful Hypercompositional Algebra, i.e., the theory of algebraic hypercompositional structures, is given by concepts that do not exist in classical Algebra, and *reducibility* is one of them.

In 1990, James Jantosciak had the idea to describe the behaviour of the elements of a hypergroup with respect to the hyperoperation by defining three equivalence relations, that emphasize the interchangeable role of the elements with respect to the hyperoperation. If two elements in a hypergroup always belong to the same hyperproducts and their hypercomposition with all the elements of the carrier set is the same, then they are called *essentially indistinguishable* [1]. A hypergroup is *reduced* if the equivalence class of each element is a singleton with respect to the essentially indistinguishable relation.

In addition, Jantosciak noticed also that factorizing the hypergroup by this equivalence one obtains a reduced hypergroup, called the *reduced form* of the initial hypergroup. Therefore, he proposed to divide into two parts the study of the hypergroups: the study of the reduced hypergroups and the study of the hypergroups having the same reduced form [1]. Due to this important property, he named as *fundamental* the three equivalences used in the definition of the concept of reducibility.

Inspired by this pioneer paper and the further results obtained by researchers on the reducibility of various types of hypergroups [2–5], we extend here this property to hyperrings. These are algebraic structures containing an additive and a multiplicative part connected by the distributivity law, where at least one of them is a hypercompositional structure. The first type of hyperring was introduced by Krasner [6] as a hypercomposi-

tional structure whose additive part is a canonical hypergroup, and the multiplicative one is a semigroup.

Currently, this structure is known as *Krasner hyperring* and considered as an additive hyperring, in order to emphasize that the addition is a hyperoperation. If one considers the multiplication to be a hyperoperation, while the addition stays an operation, the notion of *multiplicative hyperring* was introduced in 1982 by Rota [7], where the additive part is an abelian group and the multiplicative one is a semihypergroup. If both the addition and the multiplication are hyperoperations, then we talk about *general hyperrings*.

There are several types of general hyperrings: one studied by Corsini [8] in 1975 in connection with feebly hypermodules; one defined in 1973 by Mittas [9,10] and called superring, having as additive part a canonical hypergroup; another one studied in 1989 by Spartalis [11], where the additive part is a hypergroup and the multiplicative one is a semihypergroup. Expository and survey articles on this topic have been published by Nakassis [12] in 1988 and recently by Massouros [13,14].

The aim of this manuscript is to define and study the concept of reducibility in the class of hyperrings. We will do this in a very natural way, by extending the three fundamental relations defined by J. Jantosciak to both addition and multiplication. It is clear that it makes sense to do this only in a general hyperring, where the carrier set is endowed with two hyperoperations, because these fundamental equivalences are equivalent with the equality relation when they are considered with respect to an operation.

Thus, the study of the reducibility in a Krasner hyperring or in a multiplicative hyperring is not relevant since it reduces to the study of the reducibility of a hypergroup. This study, covered in Section 4, was conducted first in a general way and then for particular classes of general hyperrings, as the hyperring of formal series, or hyperrings with P-hyperoperations. Particular attention is given to the complete hyperrings and (H,R)-hyperrings. The paper ends with some conclusive remarks and ideas for future work.

2. Preliminaries on Hypergroups and Hyperrings

For a non-empty set H, we denote, by $\mathcal{P}^*(H)$, the family of all non-empty subsets of H. A binary hyperoperation, also called a hyperproduct, is an application $\circ : H \times H \to \mathcal{P}^*(H)$ and the pair $(H, \circ)$ is called a hypergrupoid. It is important to stress that, in a hypergrupoid, the hyperproduct $x \circ y$ between two arbitrary elements x and y in H is a non-empty subset of H. This is a property that we cannot find in classical algebraic structures, such as groupoids and semigroups.

The hyperoperation is extended to non-empty subsets of H as $A \circ B = \bigcup_{a \in A, b \in B} a \circ b$.

If the hyperoperation is associative, then the hypercompositional structure $(H, \circ)$ is a semihypergroup, which becomes a hypergroup when the reproducibility property also holds: $x \circ H = H \circ x = H$ for all $x \in H$.

The link between groups and hypergroups is established by the fundamental relation β defined on a semihypergroup $(H, \circ)$ as follows: $\beta = \cup_{n \geq 1} \beta_n$ where β_1 is the diagonal relation on H and for any $n > 1$, and $x, y \in H$, $x\beta_n y \Leftrightarrow \exists a_1, a_2, \ldots, a_n \in H$ such that $\{x, y\} \subseteq \prod_{i=1}^{n} a_i = a_1 \circ a_2 \circ \cdots \circ a_n$. It is clear that β is a reflexive and symmetrical relation, but generally not transitive. That is why we take its transitive closure β^*, which is an equivalence relation. Recall that, for hypergroups, we have $\beta = \beta^*$ [15,16], and the quotient $(H/\beta^*, \otimes)$ is a group with the operation $\beta^*(x) \otimes \beta^*(y) = \beta^*(z)$ for all $x, y \in H$ and $z \in x \circ y$.

Considering now the canonical projection $\varphi_H : H \to H/\beta^*$, which is a good homomorphism, i.e., $\varphi_H(x \circ y) = \varphi_H(x) \otimes \varphi_H(y)$, we may define the heart (or core) of a hypergroup H as the set $\omega_H = \{x \in H | \varphi_H(x) = 1\}$, where 1 is the identity of the group H/β^*. This set plays an important role for the structure of a hypergroup, because, if we know it, then we can determine the complete closure of a subset of H.

More exactly, if A is a non-empty subset of H, it is called a *complete part* [17] of H if for any natural number n and any elements $a_1, a_2, \ldots, a_n$ in H, the following implication holds: $A \cap \prod_{i=1}^{n} a_i \neq \emptyset \Rightarrow \prod_{i=1}^{n} a_i \subseteq A$. The intersection of all complete parts of H containing the subset A is called the *complete closure* of A in H, and it is denoted by $C(A)$. Moreover, $C(A) = \omega_H \circ A = A \circ \omega_H$. The complete closure of a set helps us to define a particular type of hypergroups, called *complete hypergroups*.

We say that a hypergroup $(H, \circ)$ is complete if $x \circ y = C(x \circ y)$ for all $x, y \in H$. Moreover, if $(H, \circ)$ is a complete hypergroup, then $x \circ y = C(\{a\}) = \beta(a)$ for every $x, y \in H$ and $a \in x \circ y$. In practice, this definition is substituted with the representation theorem, which we recall here below.

Theorem 1 ([18]). *A hypergroup $(H, \circ)$ is complete if and only if it can be partitioned as $H = \bigcup_{g \in G} A_g$, where G and the subsets A_g of H satisfy the following conditions:*

(1) $(G, \cdot)$ *is a group.*
(2) *For all $g_1 \neq g_2 \in G$, there is $A_{g_1} \cap A_{g_2} = \emptyset$.*
(3) *If $(a, b) \in A_{g_1} \times A_{g_2}$, and then $a \circ b = A_{g_1 g_2}$.*

It is clear that any group is a complete hypergroup; however, this case is not interesting for our study. This is why we will consider only proper complete hypergroups, i.e., complete hypergroups that are not groups. The heart ω_H of a complete hypergroup $(H, \circ)$ has an interesting property: it coincides with the set of identities of H. The complete hypergroups have been studied for their general properties [19], or in connection with their fuzzy grade [20], for their commutativity degree [21], or in relation with their size [22].

General hyperrings are algebraic structures equipped with two hyperoperations, i.e., hyperaddition and hypermultiplication that satisfy the distributivity condition. Here, we will recall the definitions of some particular types of general hyperrings, which will be considered further on in the paper.

Definition 1 ([23]). *A hypercompositional structure $(R, \oplus, \odot)$ is called a hyperringoid if*

1. $(R, \oplus)$ *is a hypergroup.*
2. $(R, \odot)$ *is a semigroup.*
3. *The operation "$\odot$" distributes on both sides over the hyperoperation "$\oplus$."*

This algebraic hypercompositional structure was first introduced by Massouros [24] in a study on languages and automata. If we request that both addition and multiplication are hyperoperations, then the hyperringoid becomes a general hyperring.

Definition 2 ([25]). *A triple $(R, \oplus, \odot)$ is a general hyperring if:*

1. $(R, \oplus)$ *is a hypergroup.*
2. $(R, \odot)$ *is a semihypergroup.*
3. *The multiplication is distributive with respect to the addition, i.e., for all $a, b, c \in R$ $a \odot (b \oplus c) = (a \odot b) \oplus (a \odot c)$ and $(a \oplus b) \odot c = (a \odot c) \oplus (b \odot c)$.*

The H_v-structures were introduced by Vougiouklis during the 4th AHA Congress in 1990 [26] as hypercompositional structures with weak associative hyperoperations.

Definition 3. *The hyperstructure $(H, \cdot)$ is an H_v-semigroup if $x \cdot (y \cdot z) \cap (x \cdot y) \cdot z \neq \emptyset$ for all $x, y, z \in H$. If also the eproducibility property is valid, i.e., $a \cdot H = H \cdot a = H$, then $(H, \cdot)$ is called an H_v−group.*

Definition 4. *A multi-valued system $(R, \oplus, \odot)$ is an H_v−ring if:*

1. $(R, \oplus)$ *is an H_v-group.*
2. $(R, \odot)$ *is an H_v-semigroup.*

3. The multiplication weakly distributes with respect to the addition, i.e., for all $a, b, c \in R$ $(a \odot (b \oplus c)) \cap ((a \odot b) \oplus (a \odot c)) \neq \emptyset$ and $((a \oplus b) \odot c) \cap ((a \odot c) \oplus (b \odot c)) \neq \emptyset$.

It is important to recall here one of the main properties of hypercompositional structures: the quotient of a group with respect to any of its subgroups is a hypergroup, while the quotient of a group by any equivalence relation gives birth to an H_v-group [14]. A recently published overview of the theory of weak-hyperstructures is covered in [26,27].

In the following, we will recall the construction of two types of hyperrings, which we will study in the next section. The first one leads to an H_v-ring obtained from a ring. This structure was principally studied by Spartalis and Vougiouklis [28,29], in connection with homomorphisms and numeration.

Let $(R, +, \cdot)$ be a ring and P_1 and P_2 be non-empty subsets of R. The hyperoperations defined by $xP_1^*y = x + y + P_1$ and $xP_2^*y = x \cdot y \cdot P_2$ for all $x, y \in R$ are called P-hyperoperations [30].

Theorem 2 ([29]). *Let $(R, +, \cdot)$ be a ring, $Z(R)$ be the center of the multiplicative semigroup $(R, \cdot)$ and P_1, P_2 be non-empty subsets of R. If $0 \in P_1$ and $Z(R) \cap P_2 \neq \emptyset$, then (R, P_1^*, P_2^*) is an H_v-ring.*

This kind of H_v-ring is called an H_v-ring with P-hyperoperations.

We end this section by recalling the construction of the hyperring of the formal series [31,32]. Based on this, we studied the structure of the set of polynomials over a hyperring.

Let $(R, +, \cdot)$ be a general commutative hyperring. A formal series with coefficients in R is an infinite sequence $(a_0, a_1, a_2, \ldots, a_n, \ldots)$ of elements a_i in R. The set of all such series is denoted by $R[[x]]$. We say that two series $(a_0, a_1, a_2, \ldots, a_n, \ldots)$ and $(b_0, b_1, b_2, \ldots, b_n, \ldots)$ are equal if and only if $a_i = b_i$ for all indices i.

Let define on $R[[x]]$ the addition by

$$(a_0, a_1, \ldots, a_n, \ldots) \oplus (b_0, b_1, \ldots, b_n, \ldots) = \{(c_0, c_1, \ldots, c_n, \ldots), c_k \in a_k + b_k\}$$

and the multiplication by

$$(a_0, a_1, \ldots, a_n, \ldots) \odot (b_0, b_1, \ldots, b_n, \ldots) = \{(c_0, c_1, \ldots, c_n, \ldots), c_k \in \sum_{i+j=k} a_i \cdot b_j\}.$$

The structure $(R[[x]], \oplus, \odot)$ is a general hyperring. We recall that the set of the polynomials $R[x]$ with coefficients in R is a superring with the same hyperoperations $\oplus$ and $\odot$ defined above [33]. This means that $(R[x], \oplus)$ is a canonical hypergroup, $(R[x], \odot)$ is a semi-hypergroup such that 0 is a bilaterally absorbing element and the multiplication is weakly distributive on the left side with respect to the addition, i.e., $f \odot (g \oplus h) \subseteq f \odot g \oplus f \odot h$, for $f, g, h \in R[x]$.

3. Short Review of the Reducibility in Hypergroups

In this section, we briefly recall the notion of the reducibility of hypergroups. We start with the three fundamental relations introduced by Jantosciak [1] on an arbitrary hypergroup.

Definition 5 ([1]). *Two elements x, y in a hypergroup $(H, \circ)$ are called:*

1. *operationally equivalent or by short o-equivalent, and we write $x \sim_o y$, if $x \circ a = y \circ a$ and $a \circ x = a \circ y$, for any $a \in H$;*
2. *inseparable or by short i-equivalent, and we write $x \sim_i y$, if, for all $a, b \in H$, $x \in a \circ b \iff y \in a \circ b$; and*
3. *essentially indistinguishable or by short e-equivalent, and we write $x \sim_e y$, if they are operationally equivalent and inseparable.*

Definition 6 ([1]). *A hypergroup H is called reduced if, for any $x \in H$, the equivalence class of x with respect to the essentially indistinguishable relation $\sim_e$ a singleton.*

Proposition 1 ([5]). *A total hypergroup is not reduced.*

Theorem 3 ([5]). *Any proper complete hypergroup is not reduced.*

Proposition 2. *Let ϕ be a good surjective homomorphism from the hypergroup $(R, +)$ to the hypergroup $(T, \oplus)$. If two elements are essentially indistinguishable with respect to the hyperoperation $+$, then their images are essentially indistinguishable with respect to the hyperoperation $\oplus$.*

Proof. Let x and y be elements from R such that $x + a = y + a$, where $a \in R$. This gives $\{\phi(l) | l \in x + a\} = \{\phi(k) | k \in y + a\}$, and thus $\phi(x + a) = \phi(y + a)$. From here, $\phi(x) \oplus \phi(a) = \phi(y) \oplus \phi(a)$. Denote $\phi(a) = b$ and $\phi(x) = x_1, \phi(y) = y_1$. Thus, $x_1 \oplus b = y_1 \oplus b$. If the equality $x + a = y + a$ holds for every $a \in H$, then the last equality holds for all $b \in T$ since $\{\phi(a) | a \in R\} = T$. Assuming $a + x = a + y$ for all $a \in R$, similarly, we obtain $\phi(a) \oplus \phi(x) = \phi(a) \oplus \phi(y)$ for all $a \in R$. Hence, if $x \sim_o^+ y$ then $\phi(x) \sim_o^\oplus \phi(y)$.

Let $x \sim_i^+ y$, i.e., $x \in a + b$ if and only if $y \in a + b$ for all $a, b \in R$. From this equivalence, we find that $\phi(x) \in \{\phi(l) | l \in a + b\}$ if and only if $\phi(y) \in \{\phi(k) | k \in a + b\}$, and thus $\phi(x) \in \phi(a+b)$ if and only if $\phi(y) \in \phi(a+b)$. Since ϕ is homomorphism, $\phi(x) \in \phi(a) \oplus \phi(b)$ if and only if $\phi(y) \in \phi(a) \oplus \phi(b)$. Let $\phi(x) = x_1, \phi(y) = y_1$ and $\phi(a) = a_1, \phi(b) = b_1$. Since the mapping is surjective $a_1 \oplus b_1$ covers whole set T. Hence, $x_1 \in a_1 \oplus b_1$ is equivalent to $y_1 \in a_1 \oplus b_1$, for all $a_1, b_1 \in T$. Here, $x \sim_i^+ y$ implies $\phi(x) \sim_i^\oplus \phi(y)$. The definition of the essential indistinguishability relation, together with the above implications, concludes the proof of our claim. □

4. Reducibility in Hyperrings

In a semigroup, the equivalences $\sim_o$ and $\sim_i$ coincide with the diagonal relation, i.e., $x \sim_o y \iff x \sim_i y \iff x = y$. Thus, in a Krasner hyperring or in a multiplicative hyperring (when the referential set is equipped with a hyperoperation and an operation), these two equivalences are not significant. Therefore, in this section, our first aim is to study relationships between these equivalences in a general hyperring $(R, \oplus, \odot)$, where addition and multiplication are both hyperoperations.

For any element $x \in R$, we denote by $\hat{x}_r^\oplus$ and $\hat{x}_r^\odot$, the equivalence classes of x with respect to the hyperoperations $\oplus$ and $\odot$, respectively, where $r \in \{o, i, e\}$ denotes the type of the equivalence that we consider in Definition 7. In the following, by hyperring, we mean a general hyperring.

Definition 7. *We say that two elements x and y in a hyperring $(R, \oplus, \odot)$ are operationally equivalent, inseparable or essentially indistinguishable if they have the same property with respect to both hyperoperations, i.e.,*

1. *$x \sim_o y$ if $x \oplus a = y \oplus a, a \oplus x = a \oplus y$ and $a \odot x = a \odot y, x \odot a = y \odot a$, for all $a \in R$.*
2. *$x \sim_i y$ if $x \in a \oplus b \iff y \in a \oplus b$, for all $a, b \in R$ and $x \in c \odot d \iff y \in c \odot d$, for all $c, d \in R$.*
3. *$x \sim_e y$ if $x \sim_o y$ and $x \sim_i y$.*

Definition 8. *A hyperring R is called reduced if the equivalence class of each element $x \in R$ with respect to the essentially indistinguishable relation $\sim_e$ is a singleton, i.e., $\hat{x}_e = \{x\}$ for any $x \in R$.*

The equivalence class of any element x in R with respect to the essentially indistinguishability relation $\sim_e$ is obtained as $\hat{x}_e = \hat{x}_e^\oplus \cap \hat{x}_e^\odot = (\hat{x}_o^\oplus \cap \hat{x}_i^\oplus) \cap (\hat{x}_o^\odot \cap \hat{x}_i^\odot)$. It is important to stress on the following property. If at least one of the hypergroupoids $(R, \oplus)$ or $(R, \odot)$ is reduced, then the hyperring $(R, \oplus, \odot)$ is reduced, too. Reciprocally, if $(R, \oplus, \odot)$ is reduced, then the hypergroupoids $(R, \oplus)$ and $(R, \odot)$ can be reduced or not, as one can see in the following examples.

Example 1. *Let $(R, \oplus, \odot)$ be a hyperring defined by the following Cayley tables:*

$\oplus$	e	a
e	R	R
a	R	R

$\odot$	e	a
e	e	R
a	R	a

Since $(R, \oplus)$ is a total hypergroup, based on Proposition 1, it is not reduced. Here, $\hat{a}_e^{\oplus} = \hat{e}_e^{\oplus} = \{e, a\}$. However, it is easy to check that the hypergroup $(R, \odot)$ is a reduced hypergroup, and $\hat{a}_e^{\odot} = \{a\}, \hat{e}_e^{\odot} = \{e\}$. All together, it gives that $\hat{e}_e = \{e\}$ and $\hat{a}_e = \{a\}$ which shows that $(R, \oplus, \odot)$ is a reduced hyperring.

Example 2. *Let the hyperring $(R, \oplus, \odot)$ be defined by the following Cayley tables:*

$\oplus$	x	y	z
x	x,y	x,y	R
y	x,y	x,y	R
z	R	R	R

$\odot$	x	y	z
x	R	R	R
y	R	y,z	y,z
z	R	y,z	y,z

It is elementary to check that the algebraic hyperstructure $(R, \oplus, \odot)$ is a general hyperring. Since the rows corresponding to x and y are equal in $(R, \oplus)$ and both x, y appear in the same hyperproducts $a \oplus b$, it follows that $x \sim_e^{\oplus} y$, which implies that $(R, \oplus)$ is not reduced. Similarly, $(R, \odot)$ is not a reduced hypergroup since $y \sim_e^{\odot} z$. But, $\hat{x}_e = \hat{x}_e^{\oplus} \cap \hat{x}_e^{\odot} = \{x, y\} \cap x = \{x\}$. Similarly, $\hat{y}_e = \{y\}$, and $\hat{z}_e = \{z\}$, which proves that $(R, \oplus, \odot)$ is a reduced hyperring.

4.1. Some Properties of the Reducibility in Hyperrings

In the following, subsections, we suppose that the ring $(R, +, \cdot)$ has no zero divisors.

First, we will present some relationships between the operationally equivalence (inseparability) with respect to the first hyperoperation of the hyperring and the operationally equivalence (inseparability) with respect to the second hyperoperation of the considered hyperring.

Proposition 3. *Let $(R, \oplus, \odot)$ be a general hyperring, where the hypergroup $(R, \oplus)$ contains a scalar identity. Then, the essentially indistinguishability with respect to the hyperoperation "$\oplus$" implies the essentially indistinguishability with respect to the hyperoperation "$\odot$", i.e., $x \sim_e^{\oplus} y \Rightarrow x \sim_e^{\odot} y$, for all $x, y \in R$.*

Proof. We denote by 0 the scalar identity in $(R, \oplus)$. Let x and y be two elements in R such that $x \sim_0^{\oplus} y$, i.e., $x \oplus a = y \oplus a$ and $a \oplus x = a \oplus y$, for all $a \in R$. This means that, for any $u \in R$ such that $u \in x \oplus a$, it holds $u \in y \oplus a$. Let u in $a \odot x$. Then, since $x = x \oplus 0$, it follows that $u \in a \odot (x \oplus 0)$. Now, using $x \oplus 0 = y \oplus 0$, we get $u \in a \odot (y \oplus 0) = a \odot y$. By symmetry, we can conclude that $a \odot x = a \odot y$, and $x \odot a = y \odot a$, for all $a \in R$. Hence, $x \sim_0^{\odot} y$.

Let us suppose that $x \in a \oplus b$ if and only if $y \in a \oplus b$, for any $a, b \in R$. Let c and d be elements in the hyperring such that $x \in c \odot d$. Thus, $x \in (c \oplus 0) \odot d$. Using the distributibivity, we obtain $x \in c \odot d \oplus 0 \odot d = \{m \oplus n | m \in c \odot d, n \in 0 \odot d\}$. Since x and y appear in the same hyperproducts $a \oplus b$, for any $a, b \in R$, it follows that y also belongs to the same hyperproduct, which gives $y \in c \odot d \oplus 0 \odot d$, i.e., $y \in c \odot d$. This proves the implication $x \sim_i^{\oplus} y \Rightarrow x \sim_i^{\odot} y$. Now the conclusion of the result is clear. □

Corollary 1. *Let $(R, \oplus, \odot)$ be a general hyperring such that $(R, \oplus)$ contains a scalar identity. If $(R, \oplus)$ is not a reduced hypergroup, then the hyperring $(R, \oplus, \odot)$ is not reduced, too.*

Proof. If $(R, \oplus)$ is not a reduced hypergroup, then there exist two distinct elements x and y in R such that $x \sim_e^{\oplus} y$. Based on Proposition 3, it follows that $x \sim_e^{\odot} y$, meaning that the hyperring $(R, \oplus, \odot)$ is not reduced. □

In the second part of this section, we present some particular types of general hyperrings and highlight some of their properties related to the reducibility. We start with some aspects regarding the reducibility of the hyperring of formal series.

Proposition 4. *Let $R[[x]]$ be the hyperring of the formal series with coefficients in the general commutative hyperring $(R, +, \cdot)$. The hyperring $(R, +, \cdot)$ is reduced if and only if the hyperring $(R[[x]], \oplus, \odot)$ is reduced.*

Proof. Let us suppose that the hyperring R is not reduced, i.e., there exist elements a and b such that $a + x = b + x$ and $x + a = x + b$ for all $x \in R$, and also a and b appear in the same hyperproducts $c + d$, where $c, d \in R$. As a direct consequence, the formal series $(a, a, \ldots, a, \ldots)$ and $(b, b, \ldots, b, \ldots)$ are operationally equivalent and inseparable with respect to the hyperoperation $\oplus$. Analogously, the implication holds also if we consider the multiplicative hyperoperation. Hence, if R is not reduced, then the hyperring $(R[[x]], \oplus, \odot)$ is not reduced, too.

Let us prove now that the reducibility in $(R, +, \cdot)$ implies the reducibility in $(R[[x]], \oplus, \odot)$. For that purpose, let us assume that the hyperring $R[[x]]$ is not reduced. Then, there exist two formal series $(a_1, a_2, \ldots, a_n, \ldots)$ and $(b_1, b_2, \ldots, b_n, \ldots)$, which are operationally equivalent with respect to the hyperoperation $\oplus$. This implies that:

$$(a_1, a_2, \ldots, a_n, \ldots) \oplus (x_1, x_2, \ldots, x_n, \ldots) = \tag{1}$$
$$(b_1, b_2, \ldots, b_n, \ldots) \oplus (x_1, x_2, \ldots, x_n, \ldots), \tag{2}$$

and

$$(x_1, x_2, \ldots, x_n, \ldots) \oplus (a_1, a_2, \ldots, a_n, \ldots) = \tag{3}$$
$$(x_1, x_2, \ldots, x_n, \ldots) \oplus (b_1, b_2, \ldots, b_n, \ldots), \tag{4}$$

for any formal series $(x_1, x_2, \ldots, x_n, \ldots) \in R[[x]]$. Using the definition of the hyperaddition in $(R[[x]], \oplus, \odot)$, the previous equalities give that $a_i + x_i = b_i + x_i$ and $x_i + a_i = x_i + b_i$ for any arbitrary $x_i \in R$. Hence, $a_i \sim_o^+ b_i$ for any elements $a_i, b_i \in R$, which are the coordinates of the considered formal series.

Assuming now that the series $(a_1, a_2, \ldots, a_n, \ldots)$ and $(b_1, b_2, \ldots, b_n, \ldots)$ are inseparable with respect to the hyperoperation $\oplus$, it easily follows that a_i and b_i appear in the same hyperproducts $c + d$, where $c, d \in R$, so they are inseparable with respect to the hyperproduct "$+$" on R. Similarly, we can prove that the essentially indistinguishability with respect to the hypermultiplication "$\odot$" implies essentially indistinguishability with respect to the hyperoperation "$\cdot$". We finally find that $(R, +, \cdot)$ is not reduced, which concludes the proof. □

The next part of this subsection is dedicated to the study of reducibility of the hyperrings with P-hyperoperations.

Proposition 5. *Let $(R, +, \cdot)$ be a commutative principal ideal domain with two units, i.e., 1 and -1. If $P_1 = nR$, with $n \in R$, and $P_2 = R$, then the structure (R, P_1^*, P_2^*) is a commutative H_v-ring with P-hyperoperations, which is a non-reduced hyperring.*

Proof. Any principal ideal contains 0; therefore, $0 \in P_1$. As the ring R is commutative, it coincides with its center $Z(R)$, and therefore the set $P_2 = R$ has a non-empty intersection with $Z(R)$, and thus the conditions of Theorem 2 are satisfied, proving that the hyperstructure (R, P_1^*, P_2^*) is a commutative H_v-ring.

Let x and y be distinct elements such that $xP_1^*a = yP_1^*a$ for all a in R, meaning that $x + a + P_1 = y + a + P_1$, i.e., $x + a + nR = y + a + nR$, for the fixed element $n \in R$ and any $a \in R$. Since the principal ideal nR is a subgroup, then the equality holds whenever $x - y \in nR$. Therefore, the elements x and y are operationally equivalent with respect to the hyperoperation P_1^* if and only if $x - y \in nR$.

Let x and y be two elements such that $x - y \in nR$. Let us suppose that $x \in aP_1^*b$, where $a, b \in R$. The element x belongs to $a + b + nR$, i.e., $x = a + b + n \cdot s$, with $s \in R$. Since $x = y + n \cdot k$, with $k \in R$, it follows that $y + n \cdot k = a + b + n \cdot s$, meaning that $y \in a + b + nR$. Hence, $y \in aP_1^*b$. Similarly, we can prove the other implication. Thus, $x \sim_i^{P_1^*} y$. Conversely, if $x \sim_i^{P_1^*} y$, then it is clear that $x - y \in nR = P_1$. Hence, for any two distinct elements $x, y \in R$, $x \sim_e^{P_1^*} y$ if and only if $x - y \in P_1$.

Now, suppose that x and y are operational equivalent with respect to the hyperoperation P_2^*. Thus $xP_2^*a = yP_2^*a$, i.e., $x \cdot a \cdot P_2 = y \cdot a \cdot P_2$, for any $a \in R$. Using the property that two principal ideals are equal when their generators are associated, we obtain that there exists a unit u such that $ya = uxa$, and similarly, there exists a unit v such that $xa = vya$. Both together imply that $ya = uvya$, with $uv = 1$. Since the ring R contains only two units, we have exactly two possibilities. If both units u and v are the multiplicative identity 1, then we obtain that $xa - ya = 0$, i.e., $(x - y)a = 0$, which implies that $x = y$. The second case is when $u = v = -1$ and we obtain $ya = -xa$, for any $a \in R$, thus $y = -x$.

Regarding the inseparability with respect to the hyperoperation P_2^*, we easily see that for any $x \in R$, there is $x \sim_i^{P_2^*} (-x)$ and, moreover, $x \sim_e^{P_2^*} (-x)$.

Based on these two results, it follows clearly that $x \sim_e (-x)$, for any $x \in P_1$ which says that the H_v-ring (R, P_1^*, P_2^*) is not reduced. □

Example 3. *An example of an H_v-ring with P-hyperoperations satisfying Proposition 5 can be obtained taking $R = \mathbb{Z}$, the ring of integers.*

In the following, we will construct other examples of H_v-rings with P-hyperoperations and study their reducibility.

Example 4. *Let $\mathbb{Z}$ be the ring of integers and set $P_1 = n\mathbb{Z}$ with $n \in \mathbb{Z}$ and $P_2 = \mathbb{Z}^+$, the set of positive integers. Then, the hyperstructure $(\mathbb{Z}, P_1^*, P_2^*)$ is a commutative H_v-ring with P-hyperoperations, which is reduced.*

It is easy to see that the conditions of the Theorem 2 are all fulfilled, which implies that the hyperstructure $(\mathbb{Z}, P_1^*, P_2^*)$ is an H_v-ring. Similarly, as in Example 3, we conclude that $x \sim_e^{P_1^*} y$ if and only if $x - y \in P_1$, i.e., $x - y = ns$ for some $s \in \mathbb{Z}$.

Let us suppose that $xP_2^*a = yP_2^*a$, i.e., $x \cdot a \cdot \mathbb{Z}^+ = y \cdot a \cdot \mathbb{Z}^+$, for any $a \in \mathbb{Z}$. Choosing $a = 1$, it follows that $\{xk \mid k \in \mathbb{Z}^+\} = \{yk \mid k \in \mathbb{Z}^+\}$. The equality is satisfied only in the case when $x = y$. Thus, the H_v-ring $(\mathbb{Z}, P_1^*, P_2^*)$ is reduced.

Example 5. *Let (R, P_1^*, P_2^*) be a commutative H_v – ring with P – hyperoperations such that $(R, \cdot)$ is a group and let P_1 be a subgroup of $(R, +)$ and $P_2 = R$. Then, the H_v-ring (R, P_1^*, P_2^*) is not reduced.*

It is easy to check that the hyperstructure (R, P_1^*, P_2^*) is an H_v-ring with P-hyperoperations. Let us prove its non-reducibility. Indeed, following the procedure explained in Proposition 5, we conclude that $x \sim_e^{P_1^*} y$ if and only if $x - y \in P_1$. Hence, for any two distinct elements $x, y \in R$, such that $x - y \in P_1$, there is $\hat{x}_e^{P_1^*} = \hat{y}_e^{P_1^*} \supseteq \{x, y\}$. Taking $P_2 = R$ we easily get that $xP_2^*a = yP_2^*a$, for all $a \in R$, and if x belongs to aP_2^*b, obviously also y belongs to it. Therefore, for an arbitrary element x in R, there is $\hat{x}_e^{P_2^*} = R$.

Combining the two results, we get $x \sim_e y$, whenever $x - y \in P_1$, meaning that the considered H_v-ring is not reduced.

Example 6. *Let (R, P_1^*, P_2^*) be a commutative H_v-ring with P-hyperoperations, such that $(R, +, \cdot)$ is a field and let K be a subfield of R. If $P_1 = P_2 = K$, then the H_v – ring (R, P_1^*, P_2^*) is not reduced.*

Let x and y be arbitrary elements from R. Analogously to Example 5, $x \sim_e^{P_1^*} y$ if and only if $x - y \in P_1$.

Let us suppose that the equality $xP_2^*a = yP_2^*a$ is satisfied for all $a \in R$, i.e., $xaK = yaK$ for any $a \in R$. This is equivalent to $xK = yK$, which is satisfied for any $x, y \in K$.

Merging both conclusions, we get that the hyperring (R, P_1^*, P_2^*) is not reduced, since any two elements x and y in K are essentially indistinguishable.

We conclude this subsection with the study of the reducibility of the hyperrings constructed with Corsini hypergroups. Let us recall first the definition of such a hypergroup.

Definition 9 ([34]). *A hypergroup $(H, \circ)$ is called a Corsini hypergroup, if, for any two elements $x, y \in H$, the following conditions hold:*

1. $x \circ y = x \circ x \cup y \circ y$,
2. $x \in x \circ x$,
3. $y \in x \circ x$ *if and only if* $x \in y \circ y$,
4. *for any* $(a, c) \in H^2$, $c \circ c \circ c \setminus c \circ c \subseteq a \circ a \circ a$.

Proposition 6. *Let $(H, \circ)$ be a Corsini hypergroup and $(H, \star)$ be a B-hypergroup, i.e., $x \star y = \{x, y\}$ for all $x, y \in H$. Then, the hyperring $(H, \star, \circ)$ is a reduced hyperring.*

Proof. Based on Al-Tahan and Davvaz [35], it is known that, if $(H, \circ)$ is a Corsini hypergroup and $(H, \star)$ is a B-hypergroup, then the structure $(H, \star, \circ)$ is a commutative hyperring. Kankaraš has proved in [4] that any $B-$ hypergroup is a reduced hypergroup, which easily gives that the hyperring $(H, \star, \circ)$ is reduced, too. □

Example 7. *Endow the set $R = \{x, y, z\}$ with the hyperoperations $\oplus$ and $\odot$ given by the following tables:*

$\oplus$	x	y	z
x	x, y	x, y	R
y	x, y	x, y	R
z	R	R	z

$\odot$	x	y	z
x	x	x, y	x, z
y	x, y	y	y, z
z	x, z	y, z	z

The hypergroup $(R, \oplus)$ is a Corsini hypergroup [35] and $(R, \odot)$ is a B-hypergroup. Here, $x \oplus a = y \oplus a$ for any $a \in R$. Thus, $x \sim_o^\oplus y$. x and y appear in the same hyperproducts, which gives $x \sim_i^\oplus y$. Considering the second hyperoperation, it easily follows that $\hat{x}_e^\odot = \{x\}$ for any $x \in R$. Hence, $(R, \oplus, \odot)$ is a reduced hyperring.

Remark 1. *If we consider that $(R, \oplus)$ is the hypergroup defined in Example 7 and $(R, \odot)$ is the total hypergroup, then both hypergroups are Corsini hypergroups; hwoever, the hyperring $(R, \oplus, \odot)$ is not reduced since $\hat{x}_e = \hat{y}_e = \{x, y\}$.*

4.2. Reducibility in Complete Hyperrings

The definition of complete hyperrings is based on the definition of complete hypergroups.

Definition 10 ([36]). *Let $(H, \oplus, \odot)$ be a hyperring. If $(H, \oplus)$ is a complete hypergroup, then we say that H is $\oplus$-complete. If $(H, \odot)$ is a complete semihypergroup, then we say that H is $\odot$-complete and if both $(H, \oplus)$ and $(H, \odot)$ are complete, then we say that H is a complete hyperring.*

Following the construction of complete hypergroups, De Salvo [36] proposed a method to obtain complete hyperrings starting with rings. Let us recall here this construction.

Let $(R, +, \cdot)$ be a ring, and $\{A(g)\}_{g \in R}$ be a family of nonempty sets, such that:

1. $\forall g, g' \in R, g \neq g' \Rightarrow A(g) \cap A(g') = \emptyset$
2. $g \notin R \cdot R \Rightarrow |A(g)| = 1$.

Set $H_R = \bigcup_{g \in R} A(g)$ and define on H_R two hyperoperations $\oplus$ and $\odot$ as follows: for any $a, b \in H_R$, there exist $g, g' \in R$ such that $a \in A(g), b \in A(g')$ and define

$$a \oplus b = A(g + g'), a \odot b = A(gg').$$

Lemma 1 ([36]). *Using the previous notations, for all $g, g' \in R$ and any $a \in A(g), b \in A(g')$ we have:*
$a \oplus b = A(g + g') = A(g) \oplus A(g')$,
$a \odot b = A(gg') = A(g) \odot A(g')$.

In [37] Corsini proved that $(H_R, \oplus)$ and $(H_R, \odot)$ are, respectively, a complete commutative hypergroup and a complete semihypergroup.

Remark 2. *All complete hyperrings can be constructed by the above described procedure, since it is known that any complete semihypergroup (hypergroup) can be constructed as the union of disjoint sets $A(g), g \in G$ (see Theorem 1).*

Based on Theorem 3, any complete (semi)hypergroup is not reduced; however, this property does not imply directly the non-reducibility of any complete hyperring. That's why we need to study its reducibility in a different way, as shown in the next result.

Theorem 4. *Any complete hyperring $(H_R, \oplus, \odot)$ is not reduced.*

Proof. Let $(H_R, \oplus, \odot)$ be a complete hyperring. Therefore the hypergroup $(H_R, \oplus)$ and the semihypergroup $(H_R, \odot)$ are both complete, so both are not reduced. It follows that there exist $a \neq b \in H_R$ such that $a \sim_e^{\oplus} b$. Now it is enough to prove that $a \sim_e^{\oplus} b$ implies $a \sim_e^{\odot} b$ for $a, b \in H_R$, because in this case $\hat{a}_e = \hat{a}_e^{\oplus} \cap \hat{a}_e^{\odot} \supseteq \{a, b\}$, which shows that $(H_R, \oplus, \odot)$ is not reduced.

First, we will prove that the operational equivalence relation with respect to the hyperoperation $\oplus$ implies the operational equivalence relation with respect to $\odot$. Let a, b be elements from H_R such that $a \oplus c = b \oplus c$, for all $c \in H_R$. It follows that there exist $g_a, g_b, g_c \in R$ such that $a \in A(g_a), b \in A(g_b)$ and $c \in A(g_c)$. According to Lemma 1, we have $a \oplus c = A(g_a + g_c)$ and $b \oplus c = A(g_b + g_c)$, which leads to the equality $A(g_a + g_c) = A(g_b + g_c)$, and so $g_a + g_c = g_b + g_c$ in the group $(R, +)$. Therefore, $g_a = g_b$, that implies that $g_a \cdot g_c = g_b \cdot g_c$. Therefore, $a \odot c = A(g_a \cdot g_c) = A(g_b \cdot g_c) = A(g_b) \odot A(g_c) = b \odot c$. Similarly, $c \oplus a = c \oplus b$ implies that $c \odot a = c \odot b$. This means that $a \sim_o^{\oplus} b$ implies $a \sim_o^{\odot} b$ for all $a, b \in H_R$.

Next, we will show that the indistinguishability relation with respect to $\oplus$ implies the indistinguishability relation with respect to $\odot$.

Let us suppose $a \sim_i^{\oplus} b$. This means that a and b appear in the same hyperproducts $d \oplus e$, for $d, e \in H_R$. Thus $a \in A(g_d) \oplus A(g_e) \iff b \in A(g_d) \oplus A(g_e)$, with $g_d, g_e \in R$ such that $d \in A(g_d), e \in A(g_e)$. It follows that $a \in A(g_d + g_e) \iff b \in A(g_d + g_e)$, meaning that $a, b \in A(g)$, with $g \in R$. If we consider now $a \in k \odot l$, then $a \in A(g_k \cdot g_l)$, where $k \in A(g_k), l \in A(g_l)$. Since a and b are in the same A_g, it follows that $b \in A(g_k \cdot g_l) = k \odot l$, equivalently, $b \in k \odot l$. Similarly, if $b \in k \odot l$, then $a \in k \odot l$. Hence, $a \sim_i^{\odot} b$. □

Example 8. *Let the hyperring $R = (\{a, b, c, d, e\}, \oplus, \odot)$ be defined as shown in the following tables:*

$\oplus$	a	b	c	d	e
a	a	b,c	b,c	d	e
b	b,c	d	d	e	a
c	b,c	d	d	e	a
d	d	e	e	a	b,c
e	e	a	a	b,c	d

$\odot$	a	b	c	d	e
a	a	a	a	a	a
b	a	b,c	b,c	d	e
c	a	b,c	b,c	d	e
d	a	d	d	a	d
e	a	e	e	d	b,c

The hyperring $(R, \oplus, \odot)$ is a commutative complete hyperring [38]. Since the rows corresponding to the elements b and c are exactly the same in both tables, we conclude that $b \sim_o^\oplus c$ and $b \sim_o^\odot c$, which further gives $b \sim_o c$, i.e., $\hat{b}_o = \hat{c}_o \supseteq \{b,c\}$. Furthermore, we notice that $\hat{b}_o = \hat{c}_o = \{b,c\}$. In addition, the elements b and c appear together in the same hyperproducts in $(R, \oplus)$, as well as in $(R, \odot)$, whence $b \sim_i c$, and thus $\hat{b}_i = \hat{c}_i = \{b,c\}$. Hence, $\hat{b}_e = \hat{c}_e = \{b,c\}$, which implies that the given hyperring is not reduced.

Remark 3. Since $(R, \cdot)$ is generally a semigroup, and not a group, it may happen that the operational equivalence relation with respect to the hyperoperation $\odot$ does not imply the operational equivalence relation with respect to the hyperoperation $\oplus$.

4.3. Reducibility in (H,R)-Hyperrings

(H, R)-hyperrings were introduced by De Salvo in [36], when he generalized the construction of (H,G)-hypergroups described in [39]. In the following, we will present their construction.

Let $(H, \circ, \bullet)$ be a hyperring and $\{A_i\}_{i \in R}$ be a family of nonempty sets such that:

1. $(R, +, \cdot)$ is a ring.
2. $A_{0_R} = H$.
3. For any $i \neq j \in R$, $A_i \cap A_j = \emptyset$.

Set $K = \bigcup_{i \in R} A_i$ and define on K the following hyperoperations:

$$\text{for any} \quad x, y \in H, x \oplus y = x \circ y \tag{5}$$

$$\text{and} \quad x \odot y = H \tag{6}$$

For any $x \in A_i$ and $y \in A_j$ such that $A_i \times A_j \neq H \times H$, define

$$x \oplus y = A_k \quad \text{if} \quad i + j = k, \tag{7}$$

$$x \odot y = A_m \quad \text{if} \quad i \cdot j = m. \tag{8}$$

The structure $(K, \oplus, \odot)$ is a general hyperring, called an (H, R)-hyperring. Moreover, if ω is the heart of the hypergroup $(K, \oplus)$, then $\omega = H$ and $H \odot K = K \odot H = H$ [36].

In the following, we will better describe the operational equivalence and the inseparability in an (H, R)-hyperring.

Lemma 2. *Let $(K, \oplus, \odot)$ be an (H,R)-hyperring, where $K = \bigcup_{i \in R} A_i$, with $(R, +, \cdot)$ a ring and $(H, \circ, \bullet)$ a hyperring.*

1. *Two elements x and y in $A_{0_R} = H$ are operationally equivalent with respect to the hyperoperation $\oplus$ if and only if they are operationally equivalent with respect to the hyperoperation $\circ$ on H.*
2. *Two elements x and y in $K \setminus A_{0_R}$ are operationally equivalent with respect to the hyperoperation $\oplus$ if and only if they belong to the same subset $A_i \subset K$.*
3. *Two elements x and y in K are inseparable with respect to the hyperoperation $\oplus$ if and only if they belong to the same subset $A_i \subset K$.*

Proof. 1. Let x, y be in $A_{0_R} = H$ such that $x \oplus a = y \oplus a$, for all $a \in K$. If $a \in A_{i_a}$, with $i_a \neq 0_R$, then the equality always holds. If $a \in A_{0_R}$, then $x \oplus a = y \oplus a$ whenever $x \circ a = y \circ a$, and thus the result is proved.

2. Let x and y be in $K \setminus H$, such that $x \in A_{i_x}$ and $y \in A_{i_y}$, with $i_x, i_y \in R$ and consider $x \oplus a = y \oplus a$, for all $a \in K$. If $a \in A_{0_R}$, then $x \oplus a = A_{i_x}$ and $y \oplus a = A_{i_y}$; therefore x and y are operationally equivalent if and only if $i_x = i_y$. If $a \in K \setminus A_{0_R}$, for example $a \in A_{i_a}$, then $x \oplus a = y \oplus a$ is equivalent with $i_x + i_a = i_y + i_a$, meaning again $i_x = i_y$.

3. Let us consider $x \sim_i^\oplus y$, meaning that $x \in a \oplus b$ if and only if $y \in a \oplus b$. If $a, b \in A_{0_R}$, then $a \oplus b = a \circ b$, and therefore $x \sim_i^\oplus y$ whenever $x, y \in a \circ b \subset A_{0_R}$. If $a \in A_{i_a}$ and

$b \in A_{i_b}$ with $A_{i_a} \times A_{i_b} \neq H \times H$, then $a \oplus b = A_{i_a + i_b} = A_i$, and therefore $x \sim_i^\oplus y$ whenever $x, y \in A_i$, with $i \in R$. Combining the two cases, we find that x and y are inseparable if and only if they are in the same subset A_i. □

Lemma 3. *Let* $(K, \oplus, \odot)$ *be an (H,R)-hyperring, where* $K = \bigcup_{i \in R} A_i$, *with* $(R, +, \cdot)$ *an integral domain and* $(H, \circ, \bullet)$ *a hyperring. Two elements x and y in K are essentially indistinguishable with respect to the hyperoperation $\odot$ if and only if they belong to the same subset $A_i \subset K$.*

Proof. The proof is similar to the one of Lemma 2. The only difference here is in the case of the relation "$\sim_\circ$", where the condition regarding R to be an integral domain is fundamental. □

Proposition 7. *Let* $(K, \oplus, \odot)$ *be an (H,R)-hyperring, where* $K = \bigcup_{i \in R} A_i$, *with* $(R, +, \cdot)$ *an integral domain and* $(H, \circ, \bullet)$ *a hyperring. Then, the hyperring* $(K, \oplus, \odot)$ *is not reduced if and only if there exists $i \in R, i \neq 0_R$, with $|A_i| \geq 2$, or the hypergroup $(H, \circ)$ is not reduced.*

Proof. Let us suppose that the hyperring $(K, \oplus, \odot)$ is not reduced. Then, there exist two distinct elements x and y in K such that $x \sim_e y$, i.e., $x \sim_e^\oplus y$ and $x \sim_e^\odot y$. Based on Lemma 2 and and Lemma 3, if x and y belong to the same subset A_i, with $i \neq 0_R$, we conclude that $|A_i| \geq 2$. Otherwise, if all sets $A_i, i \neq 0_R$ are singletons, then $x, y \in A_{0_R} = H$, which implies that $x \sim_\circ^\circ y$ and $x \sim_i^\circ y$, i.e., the structure $(H, \circ)$ is not a reduced hypergroup.

Conversely, suppose there exists $i \in R \setminus \{0_R\}$ such that $|A_i| \geq 2$. Then, there exist two elements x and y in the set A_i, implying that $x \sim_e^\oplus y$ and $x \sim_e^\odot y$. In other words, $x \sim_e y$, meaning that the (H, R)-hyperring $(K, \oplus, \odot)$ is not reduced. Assuming that $(H, \circ)$ is not reduced, let x and y be two elements such that $x \sim_e^\circ y$. According with Lemma 2 and and Lemma 3, we further conclude that $x \sim_e^\oplus y$. Due to the definition of the hyperoperation $\odot$, for any $x, y \in H$, it easily follows that $x \sim_e^\odot y$. Hence, $x \sim_e y$, i.e., $(K, \oplus, \odot)$ is not a reduced hyperring. □

Corollary 2. *If $(H, \circ, \bullet)$ is a not reduced hyperring, then the (H, R)-hyperring $(K, \oplus, \odot)$ is not reduced, too.*

In the following, we will give an example of an (H,R)-hyperring and show its non-reducibility.

Example 9. *Let endow the set $R = \{0, a, b, c\}$ with the following operations*

+	0	a	b	c
0	0	a	b	c
a	a	0	c	b
b	b	c	0	a
c	c	b	a	0

·	0	a	b	c
0	0	0	0	0
a	0	0	a	a
b	0	0	b	b
c	0	0	c	c

It easily follows that $(R, +, \cdot)$ is a ring. Furthermore, let $(H, \circ, \bullet)$ be a hyperring given by the tables

∘	c	d
c	c	c,d
d	c,d	c,d

•	c	d
c	c	c,d
d	c	c,d

The structure $(H, \circ, \bullet)$ is a general hyperring [36]. It is easy to check that $(H, \circ)$ is a reduced hypergroup and thus, the hyperring $(H, \circ, \bullet)$ is reduced, too.

We will endow the set $K = \{c, d, a_1, a_2, a_3, a_4, a_5, a_6\}$, where $A_0 = H$, $A_a = \{a_1, a_2\}$, $A_b = \{a_3, a_4, a_5\}$, $A_c = \{a_6\}$, with an (H,R)-hyperstructure, by defining the hyperaddition $x \oplus y = x \circ y$

if both x,y belong to H, otherwise, let $x \oplus y = A_k$, with $x \in A_i, y \in A_j$ and $k = i + j$. We define $x \odot y = H$, where $x, y \in H$ and $x \oplus y = A_k$ with $x \in A_i, y \in A_j$ and $k = i \cdot j$. Then, the structure $(K, \oplus, \odot)$ is an (H,R)-hyperring.

Let us prove that $a_1 \sim_e a_2$, i.e., $a_1 \oplus x = a_2 \oplus x$ for all $x \in K$. Indeed, if $x \in H$, $a_1 \oplus x = A_{a+0} = a_2 \oplus x$. If $x \in A_a$, $a_1 \oplus x = A_{a+a} = A_0 = a_2 \oplus x = A_0$. For $x \in A_b$, $a_1 \oplus x = A_{a+b} = A_c = a_2 \oplus x$. Finally, $a_1 \oplus x = a_2 \oplus x = A_{a+c} = A_b$ for $x \in A_c$. Due to the commutativity of the ring R, $x \oplus a_1 = x \oplus a_2$ for any $x \in K$. Similarly, $a_1 \odot x = a_2 \odot x$ and $x \odot a_1 = x \odot a_2$ for any $x \in K$. Thus, $a_1 \sim_o a_2$.

Since $x \oplus y \subseteq H$ if both $x, y \in H = A_0 = \{c, d\}$, we conclude that the elements a_1 and a_2 do not appear in such hyperproducts. All other hyperproducts $x \oplus y$ are equal to some sets A_k, where $k \in \{a, b, c\}$, with A_a, A_b and A_c being disjoint sets. Hence, a_1 and a_2 appear in the hyperproducts which are equal to A_a, so they always appear together. Analogously, a_1 and a_2 appear in the same hyperproducts $x \odot y$. Hence, $a_1 \sim_i a_2$.

Similarly, one proves that $\hat{a}_{3e} = \hat{a}_{4e} = \hat{a}_{5e} = \{a_3, a_4, a_5\}$. Thereby we conclude that the (H,R)-hyperring $(K, \oplus, \odot)$ is not reduced.

5. Conclusions

In this paper, we defined and studied the reducibility of some particular types of general hyperrings, thus, extending the concept of reducibility in hypergroups. We presented some properties of the fundamental relations in general hyperrings, and we investigated the reducibility for complete and (H, R)-hyperrings, hyperrings of formal series, and hyperrings constructed with Corsini hypergroups. In a future work, our goal is to extend this study of reducibility to the fuzzy case, i.e., to define and investigate the fuzzy reducibility in hyperrings.

Author Contributions: Conceptualization, I.C. and M.K.; methodology, I.C. and M.K.; investigation, I.C. and M.K.; writing—original draft preparation, M.K.; writing—review and editing, I.C.; funding acquisition, I.C. Both authors have read and agreed to the published version of the manuscript.

Funding: The first author acknowledges the financial support from the Slovenian Research Agency (research core funding No. P1-0285).

Institutional Review Board Statement: Not applicable.

Data Availability Statement: Not applicable.

Conflicts of Interest: The authors declare no conflict of interest.

References

1. Jantosciak, J. Reduced hypergroups. In Proceedings of the 4th International Congress Algebraic Hyperstructures and Applications, Xanthi, Greece, 27–30 June 1990; Vougiouklis, T., Ed.; World Scientific: Singapore, 1991; pp. 119–122.
2. Cristea, I. Several aspects on the hypergroups associated with n-ary relations. *An. Stiint. Univ. Ovidius Constanta Ser. Mat.* **2009**, *17*, 99–110.
3. Cristea, I.; Ştefănescu, M.; Angheluţa, C. About the fundamental relations defined on the hypergroupoids associated with binary relations. *Eur. J. Combin.* **2011**, *32*, 72–81. [CrossRef]
4. Kankaraš, M. Reducibility in Corsini hypergroups. *An. Stiint. Univ. Ovidius Constanta Ser. Mat.* **2021**, *29*, 93–109.
5. Kankaraš, M.; Cristea, I. Fuzzy reduced hypergroups. *Mathematics* **2020**, *8*, 263. [CrossRef]
6. Krasner, M. A class of hyperrings and hyperfields. *Int. J. Math. Math. Sci.* **1983**, *6*, 307–312. [CrossRef]
7. Rota, R. Sugli iperanelli multiplicativi. *Rend. Mat.* **1982**, *2*, 711–724
8. Corsini, P. Hypergroupes reguliers et hipermodules. *Ann. Univ. Ferrara-Sez.VII Mat.* **1975**, *20*, 121–135.
9. Mittas, J. Hyperanneaux canoniques. *Math. Balk.* **1972**, *2*, 165–179.
10. Mittas, J. Sur les hyperanneaux et les hypercorps. *Math. Balk.* **1973**, *3*, 368–382.
11. Spartalis, S. A class of hyperrings. *Riv. Math. Pura Appl.* **1989**, *4*, 55–64.
12. Nakasis, A. Expository and survey article: Recent results in hyperring and hyperfield theory. *Int. J. Math. Math. Sci.* **1988**, *11*, 209–220. [CrossRef]
13. Massouros, C.; Massouros, G. An overview of the foundations of hypergroup theory. *Mathematics* **2021**, *9*, 1014. [CrossRef]
14. Massouros, G.; Massouros, C. Hypercompositional Algebra, Computer Science and Geometry. *Mathematics* **2020**, *8*, 1338. [CrossRef]
15. Freni, D. Une note sur le coeur d'un hypergroupe et sur la cloture transitive β^* de β. *Riv. Mat. Pura Appl.* **1991**, *8*, 153–156.

16. Freni, D. Strongly transitive geometric spaces: Applications to hypergroups and semigroups. *Commun. Algebra* **2004**, *32*, 969–988. [CrossRef]
17. Koskas, M. Groupoides, demi-hypergroupes et hypergroupes. *J. Math. Pure Appl.* **1970**, *49*, 155–192.
18. Corsini, P. *Prolegomena of Hypergroup Theory*; Aviani Editore: Tricesimo, Italy, 1993.
19. De Salvo, M.; Lo Faro, G. On the n^*-complete hypergroups. *Discret. Math.* **1999**, *208/209*, 177–188. [CrossRef]
20. Cristea, I.; Hassani Sadrabadi, E.; Davvaz, B. A fuzzy application of the group $\mathbb{Z}_n$ to complete hypergroups. *Soft. Comput.* **2020**, *24*, 3543–3550. [CrossRef]
21. Sonea, A.C.; Cristea, I. The class equation and the commutativity degree for complete hypergroups. *Mathematics* **2020**, *8*, 2253. [CrossRef]
22. De Salvo, M.; Fasino, D.; Freni, D.; Lo Faro, G. 1-hypergroups of small sizes. *Mathematics* **2021**, *9*, 108. [CrossRef]
23. Massouros, G.G. The hyperringoid. *Multi Val. Log.* **1998**, *3*, 217–234.
24. Massouros, C.G.; Mittas, J. Languages and Automata and hypercompositional structures. In Proceedings of the 4th International Congress on Algebraic Hyperstructures and Applications, Xanthi, Greece, 27–30 June 1990; Vougiouklis, T., Ed.; World Scientific: Singapore, 1991; pp. 137–147.
25. Vougiouklis, T. The fundamental relation in hyperrings. The general hyperfield. In Proceedings of the 4th International Congress on Algebraic Hyperstructures and Applications, Xanthi, Greece, 27–30 June 1990; Vougiouklis, T., Ed.; World Scientific: Singapore, 1991; pp. 203–211.
26. Vougiouklis, T. Fundamental relation in H_v-structures. The 'Judging from the results' proof. *J. Algebr. Hyperstructures Log. Algebr.* **2020**, *1*, 21–36. [CrossRef]
27. Davvaz, B.; Vougiouklis, T. *A Walk Through Weak Hyperstructures. H_v-Structures*; World Scientific: Hackensack, NJ, USA, 2018.
28. Spartalis, S. On the number of H_v-rings with P-hyperoperations. *Discret. Math.* **1996**, *155*, 225–231. [CrossRef]
29. Spartalis, S.; Vougiouklis, T. The fundamental relations on H_v-hyperrings. *Riv. Math. Pura Appl.* **1993**, *13*, 7–20.
30. Vougiouklis, T. Isomorphisms on P-hypergroups and cyclicity. *ARS Combin.* **1990**, *29A*, 241–245.
31. Davvaz, B.; Koushky, A. On hyperrings of polynomials. *Ital. J. Pure Appl. Math.* **2004**, *15*, 205–214.
32. Jančić-Rašović, S. About the hyperring of polynomials. *Ital. J. Pure Appl. Math.* **2007**, *21*, 223–234.
33. Ameri, R.; Mansour, E.; Hoškova-Mayerova, Š. Superring of polynomials over a hyperring. *Mathematics* **2019**, *7*, 902. [CrossRef]
34. Corsini, P. Hypergraphs and hypergroups. *Algebra Univ.* **1996**, *35*, 548–555. [CrossRef]
35. Al Tahan, M.; Davvaz, B. On Corsini hypergroups and their productional hypergroups. *Korean J. Math.* **2019**, *27*, 63–80.
36. De Salvo, M. Iperanelli ed ipercorpi. *Ann. Sci. L'Université Clermont. Mathématiques* **1984**, *22*, 89–107.
37. Corsini, P. Sur les semi-hypergroupes complets et les groupoides. *Atti Soc. Pelor. Sc. Mat. Fis. Nat.* **1980**, *XXVI*, 391–398.
38. Davvaz, B.; Loreanu, V. *Hyperrings Theory and Applications*; International Academic Press: Palm Harbor, FL, USA, 2007.
39. De Salvo, M. (H,G)-hypergroups. *Riv. Mat. Univ. Parma* **1984**, *10*, 207–216.

MDPI
St. Alban-Anlage 66
4052 Basel
Switzerland
www.mdpi.com

Mathematics Editorial Office
E-mail: mathematics@mdpi.com
www.mdpi.com/journal/mathematics

Disclaimer/Publisher's Note: The statements, opinions and data contained in all publications are solely those of the individual author(s) and contributor(s) and not of MDPI and/or the editor(s). MDPI and/or the editor(s) disclaim responsibility for any injury to people or property resulting from any ideas, methods, instructions or products referred to in the content.

www.ingramcontent.com/pod-product-compliance
Lightning Source LLC
LaVergne TN
LVHW070646100526
838202LV00013B/890